CW00457946

RISK ON THE TABLE

The Environment in History: International Perspectives

Series Editors: Stefania Barca, *University of Coimbra*; Kieko Matteson, *University of Hawai'i at Mānoa*; Christof Mauch, *LMU Munich*; Helmuth Trischler, *Deutsches Museum, Munich*

ENVIRONMENT AND SOCIETY

For a full volume listing, please see the series page on our website: http://berghahnbooks.com/series/environment-in-history

Risk on the Table

*Food Production, Health,
and the Environment*

Edited by
Angela N. H. Creager and Jean-Paul Gaudillière

berghahn
NEW YORK · OXFORD
www.berghahnbooks.com

First published in 2021 by
Berghahn Books
www.berghahnbooks.com

Library of Congress Cataloging-in-Publication Data

Names: Creager, Angela N. H., editor. | Gaudillière, Jean-Paul, 1957– editor.
Title: Risk on the Table: Food Production, Health, and the Environment / edited by
Angela N. H. Creager and Jean-Paul Gaudillière.
Description: New York: Berghahn, 2021. | Series: Environment in History; 21 | Includes
bibliographical references and index.
Identifiers: LCCN 2020048764 (print) | LCCN 2020048765 (ebook) | ISBN
9781789209440 (hardback) | ISBN 9781789209457 (ebook)
Subjects: LCSH: Food—Safety measures. | Food adulteration and inspection—Risk
assessment.
Classification: LCC TX531 .R557 2021 (print) | LCC TX531 (ebook) | DDC
363.19/26—dc23
LC record available at https://lccn.loc.gov/2020048764
LC ebook record available at https://lccn.loc.gov/2020048765

British Library Cataloguing in Publication Data

A catalogue record for this book is available from the British Library

ISBN 978-1-78920-944-0 hardback
ISBN 978-1-78920-945-7 ebook

Contents

PART II. ORDERING RISKS

Figures and Tables

Figures

Table

Acknowledgments

For funding that enabled this to be a truly collective endeavor, we thank the Shelby Cullom Davis Center for Historical Studies, Program in History of Science, and Department of History at Princeton University. Alison McManus, Pallavi Podapati, and especially Gina Surita helped edit the chapters in preparation for submission. Francesca DeRosa offered her excellent skills in proofreading. Jennifer Goldman, Jennifer Loessy, and Jaclyn Wasneski provided additional support. We are grateful to Helmuth Trischler for his support and encouragement. For comments on the manuscript we thank Julia Stone and two anonymous referees. Finally, we acknowledge our copyeditor, Lynn Otto, for making the text as clear and consistent as possible, and Berghahn editors Mykelin Higham and Elizabeth Martinez for overseeing a swift and smooth publication process, even in the midst of a pandemic.

Abbreviations

ACG	African Groundnut Council
ACSH	American Council on Science and Health
ADI	Acceptable Daily Intake
AEA	Atomic Energy Authority (UK)
AEC	Atomic Energy Commission (US)
AGPs	antibiotic growth promoters
AMA	American Medical Association
AMR	antimicrobial resistance
APF	animal protein factor
ARC	Agricultural Research Council (UK)
ASARCO	American Smelting and Refining Company
AVMA	American Veterinary Medical Association
BHA	butylated hydroxyanisole
BBP	benzyl butyl phthalate
BMI	body mass index
BOF	Bureau of Food (US)
BPA	bisphenol A
BVM	Bureau of Veterinary Medicine (US)
CAFOs	concentrated animal feeding operations
CAST	Council for Agricultural Science and Technology
CBD	cannabidiol
CDC	Centers for Disease Control and Prevention, originally Communicable Disease Center (US)
CFR	Code of Federal Regulations (US)
CRISPR	clustered regularly interspaced short palindromic repeats
CSPI	Center for Science in the Public Interest
DES	diethylstilbestrol
DDT	dichloro-diphenyl-trichloroethane
DEHP	diethylhexyl phthalate
DHC	Dow Historical Collection, Science History Institute Archives, Philadelphia, PA

DINP	di-isononyl phthalate
DMAs	demethylated arsenicals
DSHEA	Dietary Supplement Health and Education Act
EEC	European Economic Community
EGCG	epigallocatechin-3-gallate
EPA	Environmental Protection Agency (US)
EU	European Union
EUROTOX	European Committee for the Protection of Populations against Risks of Chronic Toxicity
FAO	Food and Agriculture Organization (UN)
FDC	Federal Food, Drug, and Cosmetic Act (US)
FDCA	1938 Food, Drug, and Cosmetic Act (US)
FDA	Food and Drug Administration (US)
FID	Food Industry Documents, database of archival documents hosted by UCSF Library and Center for Knowledge Management
FSIS	Food Safety and Inspection Service, part of USDA
FOSHU	Food for Specific Health Uses
GABA	γ-amino butyric acid
GAO	General Accounting Office (US)
GATT	General Agreement on Tariffs and Trade
GPO	Government Printing Office (US)
GRAS	generally recognized as safe
HMSO	Her Majesty's Stationery Office (UK)
HPV	Human Papilloma Virus
iA	inorganic arsenic
IARC	International Agency for Research on Cancer
ICRP	International Commission on Radiological Protection
JAMA	*Journal of the American Medical Association*
JECFA	Joint FAO/WHO Expert Committee on Food Additives
MAFF	Ministry of Agriculture, Fisheries and Food (UK)
MDR	multiple-drug resistant
MMAs	monomethylated arsenicals
MRC	Medical Research Council (UK)
MSG	monosodium glutamate
NAS	National Academy of Science (US)
NCI	National Cancer Institute (US)
NDA	New Drug Application (US)
NCDs	noncommunicable diseases
NLEA	Nutrition Labeling and Education Act

NSDAP	*Nationalsozialistische Deutsche Arbeiterpartei*, known in English as Nazi Party
NIEO	New International Economic Order
NRC	National Research Council, arm of the NAS (US)
NRDC	Natural Resources Defense Council
NTP	National Toxicology Program (US)
NTRC	National Toxicology Research Center
OECD	Organisation for Economic Co-operation and Development
OPEC	Organization of Petroleum Exporting Countries
OSHA	Occupational Safety and Health Administration (US)
PAG	Protein Advisory Group
PFC	long-chain perfluorinated compounds
ppb	parts per billion
ppm	parts per million
POPs	persistent organic pollutants
PCBs	polychlorinated biphenyls
PCTs	polychlorinated terphenyls
PFGE	pulsed field gel electrophoresis
SCOGS	Select Committee on GRAS Substances
SMZ	sulfamethazine
STOP	Swab Test On Premise (USDA)
STS	science and technology studies
TSCA	Toxic Substances Control Act (US)
TTID	Truth Tobacco Industry Documents, database of archival documents hosted by UCSF Library and Center for Knowledge Management
UCSF	University of California, San Francisco
U.I.C.C.	Union Internationale contre le Cancer
UDMF	unsymmetrical dimethyl hydrazine
UN	United Nations
UNEP	United Nations Environmental Programme
UNICEF	United Nations Children's Fund
USDA	United States Department of Agriculture
VPHL	Veterinary Public Health Laboratory (US)
VPHS	Veterinary Public Health Section (US)
WHO	World Health Organization (UN)
WTO	World Trade Organization

Introduction

Angela N. H. Creager and Jean-Paul Gaudillière

In mid-May of 2017, the twenty-seven environment ministers of the EU met for the fifth time in less than two years in order to craft a compromise on the legal definition of endocrine disruptors. This definition is a mandatory step in the implementation of the EU-wide agreement on the regulation of chemicals (Regulation, Evaluation, and Authorization of Chemicals—better known as REACH) because the definition will determine which substances might be banned or restricted under the REACH provisions. Once again, the meeting failed. At the core of the disagreements this time was an amendment introduced by the European Commission and supported by Germany that stated that pesticides specifically designed to disrupt insect hormonal regulation should be taken out of the endocrine disruptors package, thus creating a major loophole in the law. Perhaps not surprisingly, the meeting had been preceded by a rise in lobbying and publicizing activities. Organizations funded by the chemical industry (the European Crop Protection Association, the European Chemical Industry Council, and PlasticsEurope) advocated for the amendment, whereas it was opposed by medical and scientific societies (the Endocrine Society, the European Society of Endocrinology, the European Society of Pediatric Endocrinology). Moreover, the European press had published a whole range of internal documents from PlasticsEurope acknowledging that the situation looked bleak since "current scientific evidence show that plastics display endocrine disruptors properties" and arguing for the "financing of more research."[1] Endocrine disruptors are a concern pertaining to both ecology and food quality: pesticides are not only disseminated in the environment but are also persistent contaminants in the products of industrial agriculture. An announcement in mid-June by the European Chemical Agency underlined this risk, identifying bisphenol A (also known as BPA) as an endocrine disruptor associated with increased human risks of cancer, diabetes, obesity, and neurobehavioral disorders.[2] BPA leaches from many of the plastic containers used in the food processing industry. Even as industry disputes the degree of health hazards they pose, endocrine disruptors have become a pervasive contaminant in the food supply.

The now thirty-year-old controversy about endocrine disruptors may be viewed as a paradigmatic case of the emerging downside of the industrial production of food, long hailed as a triumph of technology.[3] Technical solutions invented to increase agricultural productivity and "solve" the food security problem have become the source of new and hard-to-quantify risks, not to mention the questionable nutritional value of many processed foods. It may also be viewed as a paradigmatic case of the contemporary production of ignorance, in which vested interests obstruct or call into question scientific knowledge.[4] The European press regularly reports conflicts of interests within the national and EU regulatory agencies, as well as massive and powerful lobbying activities in Brussels. The camps seem clear: the chemical industry defends its markets against scientists, physicians, and public health authorities—the latter relying on published experimental evidence, the former pointing to knowledge gaps and uncertainty. This dispute over endocrine disruptors seems poised to be the twenty-first-century analogue of the twentieth-century "tobacco and cancer" affair, this time touching a consumable that is not voluntary—namely, food.[5]

Standing behind this one politically charged example are a host of research and regulatory complexities, which are multiplied when one adds the many other hazardous food contaminants, from *Salmonella* to mercury. While the chemical industries have many ways of constructing markets—including strategies of substitution, delocalization, and green-washing—the production of evidence regarding the thousands of chemicals employed in making food is not only meager but also multiple and often contradictory, in part due to the many levels of control as edibles circulate through the global commodities market. Why, after a century of technological progress and regulatory oversight, does the safety of industrialized food appear so elusive? To answer that question requires an in-depth look at how researchers, companies, and agencies have determined risk and safety, produced knowledge as well as commodities, and dealt with uncertainty and conflict. *Risk on the Table* tries to understand the history behind the ongoing heartburn of food worries.[6]

The purity and safety of the food supply is an old issue for ordinary people, experts, and state authorities. However, the growing use of chemicals in food production and the industrialization of agriculture catalyzed a new set of controversies in the twentieth century about the risks of food. Several developments were implicated in these debates, including the reporting of health issues in the media; the proliferation of synthetic chemicals as additives, preservatives, pesticides, drugs, and packaging; the biological selection of newly pathogenic bacteria by use of antibiotics and containment facilities in agriculture; and improved techniques for detecting minute levels of contaminants along with new understandings of health hazards for low-dose exposures. The industrialization of the food supply has been a theme in studies of business and agriculture for forty years, as illustrated by Harvey Levenstein's influential *Revolution at the Table: The Transformation of the American Diet* (1988). However, whereas this earlier scholarship tended to take for granted the role of scientific and technical expertise in the expansion and management of the food industry, including issues of its safety, our authors examine the changing nature of expert knowledge, as well as the potential conflicts among the perspectives offered by microbiologists, nutritionists, toxicologists, epidemiologists, and economists. In many cases, technological innovations designed to address the productivity and safety of the food supply—such as the tremendous postwar growth in pesticide usage—ended up generating novel hazards. This introduction briefly recounts the history of the industrialization of food and how emerging dangers from this production system prompted new scientific research and regulation. We then turn to the chapters of this book and how they contribute to understanding how science and technology have become implicated in both creating and controlling food risk.

The Industrialization of Food

The mass production in factory settings of ready-to-eat food began in the nineteenth century with the commercial canning industry.[7] As Alfred Chandler notes, American food manufacturers were among the first to take advantage of new continuous process technologies. Grain mills led the way, as exemplified by the Pillsbury brothers' automatic all-roller, gradual reduction mill. A key feature of continuous process factories was their high unit output. In the case of new industrial foodstuffs such as oatmeal, production volume required the development of new markets. Henry P. Crowell started selling Quaker Oats as a breakfast cereal, creating a now ubiquitous food product line. National companies such as Crowell's used mass advertising campaigns to promote brand name recognition and increase sales.[8] Heinz, Armour, and National Biscuit Company (NABISCO) soon joined in the mass production and distribution of food products. Quality assurance and food safety were a key part of what firms were marketing in tins and boxes. Industrial self-regulation was thus closely tied to fostering a base of customers, and their trust, nationwide.

Farms themselves did not become mechanized, large-scale elements of the US food system until the twentieth century—and later in much of Europe.[9] Vast fields of monoculture crops were appetizing to insects, plant pathogens, and vermin, spurring a growing reliance on so-called economic poisons (such as arsenic-containing chemicals) to protect the plants.[10] Industrialized farms became as dependent on agrochemicals, especially fertilizers and pesticides, as they were on the tractors, crop planes, and mechanized equipment. In 1955, the term "agribusiness" was coined by Harvard Business School's John H. Davis to describe this new system, with its heavy reliance on the chemical industry. His term was increasingly linked to the loss of small family farms to large corporate owners, as opposed to the government regulation, scientific research, labor patterns, and marketing networks that Davis saw as integral to this system—elements we seek to re-examine here.[11] Technologies for food processing were also critical to industrialization. Bacteriologists provided a scientific rationale for older food preservation practices such as conserving and canning, and contributed new methods (such as sterile processing and pasteurization) to control contamination and enable long-term preservation for the growing geographical reach of food distribution.[12]

By the 1930s, what has been called the "chemogastric revolution" introduced a wide range of new chemicals into food production, aimed at preventing spoilage, extending shelf life, making edible goods more appealing, and increasing agricultural productivity.[13] (Refrigeration and the emergence of the "cold chain" similarly transformed the storage of perishables.[14]) Food additives were not new, nor were they necessarily unnatural. Ascorbic acid, an antioxidant widely used as a preservative, remains better known by its common name, vitamin C.[15] But during the first half of the twentieth century, the chemical industry produced a host of new food chemicals, especially antioxidants and agents for stabilizing and emulsifying; five hundred were on the market by 1947, and that number nearly doubled in the following decade.[16] Many chemical additives enhanced the appearance of foods (or stopped their color from fading), making them more appetizing or even more natural-looking.[17] In fact, manufacturers of processed foods promoted the image of their products as fresh and familiar, even as they relied on an increasing array of chemicals for preservation, taste, and appearance.

By virtue of these chemicals, as well as industrial packaging and distribution systems, convenience foods multiplied on grocery store and (increasingly) supermarket shelves, including frozen foods, cake mixes, instant coffee, and also orange juice, which used 25 percent of Florida's orange crop in 1950, and 70 percent in 1960.[18] The growth of the Florida citrus juice industry provided a cautionary tale: in the 1940s, thiourea was being used as a preservative for juice oranges, but pharmacologists at the FDA subsequently documented its unexpected toxicity.[19] Beyond issues of safety, the FDA struggled to regulate standards and labeling for "fresh" orange juice, which was often reconstituted and chemically enhanced.[20]

There was also a dramatic increase after World War II in the volume of food chemicals used in agricultural production and crop storage, such as herbicides, rodenticides, insecticides, and other pesticides, especially the highly effective organochloride and organophosphate pesticides. In 1953, there were about fifty different synthetic pesticides available, but by 1964 this number had risen to five hundred.[21] In addition, the volume of pesticide production increased massively, from 100,000 tons worldwide in 1945 to near 1.5 million tons by 1970.[22] The best known organophosphate compound is DDT, but a wide range of other organic insecticides were being used in agriculture, such as dieldrin, aldrin, malathion, and parathion.[23] These compounds are generally toxic, persistent (i.e., do not break down rapidly), and also concentrate as they move up the food chain, due to their solubility in fat.[24] In addition, growth-promoting substances, such as antibiotics and diethylstilbestrol, were routinely added to livestock feed.[25] Residues of many of these chemicals began to be detected in produce, meat, and processed food that reached the market. So long as these residues were well below levels toxic to humans, the benefits of agrochemicals were seen as justifying their use. However, over the mid-twentieth century, conceptions of toxicity and tools for detecting toxins were changing.

In 1940, the food coloring agent "butter yellow" was shown by two independent scientists to be carcinogenic—after it had been used in Germany since the 1870s to tint butter and margarine.[26] This prompted political and legislative action for more stringent control of food additives in Germany.[27] Other reports about the possible carcinogenicity of food additives and chemical residues surfaced through the 1950s in Europe and in the US, alarming consumers and putting both regulatory agencies and industry on the defensive.[28] Most pesticides were known to be poisonous (they are, after all, designed to kill other living organisms), but representatives of both industry and government insisted that the so-called "tolerance" levels specified by national regulatory bodies protected consumers against hazardous pesticide exposures.[29] The so-called "cranberry scare" in the US, when the entire crop of cranberries was recalled by the government just before Thanksgiving of 1959 due to concerns about aminotriazole residues, left the public jittery about farmers' indiscriminate use of pesticides (in this case, an unauthorized use).[30]

After the publication of Rachel Carson's *Silent Spring* in 1962, consumers and environmental groups were increasingly skeptical of industry claims about the safety of chemical residues in food, as well as in the environment.[31] Since the nineteenth century, the food industry had sought to foster public trust in their products as a way to expand markets; by the 1970s, reports of the risks of synthetic chemicals in both processed and fresh foods corroded public confidence. Books such as *Eating May Be Hazardous to Your Health* and *The Chemical Feast* castigated the US government for failing to protect its citizens.[32]

Ultimately, new risks associated with these postwar food chemicals did not simply replace earlier dangers associated with pathogenic germs and toxins, but rather supplemented them.[33] While bacteriological and pasteurization techniques did enable relatively safe mass production and transportation of perishable foods (as did refrigeration), the industrialization of agriculture, including containment methods for livestock production and widespread antibiotic usage, contributed to emerging problems such as antibiotic resistance.[34] In addition to selecting for antibiotic resistance in known pathogenic bacteria, these conditions have produced new human pathogens from previously innocuous bacteria, such as the strain of *Escherichia coli* that carries the Shigella toxin (*E coli*. O157:H7). Bacterial contamination and pathogenicity continued to be problems alongside dangers from a host of new chemical residues.

In addition, reliance on agrochemicals introduced new kinds of environmental damage, in ways that manifested more slowly. The massive decrease in insect populations, including bees crucial to agricultural pollination, is linked to longstanding use of neonicotinoid insecticides.[35] Excess nitrate from fertilizer runoff in waterways has stimulated the overgrowth of algae, leading to vast "dead zones" where oxygen is insufficient for fish.[36] The agricultural herbicide atrazine is a major suspect for widespread reproductive anomalies and population declines observed in wild frogs.[37] The ecological consequences of cheap food illustrate a disconnect between, on the one hand, environmental protection and food safety and, on the other, the different kinds of regulatory science that inform them, to which we now turn.

Shifting Dangers and Regulatory Science

Early regulation of foods often focused on adulteration and quality, although there were also efforts to oversee safety in food related to public health efforts to identify disease agents and toxic substances.[38] In the late nineteenth century, the new tools of bacteriology had brought into view germs in air, water, and food responsible for many human diseases.[39] The contamination of milk with tuberculosis bacilli became a major public health concern in the early twentieth century; Upton Sinclair's 1906 novel *The Jungle* also drew attention to the unhygienic conditions in meat-packing plants.[40] In addition, microbes, now identifiable through bacteriological culturing techniques, were often found to be the cause of food spoilage. For their part, analytical chemists identified adulterants, often suspected as unsafe, as well as poisons (such as arsenic) in food. National food regulation authorities, such as the US Food and Drug Administration (FDA), relied on a staff of bacteriologists and chemists to oversee food safety and quality.[41] Below we focus on the actions of the FDA, which has been especially well studied, though the United States was certainly not the only—or even the first—nation with emerging regulatory oversight of food chemicals.[42]

Over the first half of the twentieth century, government officials and consumers became increasingly concerned about chemicals added to food in manufacturing, starting with bleaching agents in flour and caffeine in Coca-Cola.[43] Until the 1938 Food, Drug and Cosmetic Act, not even drugs needed to be tested for safety in the US.[44] The 1938 act spurred the development of testing regimes in pharmacology and toxicology, the latter of which was also stimulated by the rise of "industrial hygiene," largely concerned with occupational safety in chemical factories (and usually overseen by firms themselves).[45] Animal research was the main mechanism for assessing hazards from exposure to chemicals. Toxicologists developed standardized protocols for rodent tests (such as dose-response curves and determination of the LD-50, the dose toxic for half of an exposed animal population) used in evaluating the safety of chemicals, including food additives, colorants, preservatives, and agricultural chemicals.[46] Even before there were premarket testing requirements in the United States for such substances, the FDA's own pharmacologists began determining the toxicity of food impurities and adulterants, as well as testing drugs.[47]

In general, toxicologists assumed a level below which exposure hazards would be negligible; this threshold model put into practice the longstanding adage "the dose makes the poison."[48] However, this assumption was being called into question by scientists in two different fields. After World War II, national atomic energy agencies sponsored a great deal of genetics research on the effects of ionizing radiation. Geneticists found that radiation induced mutations even at low doses, casting doubt on the long-held assumption that exposures below a threshold were not harmful.[49] A new model of linear nonthreshold damage came to be standard for analyzing and regulating radiation exposure. This model raised questions as well about existing guidelines for low-level exposure to food chemicals, particularly carcinogens.[50] Second, pharmacologist Hermann Druckrey in Germany argued (1) that cancer was always provoked by environmental factors, (2) that there was no safe dose of a carcinogen, and (3) that carcinogens exercised cumulative and irreversible effects. The Druckrey-Küpfmüller equation, a statement of linear proportionality of carcinogenic dose to cancer incidence, became especially important in debates over food additives.[51] So while toxicology tests such as the LD-50 became widely used in screening chemicals, such laboratory experiments could not resolve whether there existed an exposure level for carcinogens low enough to be considered safe, especially for chronic, rather than one-time, exposure.[52]

Not all approaches to identifying exposure risks were laboratory based. After World War II, epidemiology became critical in tracking food-borne illnesses in populations—and, in some cases, discovering them in the first place. If invisible microbial dangers of food often made themselves known to the unwary consumer quickly, many of the newer hazards of food additives and contaminants were long-acting or latent, contributing to cancers or diseases that might appear years after exposure. Population-based studies could be used to identify long-term effects of exposure, although effects had to be very consistent to be detectable by epidemiologists. The multiplicity of scientific approaches to evaluating food safety generated the potential for conflict-ing findings and contested expertise. Was "pure" food defined by the lack of pathogenic microbes or absence of dangerous chemical contaminants? Were laboratory-based animal studies sufficient to identify toxicity and carcinoge-nicity of chemicals, especially at low doses? When epidemiological studies did not support results from toxicology tests about cancer risk, which form of sci-entific knowledge should be trusted?

In the US, congressional hearings in the late 1950s and 1960 about the safety of food additives led to new classifications of colorants by the FDA, as well as the so-called Delaney Amendment in 1958, which banned the use of carcino-genic food additives, exempting pesticides and common food compounds that were "generally recognized as safe."[53] This law reflected public perceptions of natural as healthy and artificial as dangerous, as well as a particular anxiety about residual chemical carcinogens in food, no matter how small the quanti-ties. In this sense the zero-tolerance guideline for additives aimed not only to prevent carcinogenic exposures, but also to reinforce public trust in the safety of processed foods.

In contrast to the Delaney Amendment, US pesticide oversight was not governed by a zero-tolerance principle but by the need to balance costs and benefits of these economic poisons.[54] Similar arguments were made in re-sponse to public worries about residues of hormones and antibiotics given to livestock. In the view of industrial chemists, as well as that of most farmers and regulators, the advantages of chemical technologies in food production outweighed any disadvantages to consumers and wildlife.[55] By the 1970s and 1980s, such trade-offs were generally formulated in terms of risk, as seen in the proliferation of risk assessment in environmental regulation.[56]

Even as agrochemicals became indispensable to the functioning of the late twentieth-century food system, questions remained about the adequacy of regulations in place to safeguard human health and the environment. Beyond the technical complexities in calculating safety standards for food residues was the difficulty of justifying food chemicals to a public that would prefer "pure" food. The safety issues went beyond agricultural processes, as other pollut-ants reached the food supply through the environment. The potential for com-mercial chemical production to contaminate the food supply was made vivid in Japan in 1959, when an ongoing epidemic of neurological disease, with a mortality rate of 35 percent, was finally traced to mercury contamination in seafood from industrial wastewater released into Minimata Bay.[57] Thus, by the 1970s, dangerous residues on food could include microbial contamination, pesticides and other agrochemicals, substances used in food processing and packaging, and pollutants from industry and waste disposal. While the FDA remained the central oversight body for food safety in the US, these problems also implicated the USDA, EPA, and Public Health Service.

In most European nations, analogous regulations emerged within ministerial departments. While not attempting a comprehensive global history, the chapters in this volume address some of the developments in the United Kingdom, France, and Germany. Studies of regulation have often approached the transatlantic divide as a question of delay: continental Europe was accordingly slow at adopting the FDA model of independent regulatory agencies with the consequence that mounting demands for European harmonization of the markets for drugs and food played a critical role in its generalization in the last decades of the twentieth century.[58] This volume amends this scenario in two ways. First, it shows that, from antibiotics to pesticides, issues were often similar, with numerous views on the nature of risks and the acceptable responses crossing the Atlantic.[59] Second, it suggests that the relative weakness of the agency model up to the 1970s was balanced by the strength of another "way of regulating"—that is, the regulation by professional bodies gathering veterinarians, pharmacists, or physicians whose role originated in an ancient delegation of power from the state to the professions with the granting of sales' monopolies and mandatory prescription as its core ingredients.[60]

International agencies, such as the UN's Food and Agriculture Organization or the World Health Organization, were also introduced during the period covered. The volume thus pays attention to international efforts at oversight of the food supply in what was then called the Third World. A rapidly expanding international food trade implied that similar risks could become matters of concerns in Europe and Africa or Asia, especially when exports from the latter to the former were at stake. However, the ways in which problems were framed and the responses designed and implemented rarely converged.[61] Inhabitants of the developed and developing world experienced somewhat different food risks—such as acute food scarcity in many parts of the global South as opposed to the overabundance of calorie-rich and nutrient-poor processed foods in many areas of the world—even if as the twentieth century came to a close most humans came to encounter food markets shaped by industrial capital and processing technologies. As our contributors show, this new reality did not lead to universal experiences; depending on which side of the great North/South divide one happened to live, available foods and safety regulations differed greatly.[62]

The Book

Our first section, "Objectifying Dangers," focuses on how certain scientific developments alongside the industrialization of food produced new kinds of expertise and perceptions of risk. These chapters focus on the changing tools and models that researchers used to assess the health hazards of exposure to low-level contaminants in food. They show how new knowledge challenges settled methods and assumptions for certifying safety in foods. Our authors focus on the detection of pathogenic bacteria, radioactivity, and carcinogens in food, hazards that caused widespread public concern as well as professional scrutiny—and areas bedeviled by critical scientific uncertainties.

As Anne Hardy shows in "Salad Days: The Science and Medicine of Bad Greens, 1870–2000," the popularity of watercress as a "wholesome breakfast salad" in Victorian England caused alarm among astute physicians. The streams necessary for growing watercress were often contaminated by sewage, so the greens could carry pathogens from human waste right back to the table. By the early 1900s, bacteriological methods confirmed the role of contaminated watercress in outbreaks of typhoid. Hardy notes that what we now call food poisoning became identifiable as such because enteric diseases such as cholera and typhoid declined in the early twentieth century (in the industrialized West). Moreover, the growing consumption of fresh raw vegetables over the twentieth-century, large-scale agricultural methods (including chlorine dipping of bagged greens), and increasing distance between production and market have provided new opportunities for the growth of pathogenic bacteria, including not only the well-known *Salmonella* Typhimurium and norovirus but also novel threats, such as the Shiga-toxin-producing strain of the usually harmless enteric bacterium *Escherichia coli* (*E. coli* O157:H7). Contemporary monitoring systems combine public health reporting with molecular genetic methods of typing to try to keep pace with, and contain, these new risks. Consequently, the tracking of pathogenic microbes in food now follows the genetic fingerprints of isolates, not only their species and strain.

Soraya de Chadarevian reminds us that it was the atomic age that first illustrated the potential for global contamination problems through the dispersion of fallout from atomic weapons testing. Radioactivity in the environment could make its way into the human food chain when contamination from weapons testing was taken up by fish in the ocean or by grazing cattle. Strontium-90 was one of the by-products of atomic fission that raised particular health concerns as it is chemically similar to calcium and is thus a "bone-seeker" in mammals. Moreover, it poses a long-term threat once in the body, given its half-life of over twenty-five years. In 1962, the British atomic energy plant Harwell conducted a thirty-day human experiment in which volunteer scientists were fed milk and beef contaminated with strontium isotopes. If this experiment, widely reported in the British media, was aimed at quelling apprehension about a radioactive diet, it did not succeed. Officials began monitoring milk for radioactive contamination and even working on possible methods to decontaminate affected milk. In the end, concerns about radioactive contamination making its way from the environment into food were soon extended to industrial pollutants, especially synthetic chemicals.

Looking at the postwar German debates on food additives and their putative carcinogenic potency, Heiko Stoff highlights one turning point in the assessment of cancer risk—namely, the increasing tensions between academic experts, who insisted that there was no threshold below which absolutely safe conditions existed for carcinogens, and regulatory agencies, who argued that such thresholds were necessary in order to accommodate existing food production, avoid banning too many substances, and mitigate burdensome regulation on industry. The compromises thus crafted were the "acceptable daily intake" numbers written into the 1958 German food law. In contrast to the Delaney Amendment, the German regulatory regime admitted that zero exposure to potentially hazardous food chemicals was not feasible. Yet ideals of purity remained salient to German debates on food, which in the postwar period perpetuated discourses around "poison" that had been applied to the social body during the period of National Socialism.

Complicating the growth of knowledge about food contaminants and their hazards was a growing awareness that even the natural constituents of food may not be harmless.[63] As Angela Creager's chapter details, tools developed to detect hazardous environmental chemicals were turned on dietary plants, fruits, and meats, both raw and cooked—with striking results. After developing a quick laboratory test for mutagenic chemicals (as a screen for potential carcinogens), biochemist Bruce Ames found that many foods and beverages tested positive. Epidemiologists such as Richard Doll were also analyzing the contribution of diet to human cancer, often by looking at disparities in the incidence of particular cancers (e.g., breast, stomach) across continent or country. In an influential 1981 review article, Doll and Richard Peto estimated that 35 percent of human cancer was attributable to diet. Efforts to identify which constituents of food were carcinogenic, either inherently or due to cooking processes, led to countless publications and stories in the media. Results were often contested, but behind the debate was a growing consensus that diet, including the chemical composition of foodstuffs and cooking methods, was an important aspect of environmental exposure. This chapter connects the risk from food chemicals to the confusing, ever-changing medical literature (and media coverage) on diet and disease.

Also following the line of hazards in natural foods, Lucas Mueller addresses research into one of the most potent carcinogens known, the mold-produced aflatoxin. In the early 1960s, British researchers discovered that the deaths of more than a hundred thousand turkeys were due to aflatoxin-contaminated animal feed.[64] Soon thereafter, aflatoxin ingestion was found to be associated with liver cancer in human populations, especially in Africa. Although aflatoxin is a "natural" carcinogen, its presence in human food (as well as livestock feed) is a by-product of large-scale agricultural production and crop storage. As Mueller demonstrates, the food safety controls in the first world, where aflatoxin-contaminated crops were routinely destroyed, proved too politically costly to implement in the developing world. Public health concerns about mold contamination were countered by issues of nutritional needs (especially for protein) in postcolonial nations of the global South. These considerations impacted the calculations of acceptable contamination levels, which now appear to have carried their own health costs for children. Aflatoxin thus poses a conundrum for food regulators, pitting the struggle against food scarcity directly against health safeguards.[65]

Aurélien Féron's chapter analyzes the problems that have arisen from the family of toxic chemicals called polychlorinated biphenyls, or PCBs. These synthetic substances were produced in massive quantities since the 1930s and used in a variety of industrial and consumer products, from insulating fluids for capacitors and electrical transformers to carbonless copy paper. It has now become a nearly ubiquitous global environmental contaminant, found in wildlife from remote corners of the earth as well as in human populations. Its toxicity, long observed in the realm of occupational health, was publicly demonstrated in 1968, when PCB-contaminated rice oil in Japan poisoned approximately 1,600 people, killing five. As this example illustrates, dietary exposure is the way in which PCBs most threaten people, in part because the compound is soluble in fat, bioaccumulates (concentrates in living organisms), and magnifies up the food chain. Although national and transnational polices have been implemented since the 1970s toward the phasing-out of the manufacture, use, and disposal of PCBs, contamination continues to be a serious problem. In France, fishing remains prohibited in many bays and rivers because the concentrations of PCBs in the fish exceed the thresholds set for human consumption. Féron shows that the existing frameworks designed to bring pollution levels under control have not been efficient enough to fulfill food safety requirements that motivated these regulations. PCBs, as he demonstrates, are a recalcitrant problem for both social and material reasons.

The second set of chapters, "Ordering Risks," examines how industry, government officials, and consumers have understood the costs and benefits of agricultural and food-processing chemicals. National government agencies, as well as intergovernmental organizations (such as the UN's Food and Agriculture Organization), wrestled with regulatory regimes that would reduce chemical dangers without damaging food production—or the powerful agricultural and chemical industries behind them—but the result was the complex and multilayered set of laws, agencies, and safety systems that often protect business interests more than consumer health or the environment.

We begin this section with the use of growth-promoting substances in industrial meat production. Arsenicals, hormones, and antibiotics were widely used in livestock feed before their potential hazards to consumers were recognized. However, the thin line between growth-promoting and therapeutic uses of these substances in massive livestock containment lots complicated efforts at effective regulation. In "Trace Amounts at Industrial Scale: Arsenicals and Medicated Feed in the Production of the 'Western Diet,'" Hannah Landecker examines the rise of medicated animal feed and its close connection to the chemical industry. While the (over)use of diethylstilbestrol and antibiotics in livestock farming are commonly acknowledged, the first growth-promoting chemicals to be used in commercial animal feed were arsenic-based compounds. Organoarsenicals were first explored for industrial farming as a way to control diseases common in large chicken lots, but their surprising growth-promoting properties soon took center stage. By the mid-1940s they became a standard and widely advertised ingredient in chicken feed. Landecker shows how the attempt to chemically define and control growth, both in animals and humans, reflected a new understanding of animal economy, one that was literally fed by industrial production of chemicals. Inconsistent evidence that arsenicals were carcinogenic led to the relative neglect of this concern until the late-twentieth century, when the metabolism of arsenic in mammals was more thoroughly investigated. In turn, new conceptions of food and metabolism have shown the industrial framework that still undergirds livestock agriculture to be dangerously misguided, and, in the meantime, environmental arsenic pollution has also emerged as a major concern—one that has literally changed the living world and health risks for both humans and nonhumans.

Claas Kirchhelle's chapter examines why antibiotics proved so difficult to manage in the United States, focusing on how public health, agriculture, and drug regulation are separated into different government agencies. Introduced on an industrial scale to US agriculture in 1949, antibiotics were soon routinely given to animals to boost weight gain and combat disease, sprayed on plants, and used to preserve fish and poultry. Concerned scientists and consumers soon accused livestock farmers of leaving hazardous residues in food and of selecting for antimicrobial resistance (AMR). Regulators struggled to reassure critics. Traditionally, US drug regulation had been geared to regulate substances at the point of licensing. After licensing, it was nearly impossible to control drug use by lay users like farmers. The chapter reveals how officials in the FDA and the Centers for Disease Control (CDC) attempted to reconcile new risk scenarios such as horizontal AMR proliferation with classic regulatory protocols centering on the establishment of thresholds for hazardous substances and the containment of bacterial organisms. Kirchhelle argues that the US regulatory system's focus on proof of harm and preoccupation with toxic and carcinogenic substances repeatedly impaired its ability to recognize emerging nontraditional threats like AMR and to control food-related consumption of antibiotics, which remains unabated.

The remaining three chapters in this section further explore the conflicting demands made on the regulatory systems for food safety. Jean-Paul Gaudillière's contribution provides a fresh look at the "diethylstilbestrol and meat affair," which in the early 1970s focused on the carcinogenic effects of this analog of estrogen, especially its uses as a growth enhancer in agriculture. Concerns about residues of this synthetic hormone in meat led to important technical debates over the possibility of carcinogenicity at low doses, the presence of residues in consumed products, and the benefits of increased productivity. Using the records of the trial, in which the FDA—which had banned agricultural uses of DES in beef in 1973—had to defend its rule in the face of a challenge from industry, the chapter focuses on the influence corporations exerted on the production of knowledge and the process of expertise—that is, the conjunction of commissioned research, assessment of the published literature, and political action before and during the controversy. Discussing the merits and limitations of three main categories (conflict of interest, ignorance, and capture) that historians and social scientists have used for understanding influence, the chapter concludes with a plea for including *hegemony* in the palette.

In their chapter, Maricel Maffini and Sarah Vogel look back at the FDA's Delaney clause, which (as mentioned above) banned the addition to food of any substance that had proven carcinogenic in animals and/or humans. This decision stands out as exceptionally precautionary in a country whose regulatory frameworks more often accommodate industry, especially in the realm of agriculture. Why was this rule enacted and how did it last so long before being revamped in the 1990s? The authors explain that the exemption of "generally recognized as safe" (so-called GRAS) substances, which included many chemicals already used by the food industry, limited the actual effects of the law, and the slow pace of animal testing for carcinogenicity meant there were few additions to the list of banned chemicals. Thus even as Congress was credited with redressing known inadequacies of the 1938 food law, the GRAS clause created a self-certification mechanism that effectively enabled the food industry to by-pass agency oversight in most cases. In addition, the law provided no directives or incentives for academic or government scientists to engage in laboratory testing that would enable scrutiny of carcinogenicity data provided by industry. Such structural asymmetry is more often the rule than the exception and, in this case, explains why the number of chemicals tested and evaluated actually declined after the 1950s.

Xaq Frohlich examines another aspect of the FDA's regulation, namely the way in which the agency has created and maintained a distinction between food and drugs. As the regulation of drugs shifted from safety and standard dosages to utility, as seen in the 1962 reform and the introduction of clinical trials, the regulation of food not only remained solely a matter of safety but also became increasingly rooted in a paradigm of consumers' individual informed choices. Yet this distinction often breaks down in practice, both for consumers and producers, as illustrated by the growing "functional foods" market or the nutritional supplements industry. While scholars often point to examples of inadequate standards for food safety based on out-of-date science, Frohlich shows that regulation can be a source of innovation. The comparison between foods and drugs illuminates another point that is reflected by many other chapters. From the 1930s through the 1960s, manufacturers and FDA regulators generally tried to emphasize the inherent differences between food and drugs, drawing on the commonplace notion that food is familiar and self-evident in contrast to pharmacologically engineered drugs, whose manufacture is more complex. In fact, however, chemical technology is central to both the food and drug industries, which also share many toxicological practices and assumptions. Yet their regulatory regimes remain distinct. For one thing, whereas drugs must be shown to be efficacious as well as safe in order to enter the market, the nutritional value of food is not relevant to whether it can be sold, but only to how it can be labeled.

Legal differences between drug regulation and oversight of food production clearly guide industrial testing regimes. The burden of proof for the safety and efficacy of drugs is on industry, even as there are firm guidelines for what constitutes acceptable data. The pharmaceutical industry often complains that this burden, along with the expense of new research, is what keeps prices of prescription drugs so high. By contrast, firms in the agro-industry do not have to integrate (or even know about) health-related "use value" in their evaluation of products and their market potential. Safety matters, but the state's priority in assuring inexpensive foodstuffs means that certain levels of pesticides (still termed "economic poisons" in law) and other food chemicals are tolerated. When it comes to regulatory mechanisms, this translates into the infeasibility of instituting marketing authorization procedures for products from the food industry that would require data comparable to that produced by clinical trials—that is, involving medical scientists who are looking for health-related outcomes. It also perpetuates the fiction that food is "natural" and "traditional" while drugs are innovative. Additives have been the exception that proves the rule in terms of safety testing.

Taken together, the chapters in this section document two aspects of food regulation that might seem contradictory. On the one hand, we see the relative openness of debates, the permanent contestation of the regulatory proposals by industry, and the contingency of the legislation and rules adopted at some point. On the other hand are the pervading, almost permanent, structural asymmetries that were built in the regulatory machine accounting for the very limited impact new scientific knowledge has had on the practices of food production and food processing.

Considering this section of the book alongside the essays in part I leads to two further observations. First, many of the chemicals of concern were introduced to address other food risks, such as microbial contamination and crop failures. In this sense, chemical food risks arose in part due to the systematic efforts to make agriculture as economically productive as possible, and in part as a by-product of technological solutions to ensure sanitary food. Scientific and economic solutions to some kinds of risk—safety and scarcity—have, in turn, generated new risks of their own. Second, consequent to the massive reliance on agrochemicals in the mid-twentieth century, there have been significant changes in the technical knowledge and methods used to assess exposure risk, such as in the field of toxicology. This means that earlier standards for safety were based on what is now regarded as obsolete science—and in many cases, these older guidelines persist in the regulatory framework. Such technical obsolescence is a recurrent feature of high-tech societies, which must rely on the state of knowledge at a given moment for oversight of food safety. Yet this observation should not lead us to conclude that problems with existing regulatory regimes derive solely from scientific and technical inadequacies. To put it another way, the perceived "lag" between law and science draws attention away from the actual aims of existing regulation, including mollifying public distrust, providing industry with reliable standards and review processes, and not interfering with economic growth or innovation. Indeed, the food sector is a major market for the chemical industry, and safety standards generally reinforce mass production and heavy reliance on agrochemicals.

Conclusions

As a whole, *Risk on the Table* speaks to three important issues in the social science and environmental studies literature. First, as the title suggests, the book is in dialogue with a literature on risk that goes back to the 1986 publication of Ulrich Beck's *Risk Society*.[66] Pointing to risks in the nuclear industry as emblematic, Beck depicts technology's dangers as outpacing the ability of experts to control them. The history of food risk is in some ways a confirmation of Beck's pessimism. The turn to agro- and food chemicals in the twentieth century was aimed at increasing agricultural productivity and protecting food from spoilage and pathogens (as well as pests more generally). As we have highlighted, residues of these chemicals in food turned out to have their own hazards, and large-scale containment facilities for livestock have contributed to a resurgence of pathogens in food as well as antibiotic resistance in the clinic. That said, the cultural interpretation of risk by Mary Douglas is also especially salient for food.[67] Douglas argued that risk is a matter of cultural perception, not simply technological hazards. Notions of purity in food are deeply cultural, from bacteria-free milk to wholesome pesticide-free produce. Part of what has made consumers so suspicious of food chemicals is the perception that they are contaminating, impure. In the case of pre–World War II Germany, these cultural perceptions around food purity overlapped with ideals of a "pure" body politic, reinforcing anti-Semitism and the horrific policies of National Socialism. More recent politics of food purity have led to green consumerism without addressing the continuing massive reliance on agrochemicals in most food production. In sum, cultural frames about risk and purity complicate any simple critique of technology as endangering the food supply.[68]

Second, we ask what we can learn by comparing food and drugs when it comes to mass production and regulation. The social worlds of food and drugs share many features. Both were once in the realm of small-scale craft production before being massively industrialized, which has given marketing a central role in the making of value; both involve not only engineers and production plants but also older professionals with their own training and organization; and both are research-intensive domains giving the chemical sciences a fundamental place in the innovation machinery. Yet historiographies of food and drugs in the twentieth century have—in spite of a common interest by business historians—developed in rather different directions. The recent historiography of drugs has been dominated by three issues: the "molecularization" of therapeutic agents after World War II and its relationship with the rise of biomedicine; regulation and its relationship to clinical practice with the emergence of statistically based clinical trials as a gate-keeping mechanism; and the changing relationship between firms and physicians and the former's role in framing prescriptions and uses. By contrast, until recently the historiography of food placed less emphasis on the industries, focusing on farmers, local systems of production and their ecology, the diversity of their knowledge basis, and the inequalities associated with international trade and mass processing.[69] As a consequence, its scholarship often overlaps with environmental history—for instance, in studies of traditional food products, from rice to cheese.[70] By looking at food risks, the science used to define them, and associated regulations, this volume opens a path for a more systematic engagement between histories of food and drugs. As argued above, the comparison helps us understand the constraints that shape their regulatory patterns. Rather than offering a straightforward comparison, our collection interrogates the existence of the divide between food and drugs, shedding light on ways in which the boundary has been both crossed and (re)constructed during the past century.

Third, the meaning of "environment" is very much at stake in debates over food risk. For the last quarter century, environmental historians have been challenging scholars to rethink what they mean by "nature." Among others, William Cronon showed "wilderness" to be a romantic, sometimes nationalistic construct, and there is now a substantial literature on urban environments.[71] Our collection reinforces an observation made in recent scholarship that the distinction between "natural" and "artificial" is unstable and often symbolic.[72]

Even the whole foods that arrive on our tables are the products of complex technological systems. We worry about the hazards of added and contaminating chemicals in the food supply, even as researchers document the "natural" hazards of foods and traditional preparation techniques.[73] Moreover, standards of food safety tend to be based on laboratory studies dealing with single substances, pure chemicals, and controlled doses. By contrast, food consists of complex mixtures, whose nutritional content seems nearly irreducible to known constituents. Perhaps most significantly, eating is a major way in which we are exposed to our environment—by making it part of us. This collection helps us rethink longstanding ontologies of natural/artificial, pure/mixture, and outside/inside in an attempt to see food—its healthiness and hazards—in new ways.

Angela N. H. Creager is the Thomas M. Siebel Professor in the History of Science at Princeton University, where she specializes in history of biology and biomedicine. She has published two books with University of Chicago Press, *The Life of a Virus: Tobacco Mosaic Virus as an Experimental Model, 1930–1965* (2002) and *Life Atomic: A History of Radioisotopes in Science and Medicine* (2013), which was awarded the Patrick Suppes Prize in the History of Science by the American Philosophical Society. Her current work focuses on the role of genetic tests in environmental science and regulation.

Jean-Paul Gaudillière is a historian at Cermes3 and professor at École des Hautes Études en Sciences Sociales, Paris. His research explores the history of the life sciences and medicine during the twentieth century. His recent work focuses on the history of pharmaceutical innovation and the uses of drugs on the one hand, and the dynamics of health globalization after World War II on the other. He is coordinator of the European Research Council project From International to Global: Knowledge, Diseases, and the Post-war Government of Health.

Notes

1. Foucart and Horel, "Le Bisphénol A"; Foucart and Horel, "Perturbateurs endocriniens."
2. European Chemicals Agency, "MSC Unanimously Agrees"; Vogel, *Is It Safe?* On the longer history of endocrine disruptors, see also Langston, *Toxic Bodies.*
3. After the events mentioned above and a negative vote on the proposed definition by the European Parliament, the European Commission revised its position, and the amendment Germany had proposed was deleted, thus clearing the way for a qualified vote on endocrine disruptors identification criteria, which took place in December 2017. Most scientists nonetheless consider the definition passed to be very restrictive and drastically limiting the number of chemicals falling under the REACH-based regulation of endocrine disruptors.

4. Proctor and Schiebinger, *Agnotology*; Oreskes and Conway, *Merchants of Doubt*; Nestle, *Unsavory Truth*.

5. For the tobacco story: Michaels, *Doubt Is Their Product*; Proctor, *Golden Holocaust*, who questions the voluntary nature of tobacco use for regular smokers.

6. As compared with earlier work (e.g., Levenstein's more comprehensive *Fear of Food*), we focus on worrisome contaminants associated with industrialization and agrochemicals in the twentieth century, as risk became a key regulatory category.

7. Petrick, "Industrialization of Food"; Zeide, *Canned*.

8. Chandler, *Visible Hand*, especially 287–314.

9. Stoll, *Fruits of Natural Advantage*; Fitzgerald, *Every Farm*; Harwood, *Europe's Green Revolution*.

10. Whorton, *Before Silent Spring*. In 1910, an Insecticide Act in the US was passed that set labeling standards for insecticides. But the board that enforced this act was situated within the USDA, whose mission was to promote American agriculture, including use of so-called economic poisons. Davis, *Banned*, 5–6.

11. Hamilton, "Analyzing Commodity Chains," 23.

12. Petrick, "Ambivalent Diet"; Petrick, "Like Ribbons."

13. White, "Chemogastric Revolution." She borrows the term from historian James Harvey Young, "Oral History of the U.S. Food and Drug Administration: Pharmacology, 1980," transcript, History of Medicine Manuscripts Division, National Library of Medicine, Bethesda, MD, on 123.

14. Freidberg, *Fresh*; Rees, *Refrigeration Nation*.

15. White, "Chemistry and Controversy," 225.

16. Ibid., 227; Levenstein, *Paradox of Plenty*, 109.

17. Hisano, *Visualizing Taste*.

18. White, "Chemistry and Controversy," 223.

19. Ibid., 230.

20. Hamilton, *Squeezed*.

21. Jas, "Public Health and Pesticide Regulation," 377.

22. Tilman et al., "Agricultural Sustainability."

23. Dunlap, *DDT*; Nash, *Inescapable Ecologies*, 127–69.

24. McGinn, "POPs Culture."

25. Marcus, *Cancer from Beef*; Landecker, "Metabolic History."

26. Schwerin, "Vom Gift im Essen"; Stoff, "Zur Kritik der Chemisierung"; Stoff, *Gift in der Nahrung*. For nineteenth-century uses of dyes from the chemical industry in food production, see Cobbold, "Introduction of Chemical Dyes"; for the growing use of food colors in the twentieth-century US, and their regulation, see Hisano, *Visualizing Taste*.

27. Stoff, *Gift in der Nahrung*.

28. Gaudillière, "Food, Drug and Consumer Regulation."

29. Wargo, *Our Children's Toxic Legacy*.

30. White, "Chemistry and Controversy," 362–409; Bosso, *Pesticides and Politics*.

31. Dunlap, *DDT*; Bosso, *Pesticides and Politics*; Marcus, *Cancer from Beef*.

32. Verrett and Carpet, *Eating May Be Hazardous*; Turner, *Chemical Feast*; see also Wellford, *Sowing the Wind*.

33. Hardy, *Salmonella Infections*.

34. Podolsky, *Antibiotic Era*; Kirchhelle, "Pharming Animals"; Kirchhelle, *Pyrrhic Progress*.

35. Suryanarayanan and Kleinman, *Vanishing Bees*.

36. Biello, "Fertilizer Runoff."
37. Helgen, *Peril in the Ponds*.
38. Wilson, *Swindled*; French and Phillips, *Cheated Not Poisoned*.
39. Sturdy and Cooter, "Science, Scientific Management."
40. Smith-Howard, *Pure and Modern Milk*; Young, *Pure Food*; Cohen, *Pure Adulteration*; Sinclair, *The Jungle*. As is often noted, Sinclair's book was meant to draw attention to the appalling conditions for stockyard and meat packaging workers (in the name of promoting socialism), but it was the unsanitary food that horrified the public.
41. Young, *Pure Food*; White, "Chemistry and Controversy"; Cohen, *Pure Adulteration*.
42. E.g., French and Phillips, *Cheated Not Poisoned*; Jas, "Public Health and Pesticide Regulation"; Ramsingh, "History of International Food Safety Standards."
43. White, "Chemistry and Controversy."
44. Parascandola, "Historical Perspectives," 88.
45. Davis, *Banned*; Christopher C. Sellers, *Hazards of the Job*.
46. Lehman et al., "Procedures."
47. White, "Chemistry and Controversy," 194; Davis, *Banned*, 27–28. In 1947, federal pesticides regulation was strengthened to require premarket registration and authorization of fungicides, insecticides, and rodenticides. However, the US Department of Agriculture, which administered the law, was primarily committed to protecting farmers rather than consumers or wildlife. Bosso, *Pesticides and Politics*, 45–60.
48. Grandjean, "Paracelsus Revisited."
49. Creager, "Radiation, Cancer, and Mutation"; Schwerin, "Low Dose Intoxication"; Schwerin, "Vom Gift im Essen."
50. Boudia, "From Threshold to Risk."
51. Druckrey developed his theory in 1948 with Karl Küpfmüller while they were in an internment camp. See Stoff, this volume; Stoff, *Gift in der Nahrung*; Wunderlich, "Zur Entstehungsgeschichte"; Wunderlich, "'Mit Papier, Bleistift und Rechenschieber.'" For evidence of how Druckrey's views influenced discussions of carcinogens in food additives, see International Union Against Cancer, "Report of Symposium."
52. Pharmacologists at the FDA from the late 1930s through the 1950s led the development of toxicity testing, especially for chronic effects. White, "Chemistry and Controversy," 211–14; Davis, *Banned*.
53. National Research Council, *Regulating Pesticides in Food*; Maffini and Vogel, this volume.
54. Bosco, *Pesticides and Politics*.
55. See, for example, this book by a Dow chemist, whose rhetorical question he answers affirmatively: Barron, *Are Pesticides Really Necessary?*
56. Boudia, "From Threshold to Risk"; Demortain, *Science of Bureaucracy*.
57. Walker, *Toxic Archipelago*, 137–75.
58. Smith and Phillips, *Food, Science, Policy and Regulation*; Vogel, *The Politics of Precaution*; Daemmrich, *Pharmacopolitics*; Demortain, *Science of Bureaucracy*.
59. See also Kirchhelle, *Pyrrhic Progress*; Kirchhelle, "Toxic Confusion"; Stoff and von Schwerin, "Eine Geschichte gefährlicher Dinge"; Stoff, *Gift in der Nahrung*; Jas, "Adapting to Reality."
60. See also Gaudillière and Hess, *Ways of Regulating Drugs*; Anderson, "Drug Regulation and the Welfare State"; Corley and Godley, "The Veterinary Medicines Industry"; Jas,

"Public Health and Pesticides Regulation"; Bonnaud and Fortané, "L'état sanitaire de la profession vétérinaire."

61. Staples, *The Birth of Development*; Winickoff and Bushey, "Science and Power in Global Food Regulation"; Cornilleau and Joly, "La revolution verte."
62. See especially the chapter by Mueller, "Risk on the Negotiating Table."
63. The potential hazards of food's natural constituents also became visible in connection with food allergies, which generated intense medical and social controversy. Smith, *Another Person's Poison.*
64. The outbreak was referred to as "Turkey X," though it quickly became clear that it affected other animals and was responsible for trout hepatoma. Goldblatt, *Aflatoxin*; Linsell and Peers, "Field Studies."
65. Aflatoxin posed a conundrum for mold classification as well, since the toxin-producing strain is so similar to that used by Japanese to ferment foods. Lee, "Wild Toxicity."
66. Beck, *Risikogesellschaft*, translated as *Risk Society.* For a useful guide, see Boudia and Jas, "Introduction."
67. Douglas and Wildavsky, *Risk and Culture*; Douglas, *Risk and Blame.*
68. Along related lines: Shotwell, *Against Purity*; Tsing, *Mushroom at the End of the World.*
69. Excellent examples of the recent focus on food industry are Cohen, *Pure Adulteration*; Hamilton, *Squeezed*; Hisano, *Visualizing Taste*; Zeide, *Canned.*
70. Ceccarelli, Grandi, and Magagnoli, *Typicality in History*; Paxson, *Life of Cheese.*
71. For example, Cronon, *Uncommon Ground*; Ammon, *Bulldozer.*
72. Berenstein, "Making a Global Sensation"; Levinovitz, *Natural.*
73. Nash, *Inescapable Ecologies.*

Bibliography

Ammon, Francesca Russello. *Bulldozer: Demolition and Clearance of the Postwar Landscape.* New Haven: Yale University Press, 2016.

Anderson, Stuart. "Drug Regulation and the Welfare State: Government, the Pharmaceutical Industry and the Health Professions in Great Britain, 1940–80." In *Medicine, the Market and the Mass Media: Producing Health in the Twentieth Century*, edited by Virginia Berridge and Kelly Loughlin, 192–217. London: Routledge, 2005.

Barron, Keith C. *Are Pesticides Really Necessary?* Chicago: Regnery Gateway, 1981.

Beck, Ulrich. *Risikogesellschaft auf dem Weg in eine andere Moderne.* Frankfurt am Main: Suhrkamp, 1986.

———. *Risk Society: Towards a New Modernity.* Translated by Mark Ritter. London: Sage, 1992.

Belasco, Warren J. *Appetite for Change: How the Counterculture Took on the Food Industry, 1966–1988.* New York: Pantheon, 1989.

Belasco, Warren, and Roger Horowitz, eds. *Food Chains: From Farmyard to Shopping Cart.* Philadelphia: University of Pennsylvania Press, 2009.

Berenstein, Nadia. "Making a Global Sensation: Vanilla Flavor, Synthetic Chemistry, and the Meanings of Purity." *History of Science* 54 (2016): 399–424.

Biello, David. "Fertilizer Runoff Overwhelms Streams and Rivers—Creating Vast 'Dead Zones.'" *Scientific American* (online), 14 March 2008. Accessed 26 August 2020, https://www.scientificamerican.com/article/fertilizer-runoff-overwhelms-streams/.

Bonnaud, Laure, and Nicolas Fortané. "L'état sanitaire de la profession vétérinaire. Action publique et régulation de l'activité professionnelle." *Sociologie* 9 (2018): 253–68.

Bosso, Christopher. *Pesticides and Politics: The Life Cycle of a Public Issue*. Pittsburgh: University of Pittsburgh Press, 1987.

Boudia, Soraya. "From Threshold to Risk: Exposure to Low Doses of Radiation and Its Effects on Toxicants Regulation." In *Toxicants, Health and Regulation since 1945*, edited by Soraya Boudia and Nathalie Jas, 71–88. London: Pickering & Chatto, 2013.

Boudia, Soraya, and Nathalie Jas. "Introduction: Risk and 'Risk Society' in Historical Perspective." *History and Technology* 23 (2007): 317–31.

Carpenter, Daniel. *Reputation and Power: Organizational Image and Pharmaceutical Regulation at the FDA*. Princeton: Princeton University Press, 2010.

Ceccarelli, Giovanni, Alberto Grandi, and Stefano Magagnoli, eds. *Typicality in History: Tradition, Innovation, and Terroir*. European Food Issues no. 4. Brussels: Peter Lang, 2013.

Chandler, Alfred D., Jr. *The Visible Hand: The Managerial Revolution in American Business*. Cambridge: Belknap/Harvard University Press, 1977.

Cobbold, Carolyn. "The Introduction of Chemical Dyes into Food in the Nineteenth Century." *Osiris* 35 (2020): 142–61.

Cohen, Benjamin R. *Pure Adulteration: Cheating on Nature in the Age of Manufactured Food*. Chicago: University of Chicago Press, 2019.

Corley, Tony A. B., and Andrew Godley. "The Veterinary Medicines Industry in Britain in the Twentieth Century." *Economic History Review* 64 (2011): 832–54.

Cornilleau, Lise, and Pierre-Benoît Joly. "La révolution verte, un instrument de gouvernement de la 'faim dans le monde.' Une histoire de la recherche agronomique international." In *Le Gouvernement des technosciences. Gouverner le progrès et ses dégâts*, edited by Pestre Dominique, 171–201. Paris: La Découverte, 2014.

Creager, Angela N. H. "Radiation, Cancer, and Mutation in the Atomic Age." *Historical Studies in the Natural Sciences* 45 (2015): 14–48.

Cronon, William, ed. *Uncommon Ground: Toward Reinventing Nature*. New York: W. W. Norton, 1995.

Daemmrich, Arthur A. *Pharmacopolitics: Drug Regulation in the United States and Germany*. Chapel Hill: University of North Carolina Press, 2004.

Davis, Frederick Rowe. *Banned: A History of Pesticides and the Science of Toxicology*. New Haven: Yale University Press, 2014.

Demortain, David. *The Science of Bureaucracy: Risk Decision-Making and the US Environmental Protection Agency*. Cambridge: MIT Press, 2020.

Douglas, Mary. *Risk and Blame: Essays in Cultural Theory*. London: Routledge, 1992.

Douglas, Mary, and Aaron Wildavsky. *Risk and Culture: An Essay on the Selection of Technological and Environmental Dangers*. Berkeley: University of California Press, 1983.

Dunlap, Thomas. *DDT: Scientists, Citizens, and Public Policy*. Princeton: Princeton University Press, 1981.

European Chemicals Agency. "MSC Unanimously Agrees That Bisphenol A Is an Endocrine Disruptor." ECHA, published 16 June 2017. Accessed 26 August 2020, https://echa.europa.eu/-/msc-unanimously-agrees-that-bisphenol-a-is-an-endocrine-disruptor.

Fitzgerald, Deborah. *Every Farm a Factory: The Industrial Ideal in American Agriculture*. New Haven: Yale University Press, 2003.

Foucart, Stéphane, and Stéphane Horel. "Le Bisphénol A considéré 'extrêmement préoccupant' par l'Europe." *Le Monde*, 16 June 2017. Accessed 20 August 2020, https://www.lemonde.fr/planete/article/2017/06/16/le-bisphenol-a-classe-extremement-preoccupant-par-l-europe_5145840_3244.html.

———. "Perturbateurs endocriniens : les scientifiques alertent sur le laxisme de Bruxelles." *Le Monde*, 17 June 2017. Accessed 20 August 2020, https://www.lemonde.fr/planete/article/2017/06/17/perturbateurs-endocriniens-les-scientifiques-alertent-sur-le-laxisme-de-bruxelles_5146113_3244.html.

Freidberg, Susanne. *Fresh: A Perishable History*. Cambridge: Belknap/Harvard University Press, 2009.

French, Michael, and Jim Phillips. *Cheated Not Poisoned? Food Regulation in the United Kingdom, 1875–1938*. Manchester: Manchester University Press, 2000.

Gaudillière, Jean-Paul. "Food, Drug and Consumer Regulation: The 'Meat, DES and Cancer' Debates in the United States." In *Meat, Medicine, and Human Health in the Twentieth Century*, edited by David Cantor, Christian Bonah, and Matthias Dörries, 179–202. London: Pickering & Chatto, 2010.

Gaudillière, Jean-Paul, and Volker Hess, eds. *Ways of Regulating Drugs in the 19th and 20th Centuries*. Basingstoke: Palgrave Macmillan, 2013.

Goldblatt, Leo A., ed. *Aflatoxin: Scientific Background, Control, and Implications*. New York: Academic Press, 1969.

Grandjean, Philippe. "Paracelsus Revisited: The Dose Concept in a Complex World." *Basic & Clinical Pharmacology & Toxicology* 119 (2016): 126–32.

Hamilton, Alissa. *Squeezed: What You Don't Know about Orange Juice*. New Haven: Yale University Press, 2009.

Hamilton, Shane. "Analyzing Commodity Chains: Linkages or Restraints?" In *Food Chains: From Farmyard to Shopping Cart*, edited by Warren Belaso and Roger Horowitz, 16–25. Philadelphia: University of Pennsylvania Press, 2009.

Hardy, Anne. *Salmonella Infections, Networks of Knowledge, and Public Health in Britain, 1880–1975*. Oxford: Oxford University Press, 2015.

Harwood, Jonathan. *Europe's Green Revolution and Others Since: The Rise and Fall of Peasant-Friendly Plant Breeding*. London: Routledge, 2012.

Helgen, Judith Cairncross. *Peril in the Ponds: Deformed Frogs, Politics, and a Biologist's Quest*. Amherst: University of Massachusetts Press, 2012.

Hisano, Ai. *Visualizing Taste: How Business Changed the Look of What You Eat*. Cambridge: Harvard University Press, 2019.

International Union Against Cancer. "Report of Symposium on Potential Cancer Hazards from Chemical Additives and Contaminants to Foodstuffs." *Acta—Unio Internationalis Contra Cancrum* 13 (1957): 179–363.

Jas, Nathalie. "Adapting to 'Reality': The Emergence of an International Expertise on Food Additives and Contaminants in the 1950s and Early 1960s." In *Toxicants, Health and Regulation Since 1945*, edited by Soraya Boudia and Nathalie Jas, 47–69. London: Pickering and Chatto, 2013.

———. "Public Health and Pesticide Regulation in France before and after *Silent Spring*." *History and Technology* 23 (2007): 369–88.

Kirchhelle, Claas. "Pharming Animals: A Global History of Antibiotics in Food Production." *Palgrave Communications* 4 (2018): 96.

————. *Pyrrhic Progress: The History of Antibiotics in Anglo-American Food Production*. New Brunswick: Rutgers University Press, 2020.

————. "Toxic Confusion: The Dilemma of Antibiotic Regulation in West German Food Production." *Endeavor* 40, no. 2 (2016): 114–27.

Landecker, Hannah. "A Metabolic History of Manufacturing Waste: Food Commodities and Their Outsides." *Food, Culture & Society* 22 (2019): 530–47.

Langston, Nancy. *Toxic Bodies: Hormone Disruptors and the Legacy of DES*. New Haven: Yale University Press, 2010.

Lee, Victoria. "Wild Toxicity, Cultivated Safety: Aflatoxin and Kōji Classification as Knowledge Infrastructure." *History and Technology* 35 (2019): 405–24.

Lehman, Arnold J., Wilbur I. Patterson, Bernard Davidow, Ernest C. Hagan, Geoffrey Woodard, Edwin P. Laug, John P. Frawley, O. Garth Fitzhugh, Anne R. Bourke, John H. Draize, Arthur A. Nelson, and Bert J. Vos. "Procedures for the Appraisal of the Toxicity of Chemicals in Foods, Drugs and Cosmetics." *Food, Drug, and Cosmetic Law Journal* 10 (1955): 679–747.

Levenstein, Harvey. *Fear of Food: A History of Why We Worry About What We Eat*. Chicago: University of Chicago Press, 2012.

————. *Paradox of Plenty: A Social History of Eating in Modern America*. New York: Oxford University Press, 1993.

————. *Revolution at the Table: The Transformation of the American Diet*. New York: Oxford University Press, 1988.

Levinovitz, Alan. *Natural: How Faith in Nature's Goodness Leads to Harmful Fads, Unjust Laws, and Flawed Science*. Boston: Beacon, 2020.

Linsell, C. Allen, and Frank G. Peers. "Field Studies on Liver Cell Cancer." In *Incidence of Cancer in Humans*, Book A of *Origins of Human Cancer*, edited by Howard H. Hiatt, James D. Watson, and Jay A. Winsten, 549–56. Cold Spring Harbor: Cold Spring Harbor Laboratory, 1977.

Marcus, Alan I. *Cancer from Beef: DES, Federal Food Regulation, and Consumer Confidence*. Baltimore: Johns Hopkins University Press, 1994.

McGinn, Anne Platt. "POPs Culture." *World Watch* 13, no. 2 (2000): 26–36.

Michaels, David. *Doubt Is Their Product: How Industry's Assault on Science Threatens Your Health*. New York: Oxford University Press, 2007.

Nash, Linda. *Inescapable Ecologies: A History of Environment, Disease, and Knowledge*. Berkeley: University of California Press, 2006.

National Research Council, Commission on Scientific and Regulatory Issues Underlying Pesticide Use Patterns and Agricultural Innovation. *Regulating Pesticides in Food: The Delaney Paradox*. Washington, DC: National Academy Press, 1987.

Nestle, Marion. *Unsavory Truth: How Food Companies Skew the Science of What We Eat*. New York: Basic Books, 2018.

Oreskes, Naomi, and Erik M. Conway. *Merchants of Doubt: How a Handful of Scientists Obscured the Truth on Issues from Tobacco Smoke to Global Warming*. New York: Bloomsbury Press, 2010.

Parascandola, John. "Historical Perspectives on *In Vitro* Toxicology." In *In Vitro Toxicology: Mechanisms and New Technology*, edited by Alan M. Goldberg, 87–96. New York: Mary Ann Liebert, 1991.

Paxson, Heather. *The Life of Cheese: Crafting Food and Value in America*. Berkeley: University of California Press, 2013.

Petrick, Gabriella M. "An Ambivalent Diet: The Industrialization of Canning." *OAH Magazine of History* 24, no. 3 (2010): 35–38.

———. "The Industrialization of Food." In *The Oxford Handbook of Food History*, edited by Jeffrey M. Pilcher, 258–78. New York: Oxford University Press, 2012.

———. "'Like Ribbons of Green and Gold': Industrializing Lettuce and the Quest for Quality in the Salinas Valley, 1920–1965." *Agricultural History* 80 (2006): 269–95.

Podolsky, Scott. *The Antibiotic Era: Reform, Resistance, and the Pursuit of Rational Therapeutics.* Baltimore: Johns Hopkins University Press, 2015.

Proctor, Robert N., and Londa Schiebinger, eds. *Agnotology: The Making and Unmaking of Ignorance.* Stanford: Stanford University Press, 2008.

Proctor, Robert N. *The Golden Holocaust: Origins of the Cigarette Catastrophe and the Case for Abolition.* Berkeley: University of California Press, 2011.

Ramsingh, Brigit. "The History of International Food Safety Standards and the Codex Alimentarius (1955–1995)." Ph.D. diss., University of Toronto, 2011.

Rees, Jonathan. *Refrigeration Nation: A History of Ice, Appliances, and Enterprise in America.* Baltimore: Johns Hopkins University Press, 2013.

Schwerin, Alexander von. "Low Dose Intoxication and a Crisis of Regulatory Models: Chemical Mutagens in the Deutsche Forschungsgemeinschaft (DFG), 1963–1973." *Berichte zur Wissenschaftsgeschichte* 33 (2010): 401–18.

———. "Vom Gift im Essen zu chronischen Umweltgefahren. Lebensmittelzusatzstoffe und die risikopolitische Institutionalisierung der Toxikogenetik in der Bundesrepublik, 1955–1964." *Technikgeschichte* 81 (2014): 251–74.

Sellers, Christopher C. *Hazards of the Job: From Industrial Disease to Environmental Health Science.* Chapel Hill: University of North Carolina Press, 1997.

Shotwell, Alexis. *Against Purity: Living Ethically in Compromised Times.* Minneapolis: University of Minnesota Press, 2016.

Sinclair, Upton. *The Jungle.* New York: Doubleday, 1906.

Smith, David F., and Jim Philips, eds. *Food, Science, Policy and Regulation in the Twentieth Century: International and Comparative Perspectives.* London: Routledge, 2000.

Smith, Matthew. *Another Person's Poison: A History of Food Allergy.* New York: Columbia University Press, 2015.

Smith-Howard, Kendra. *Pure and Modern Milk: An Environmental History Since 1900.* Oxford: Oxford University Press, 2014.

Staples, Amy L. S. *The Birth of Development: How the World Bank, Food and Agriculture Organization, and World Health Organization Changed the World, 1945–1965.* Kent: Kent State University Press, 2006.

Stoff, Heiko. *Gift in der Nahrung. Zur Genese der Verbraucherpolitik Mitte des 20. Jahrhunderts.* Stuttgart: Franz Steiner Verlag, 2015.

———. "Zur Kritik der Chemisierung und Technisierung der Umwelt: Risiko- und Präventionspolitik von Lebensmittelzusatzstoffen in den 1950er Jahren." *Technikgeschichte* 81 (2014): 229–49.

Stoff, Heiko, and Alexander von Schwerin. "Einleitung—Lebensmittelzusatzstoffe. Eine Geschichte gefährlicher Dinge und ihrer Regulierung 1950–1970." *Technikgeschichte* 81 (2014): 215–28.

Stoll, Steven. *The Fruits of Natural Advantage: Making the Industrial Countryside in California.* Berkeley: University of California Press, 1998.

Sturdy, Steve, and Roger Cooter. "Science, Scientific Management, and the Transformation of Medicine in Britain, c. 1870–1950." *History of Science* 36 (1998): 421–66.

Suryanarayanan, Sainath, and Daniel Lee Kleinman. *Vanishing Bees: Science, Politics, and Honeybee Health.* New Brunswick: Rutgers University Press, 2017.

Tilman, David, Kenneth G. Cassman, Pamela A. Matson, Rosamond Naylor, and Stephen Polasky. "Agricultural Sustainability and Intensive Production Practices." *Nature* 418 (2002): 671–77.

Tsing, Anna Lowenhaupt. *The Mushroom at the End of the World: On the Possibility of Life in Capitalist Ruins.* Princeton: Princeton University Press, 2015.

Turner, James S. *The Chemical Feast: The Ralph Nader Study Group Report on Food Protection and the Food and Drug Administration.* New York: Grossman Publishers, 1970.

Verrett, Jacqueline, and Jean Carper, *Eating May Be Hazardous to Your Health: How Your Government Fails to Protect You from the Dangers in Your Food.* New York: Simon & Schuster, 1974.

Vogel, David. *The Politics of Precaution: Regulating Health, Safety, and Environmental Risks in Europe and the United States.* Princeton: Princeton University Press, 2012.

Vogel, Sarah A. *Is It Safe? BPA and the Struggle to Define the Safety of Chemicals.* Berkeley: University of California Press, 2013.

Walker, Brett. *Toxic Archipelago: A History of Industrial Disease in Japan.* Seattle: University of Washington Press, 2010.

Wargo, John. *Our Children's Toxic Legacy: How Science and Law Fail to Protect Us from Pesticides.* New Haven: Yale University Press, 1996.

Wellford, Harrison. *Sowing the Wind: A Report from Ralph Nader's Center for Study of Responsive Law on Food Safety and the Chemical Harvest.* New York: Grossman Publishers, 1972.

White, Suzanne Rebecca. "Chemistry and Controversy: Regulating the Use of Chemicals in Foods, 1883–1959." Ph.D. diss., Emory University, 1994.

———. "The Chemogastric Revolution and the Regulation of Food Chemicals." In *Chemical Sciences in the Modern World,* edited by Seymour H. Mauskopf, 322–55. Philadelphia: University of Pennsylvania Press, 1993.

Wilson, Bee. *Swindled: The Dark History of Food Fraud, from Poisoned Candy to Counterfeit Coffee.* Princeton: Princeton University Press, 2008.

Winickoff, David E., and Douglas Bushey. "Science and Power in Global Food Regulation: The Rise of the *Codex Alimentarius.*" *Science Technology & Human Values* 35 (2010): 356–81.

Whorton, James. *Before Silent Spring: Pesticides & Public Health in Pre-DDT America.* Princeton: Princeton University Press, 1974.

Wunderlich, Volker. "'Mit Papier, Bleistift und Rechenschieber.' Der Krebsforscher Hermann Druckrey im Internierungslager Hammelburg (1946–1947)." *Medizinhistorisches Journal* 43 (2008): 327–43.

———. "Zur Entstehungsgeschichte der Druckrey-Küpfmüller-Schriften (1948–1949). Dosis und Wirkung bei krebserzeugenden Stoffen." *Medizinhistorisches Journal* 40 (2005): 369–97.

Young, James Harvey. *Pure Food: Securing the Federal Food and Drugs Act of 1906.* Princeton: Princeton University Press, 1989.

Zeide, Anna. *Canned: The Rise and Fall of Consumer Confidence in the American Food Industry.* Oakland: University of California Press, 2018.

 PART I

Objectifying Dangers

 CHAPTER 1

Salad Days

The Science and Medicine of Bad Greens, 1870–2000

Anne Hardy

The history of salad, and even more of what we can call salad science, has received remarkably little historical attention.[1] In the 2010s, however, numbers of incidents of food poisoning from eating bagged, prewashed salad leaves attracted widespread attention in Britain. The issue of salad has not featured largely in the existing history of food poisoning, but the apparently new phenomenon of salad-related incidents suggested an exploration of its prehistory, if any, and of the place of salad eating in the British diet.[2] How widely was salad consumed by the British in the past, and when and how did it become associated with food poisoning? As a dish, the salad has an ancient history, but salad science itself is a twentieth-century phenomenon, which only began to gather momentum in the 1960s, in that prosperous postwar, postrationing decade when the transport revolution enabled people to travel more freely than ever before, and a global expansion of trade brought new and tantalizing salad vegetables within the reach of populations who had never considered them before. Among those populations were the British, who, despite a vigorous and enterprising market garden sector in the nineteenth century, and a national tradition of "meat and two veg" (one of them the potato) dinners, were singularly uninterested in vegetable, let alone salad, consumption. To this generalization, the capital was perhaps an exception. In 1970s London, it was possible to buy globe artichokes, French beans, eggplants, zucchinis, and red peppers, originally, no doubt, to meet demand from the city's Italian and French immigrant communities. In provincial cities like Oxford and Cambridge, such luxuries were unheard of.[3]

The British Salad Tradition

The British have, in fact, a proud salad heritage dating back to the seventeenth century, championed by the diarist John Evelyn, who discovered the charms of salad while self-exiled to Europe during the English Civil War.[4] The great and robust early nineteenth-century gourmand, the Reverend Sydney Smith, was also an enthusiast.[5] Nonetheless, the British generally were not good on salads. Samuel Pepys, although he enjoyed a "fine discourse" with Evelyn on trees and the nature of vegetables while traveling to Greenwich in 1665, admits only twice in all his voluminous diaries to eating salad.[6] The playwright Ben Jonson (1572–1637) referred to "coarse, cold salad," and the eighteenth century diarist James Woodforde made no mention of it at all.[7] Mrs. Beeton showed very little interest in salads, except perhaps when they included lobster.[8] An American observer, circa 1900, noted "the primitive notion of salad preparation in England."[9] As recently as 1965 the English food writer Elizabeth David, "dreaded" the English "season of salads": "What becomes of the hearts of lettuces?" she demanded.

> What makes an English cook think that beetroot spreading its hideous purple dye over a sardine and a spoonful of tepid baked beans makes an hors d'oeuvre? What is the object of spending so much money on cucumber, tomatoes and lettuce because of their valuable vitamins, and then drowning them in vinegar and chemical salad cream?[10]

The Americans, however, had long been keener than the British on lettuce and its companions, and already in 1900 considered their country to be "the land of salads."[11]

Yet this does not mean that salad vegetables were not available in Britain if wanted. John Evelyn wrote up the many varieties of salad vegetables he himself cultivated, many originating on the continent of Europe.[12] It may be that the preparation put people off: Mrs. Beeton's single specific recipe for a "summer salad" of lettuce, mustard, cress, and "a few radishes," gave instructions to wash carefully, pick over and drain thoroughly, presumably to eliminate slugs, snails, and grit.[13] Some ten years after Elizabeth David complained about English salad habits, another English food writer, Elizabeth Ayrton, devoted a whole chapter to the English use of vegetables and salads in her book *The Cookery of England* (1974). Ayrton noted "the fairly general belief" that vegetables were not appreciated in England until the last two decades of the twentieth century, but she argued for, and provided evidence of, modest but general vegetable consumption among both rich and poor in the past—the rural poor, at least. "It is obvious" she wrote, "that root vegetables and leaves, shoots and fruits which could easily be grown or gathered wild were of a greater and more desperate importance to the poor." She noted that carrots, kale, and onions were grown by the poor, and she also cited William Langland's vision of Piers the Plowman (thought to have been written 1360–69), for whom parsley, cabbages, and leeks helped to sustain a meager existence until harvest time. But, as Ayrton also noted, "the Englishman has always been more then usually carnivorous when he could afford to be. Meat . . . was what everyone . . . wanted for the dinner and the supper table."[14]

It becomes evident that by the later nineteenth century, salad consumption was sustained among certain of the populace at least, for the Christmas market of 24 December 1876 at Covent Garden included lettuce, celery, spinach, coleworts, radishes, cucumber, asparagus, green peas, "salads," Batavian endive, and "mustard and cress and herbs very plentiful."[15] Before the days of international air freight and chilled compartments, the Christmas selection at Covent Garden, the product of British greenhouses and market gardens, was deeply impressive. And there must have been a market for this produce, if not among the English, then perhaps among the city's expatriate communities—the French, Italians, and Germans, the Europeans from whom John Evelyn had learned the charms of salad.

The one salad vegetable for which there is solid evidence of consumption in Victorian and Edwardian England—well, London, at least—is watercress. Henry Mayhew's classic account of the rough lives of London's child watercress sellers around Farringdon Market dates from 1860–61. Watercress seems to have been a key part of the mechanics' breakfast—the breakfasts of bricklayers, carpenters, blacksmiths, and plumbers.[16] Mayhew calculated that some 6.5 million bunches of watercress were supplied by the child sellers every year, with a similar quantity sold through the retail shops. The Covent Garden market, however, sold far less than Farringdon, with a mere 1,578,000 bunches sold wholesale, the difference perhaps due to the different social groups catered to.[17] Andrew Wynter, essayist and first editor of the *British Medical Journal*, described the London watercress market in more detail, again giving primacy to Farringdon, but also giving some account of where what he called "this wholesome breakfast salad" was grown.[18] In 1854, cresses were mainly grown west of London, at Walthamstow, Cookham, Shrivenham, and Farringdon, from where the Great Western Railway shipped them up to London. Cresses, Wynter noted, were also grown at Camden Town. Published in 1854, the very year in which John Snow took action against the infamous Broad Street Pump, Wynter's remarks on Camden Town show no consciousness of any damaging effects of sewage pollution. "Most people," he wrote, "fancy that clear purling streams are necessary for [the] production [of watercress] . . . [but] The Camden Town beds are planted in an old brick field watered by the Fleet Ditch; and though the stream at this point is comparatively pure, they owe their unusually luxuriant appearance to a certain admixture of sewage."[19] Wynter offered no further comment, but had, perhaps, issued a covert warning to his middle-class readers. His lack of explicit sewage consciousness is interesting, however, because when a sanitary storm broke around watercress some twenty years later, the basis for its scientific indictment was the fact that the East London cholera outbreak of 1866 had been shown to have been caused by sewage-polluted water.[20]

Watercress, that "wholesome breakfast salad," thus marks the point at which health hazard enters into the balance against wholesomeness in matters concerning salad, and, arguably, the science of salad takes its first faltering steps. From 1866, the possibility of watercress being contaminated with sewage-tainted water began a gradual process of realization of hidden—or possibly not so hidden—dangers in salad vegetables. Watercress became an intermittent public health issue in England between 1873 and 1903, resurfacing briefly in 1912, and again in 1953 around an outbreak of Sonne dysentery at Papworth Village Settlement.[21] Predictably, however, it was not as a health risk to the urban working class that it initially attracted attention. The plant was also widely eaten by the middle and upper classes, esteemed for its taste, for its value as an antiscorbutic, and for its sulfur and iodine content—"the most important of all known minerals that act as correctives" according "A Physician" in 1876.[22] The eminent physician and chemist Sir Alfred Smee, a great amateur gardener, advised that "a salad of some kind should be grown for every day of the year," and selected watercress as "of all salading plants . . . the most valuable." It was available all year round, could be eaten with every meal, was "warm and grateful to the stomach," and there were "very few persons to whom it is distasteful."[23] According to Smee (who had grounds to know), it even graced the tables of the aristocracy.[24]

Watercress became topical in November 1875, when a former gardener to the distinguished surgeon Sir Benjamin Brodie reported Brodie's horrified reaction to the prospect of any of his household eating watercress. Watercress, replied a medical correspondent, was "commonly believed" to be grown in sewage, and "for *that* reason" might be thought unwholesome and "rejected as dangerous."[25] Brodie—who died in 1862—was ahead of his time, but had clearly taken due note of William Budd and John Snow's findings on the transmission of typhoid and cholera. It was Smee's daughter, Elizabeth Odling, who dated the more general realization to the East London cholera outbreak of 1866.[26] Smee himself engaged in an epic tussle with Croyden Board of Health when their new sewage system, dependent on the river Wandle for outlet, ravaged his cherished garden and destroyed his trout stream and watercress beds on that river, around 1870.[27]

Smee was well aware of the health hazards posed by polluted watercress and had clearly made some study of the subject. "Watercresses," he wrote in 1872, "should be thoroughly cleaned before they are eaten, and should never be used where the stream has any sewage contamination."[28] Following a two-year battle with the Croydon Board of Health, Smee emerged victorious in forcing them to make other arrangements for their sewage, and turned his hand to restoring the status quo ante in his garden. But the episode turned him into an active and vociferous critic of boards of health and sewage schemes, and an anti-sewage pollution advocate. He launched a forceful but unsuccessful campaign for regulations to impose severe fines on those polluting land or water courses with sewage and produced a memorable description of the effects of sewage on watercress, which was still being quoted by the London County Council in 1904.[29]

Smee's description of sewage-polluted watercress was, indeed, decidedly memorable. "Water-cresses," he wrote, "act as a scrubbing brush to the sewage, and remove all the solid flocculi from the water which adhere to the stalks."[30] It was quite possibly Smee who was the "sensitive friend" who in questioning the provenance of cresses presented on the supper table, alerted Shirley Hibberd, the most popular and successful of Victorian gardening writers, to the dangers of sewage-polluted watercress, and inspired him to develop a system of growing the plants in pots irrigated by water from a known safe source.[31] Hibberd himself, by his own account, was in no way troubled by the "possible taint the water might acquire in passing the piggeries and poultry yards half a mile upstream" from his own watercress beds. But as a good host he could not possibly have the question raised a second time, and so sought alternative strategies for producing his favorite salad.[32]

Despite this small squall in the gardening world in the 1870s, there is no positive evidence of widespread evil caused by eating watercress, although the controversy might have been one of the factors slowing the decline in typhoid death rates in the 1890s.[33] Overt public health suspicions of watercress only emerged in that decade, at a time when cholera and typhoid were being linked to the consumption of sewage-polluted shellfish.[34] The Medical Officer of Health for the London district of Marylebone analyzed sixty-four cases of typhoid occurring in the district in 1894, finding that ten had eaten both watercress and oysters, eight had eaten cress but no oysters, and four had eaten oysters only. The numbers involved were not, he suggested, sufficient to warrant any general conclusion, and there the matter rested.[35] In 1900, watercress was the suspected vehicle of a typhoid outbreak in a very poor area of Southwark, until several victims declared that watercress was a luxury they could not afford.[36] By the time watercress was implicated in two successive outbreaks in a small area of Hackney in the summer of 1903, with watercress eaters several times more likely to suffer, bacteriology was sufficiently advanced to confirm that all samples of watercress supplied to the area contained "organisms of the intestinal type."[37] On this basis, and on the record going back to 1894, the London County Council organized an investigation of all watercress beds in the county, but decided, in view of the limited nature of the outbreaks and the resumed fall in the city's typhoid death rates, to take no further action.[38]

In the years around 1900, however, the English gastronomic scene does not seem to have been stirred by issues around the cleanliness or otherwise of salad vegetables. Indeed, further evidence of the English lack of interest in salad can be found in culinary writings of the time. Thus the wonderfully named Maximilian de Loup, whose *American Salad Book* was published by William Heinemann in 1901, noted that England was "still barbaric in much of her salad serving and eating"—a judgment which may call to mind Elizabeth David's strictures from the 1960s.[39] De Loup made no mention of salad hygiene, but the French chef Alfred Suzanne, with a forty-year career in London behind him, throws some light on contemporary issues around salad: "Pretend gourmands," he wrote, not mincing words, "contend that certain salads (unspecified) should not be washed, but that the leaves should simply be wiped." This was "a misguided instruction"—"it must be remembered," Suzanne admonished his readers, "that the plant has been in contact with garden compost, and it has sheltered myriads of microbes, from which the leaves can only be freed by washing."[40]

The issue of garden compost, indeed of agricultural manuring in general, was to loom increasingly large in the decades that followed. It is in this period, between circa 1910 and 1920, that we can see microbiology taking off as a science through the lens of salad. Edwin Chadwick's original ambition for the disposal of sewage waste had been to utilize it as agricultural manure, and such practices had indeed been adopted in parts of England, such as Nottinghamshire.[41] But the issue of how to dispose of sewage wastes remained, and the lengthy explorations of the Royal Commission on Sewage Disposal (1898–1912) survive as testimony to the grave attention the topic was afforded.[42] The fifth (1908) report of the commission, for example, gave studied attention to the "manurial value of sewage sludge."[43] At the same time, the Medical Department of the Local Government Board was commissioning research into the survival of typhoid bacilli in water and soil, and into the possibilities of animal intestines harboring pathogenic gastroenteric bacilli.[44] Typhoid's lesser cousins, the food-poisoning salmonellas, had begun to be identified in the late 1890s as serological methods for the differentiation of different types of bacteria came into use. Whereas it had originally been assumed that a single unique bacillus would be responsible for each individual disease, it became apparent that this was not the case.[45] Between 1900 and 1918, considerable progress was made in the study of bacterial variation, establishing that bacilli like the streptococci, the dysenteries, and the salmonellae existed as tribes rather than as individual, unrelated entities.[46] It was in these years that a science of salad began to emerge in America.

The Emergence of Salad Science

Food poisoning as a public health problem—as a problem distinct from typhoid and cholera infections—had been discovered in the 1880s, but was long almost entirely associated with meat, milk, and shellfish, and with duck eggs from the later 1920s.[47] It is caused by ingestion, in food or drink, of pathogenic bacteria or viruses, and is generally characterized by acute abdominal pain or cramps, vomiting and diarrhea, or plain diarrhea, and profound exhaustion. It does not last long but is a far-from-pleasant experience. That it began to loom large in British public health concerns in the twentieth century was partly because the virtual disappearance of typhoid and cholera allowed this lesser genre of gastric infection to command attention in a public health void. Further, early twentieth-century bacteriological investigations into the residual reservoirs of typhoid infection in nature indicated that fecally contaminated soil enabled the typhoid bacillus to survive for long periods outside the human body. As the relationship between typhoid and the lesser salmonellas became established, and it was discovered that some of these did indeed inhabit animal guts, the family's ability to survive in the wider environment became a matter of concern.

The uncertainties of early bacteriology meant that many of the first studies of bacterial habitat were flawed and unreliable. As noted by the American R. H. Creel in 1912, "The diversity of results has been in direct proportion to the number of investigators," probably because the typhoid bacillus itself was constantly being confused with its cousins.[48] Creel's own investigation, at the behest of the US Public Health and Marine Hospital Service, focused on vegetables as a possible factor in spreading typhoid. The most reliable of existing studies, Creel noted, suggested that typhoid bacteria survived in soil for longer than it took to grow many common garden vegetables. He himself was able to isolate typhoid bacilli from the leaves and upper stems of lettuce and radishes grown in glass jars of typhoid-inoculated soil under hothouse conditions after twenty-five days, while plants grown in the open yielded isolates for up to thirty-one days.[49]

Creel's work was followed up by Melick in Chicago, who found that depending on exposure, soil, and the strain of typhoid involved, the bacteria survived for between twenty-nine and seventy-four days on radishes and lettuce grown in the open.[50] Melick's findings were worrying. He had experimented with both old and new cultures of the bacillus, and his results indicated not only that survival time varied between different strains of the bacillus, but that "long-continued growth outside the human body" increased their resilience in soil.[51] The bacilli also demonstrated "extreme tenacity" in surviving on plant surfaces, and where roots such as radishes were planted in infected soil, they carried the organisms on their surfaces. Even thorough washing did not free these leaves and roots of their microbial load. For as long as their soil contained typhoid bacilli, Melick concluded, such plants were not safe to eat.[52]

These American studies marked the beginning of a slow and geographically limited concern around the ability of fruit and salad vegetables to spread bacterial infections and parasites such as flukes and intestinal worms. The progress of this interest can be charted from a literature review published in 1971 by Edwin Geldreich and Robert Bordner of the US Environmental Protection Agency in Ohio.[53] Geldreich and Bordner cited no relevant literature from the 1920s—just two studies from the mid-1930s, and four from the 1940s, with a jump to sixteen in the 1950s. From a slow start between 1949 and 1953, literature on the subject burgeoned, reaching a total of forty-four cited studies for the 1960s. Much of the drive in this production came from America, but the 1950s saw five citations from Austria/Germany, where typhoid outbreaks due to sewage-irrigated vegetable and salad crops had occurred in Lanceburg, Stuttgart, and Vienna, compounded by the new practice of spray irrigation. Very little British literature is cited on the topic, perhaps because British typhoid incidence had fallen away to practically nothing in Scotland, England, and Wales, following the introduction of mains drainage, effective sewage treatment plants, and cesspits in rural areas, or perhaps because the British were bad salad eaters and were in the habit of boiling all vegetables to death.[54] Food poisoning as distinct from typhoid had been on the British public health agenda since 1880, but preventive concern was by the 1950s directed to food handler practices rather than agricultural practices.[55]

It was in America, which in the interwar period was still wrestling with a considerable endemic typhoid problem, that the devastating health effects of sewage pollution of vegetable and salad crops became apparent in the 1930s.[56] A nationally and locally coherent and effective public health administration began to emerge in America only in the early twentieth century, and, as such systems do, took time to find its feet. By the 1930s, however, a coherent system of death registration on a state-by-state basis had been introduced, and local public health services instituted at state level. As late as 30 June 1941, however, only 54 percent of the 3,070 counties within the boundaries of the continental United States had a full-time local health organization, and these were heavily concentrated on the eastern side of the country and along the Pacific Rim in the West.[57] A great block of states in the center of the country from Montana, North Dakota, and Minnesota south to Texas was largely without full-time medical officers at the county level.

In the early 1930s, with death registration statistics now available, the Medical Officer for Colorado, Edward Chapman, concerned by the statistics and anxious for his state's reputation as a "healthy" recreational resort, set out to discover the causes of a summer and autumn diarrhea epidemic that occurred regularly in certain sections of the state.[58] Colorado was still cheerfully pouring untreated sewage into its rivers and streams, and those water courses also supplied the irrigation water for the truck gardens (market gardens) that supplied its denizens with vegetables. Although Americans were far more sanitary-conscious than the British, compliance with public health standards by poorly paid food handlers was (and is) far from perfect in either nation. Staff turn-over in this sector was generally high, meaning that too little attention could be given to proper training in hygiene.[59] Even in excellent restaurants in the 1930s, lettuce had been known to be served up with what was unquestion-ably fecal material still clinging to its leaves.[60] As Chapman pointed out, the handling of such vegetables preparatory to cooking could lead to the contam-ination of other foods, and competent bacteriologists had found that even thorough washing did not make them safe to eat.

Chapman reminded his readers that sewage was not simply a matter of storm and dishwater, as was too often generally assumed. In reality it con-tained "urine, sputum, feces, disintegrated toilet paper, disintegrating sanitary pads, contraceptive devices," all of which could carry disease.[61] He executed an elaborate comparison of Colorado death rates for diarrheal disease with those for neighboring Nebraska and Wyoming, which had no sewage-irrigated areas; Utah and Kansas, which each had one sewage-irrigated county; and California—not a neighbor but like Colorado recently rated among the top ten "intelligent states" by Dr. Frederick Henry Osborn, a research associate at the American Museum of Natural History, and with a Mexican population of 6.5 percent, similar to Colorado's 5.6 percent.[62] Crucially, California also had extensive truck gardening areas under irrigation but forbade the use of con-taminated water in cultivating all market produce—a prohibition which the state enforced, as Colorado did not. Between 1923 and 1931, Colorado's typhoid death rate was 21 percent higher than that for the USA as a whole, and 175 percent higher than that for California.[63]

Chapman went on to analyze rates for amoebic dysentery, a recent arrival in Colorado, and diarrhea among the under twos. The average Colorado death rate for the latter in 1923–31 was 35.3 per 100,000 compared to 23.8 for the US as a whole and 20.4 for California, and was the highest among the com-parable state group.[64] He mapped dumping points for raw sewage and grossly polluted water courses across the state, showing a dramatic contrast in high diarrhea death rates for "sewage" counties and minimal death rates in non-irrigation counties. His paper appeared in the *Colorado Medical Journal* for January 1934, and in October that year he followed it up with a lengthier paper on sewage disposal as a major public health problem in the same journal.[65] Despite hearty endorsement from his peers in the Colorado Medical Society, Chapman's work seems to have gone unnoticed in the larger medical commu-nity at the time. It was not until 1946 that W. H. Gaub, director of the Colorado Division of Public Health Laboratories and a fellow of the American Public Health Association, revisited the question with an updated analysis of data from the years since 1933, apparently stimulated thereto by an outbreak of dysentery at a military barracks in Denver that had been caused by the con-sumption (as coleslaw) of raw cabbage grown in sewage-irrigated soil.[66] By this time, half the state's population had access to treated water and "sewage infections" were declining. The critical finding of this study was, however, that the three Colorado counties that had introduced their own full-time "health units" in these years had done markedly better than all the others.[67] Sanitary science, applied by qualified practitioners, worked.

Although Gaub's study was confined to the problems of Colorado and published in the *Rocky Mountain Medical Journal*, sewage irrigation and its hazards were becoming topical in the late 1940s. The dangers inherent in such practices were graphically described in Gerald Winfield's 1948 publication *China: The Land and the People*.[68] China's agriculture and economy had become a subject of interest to many academics, including R. H. Tawney, in the 1930s, many of whom (Tawney was not one) waxed lyrical over the wonderful ingenuity and infinite patience with which Chinese farmers utilized all human excreta in their agriculture.[69] Winfield, an American who had been a professor of biology at Cheeloo University, Shantung Province, since 1932, was very explicit on the devastating public health consequences of this style of farming across China in terms of fecal-borne infection—enteric fevers and bacillary dysentery, infestations of hookworm, liver fluke, the intestinal worms *Fasciolopsis buskii* and *Ascaris lumbricoides*, and—"the most widespread and important of all"—Japanese blood fluke.[70] Of China's annual death rate of fifteen million persons, Winfield calculated, nearly a third was due to fecal contamination created by traditional manuring practices.[71] In the following year, 1949, Lloyd Falk of the New Jersey Agricultural Experiment Station warned that the potential transmission of human bacterial diseases through the consumption of night-soil fertilized vegetables should also be a concern in America, "where irrigation of crops with polluted stream water is becoming increasingly common."[72] Similar problems were identified in Great Britain. In the rural village of Sampford Peverell, a rare type of *Salmonella enterica* Typhi was identified on watercress that had been grown in beds on a stream contaminated with sewage. Application of new phage-typing tools in bacteriology enabled researchers to reconstruct the chains of infection associated with this classic English green.[73]

Post–World War II: Pollution by Wildlife and Others

These studies, coming as they did on the heels of World War II, and at the beginning of a period of unprecedented government interest in and support for a broad range of scientific enterprises, seem to have kick-started a fresh interest in the sanitary aspects of public health. New journals emerged in America—the short-lived *Sewage and Industrial Wastes* (1950–59), for example, and the longer-lived *Journal of Milk and Food Technology*, now the *Journal of Food Protection* (1947–). At the same time, new research into the role of salmonella infections as zoonoses, springing from the discovery that domestic cattle, ducks, and pigs harbored organisms also infecting humans, extended the range of inquiries associated with salad fruit and vegetables.[74] Studies of wild birds and of insects as disseminators of fecal pathogens in the field suggested multiple sources of disease-producing environmental pathways of human infection.[75] Yet another study from Colorado, again reflecting concern with its health status as a recreational resource, demonstrated that small wild animals and wild birds carried sufficient loads of gastroenteritis-producing organisms to be of public health significance.[76]

By 1971, Geldreich and Bordner could report that wild fauna was "probably the most significant natural means of disease transmission." Leafy crops, they noted, attract mice, rats, and rabbits, and such crops can support one hundred-plus rodents per acre. *Salmonella* isolates from wild animals on Illinois farms suggested that 7.5 percent of the total population carried residual infections, while a Colorado study showed a 6.3 percent incidence of *Salmonella* in the intestinal flora of wild birds. Fecal contamination of crops by wild creatures, they concluded, existed at a residual level that could endanger human populations.[77] This was not a problem that was going to go away any time soon. As recently as 2012 it was recorded that fecal contamination of the growing environment by wildlife can be "an on-going source of contamination" for field-grown vegetables.[78] It is very difficult to prevent the pollution of vegetables during growth and, since the introduction of techniques such as artificial rain, impossible to prevent the use of surface water for irrigation or refreshing cut produce.[79]

Soon after its creation in 1970, the US Environmental Protection Agency, following the survey conducted by its agents Geldreich and Bordner, reached certain drastic conclusions on the health risks of fresh fruit and vegetables. The microbiological information relating to crops, the soils in which they were grown, and the water with which they were irrigated or processed, acquired through a wide range of field studies on salad vegetable crops, "correlated to demonstrate the magnitude of fecal contamination on raw food products." Other studies pointed to many different sources of contamination both in the field and during handling and marketing.[80] The development of methods for measuring fecal coliforms in water had enabled the National Technical Advisory Committee to establish a valid water quality standard, but Geldreich and Bordner emphasized that yet further safeguards were needed against the infection of salad vegetables and fruits during cultivation, handling, and processing.[81] The development and application of such safeguards proved an almost impossible task, however. Despite growing international engagement with these issues, especially from the 1980s, and the rapid development of scientific interest in microbiological standards and guidelines, the task of coordinating laboratory findings with workable standards and achieving their application in the field was unmanageable. In America, where the modernization of the food industries predated that in Europe, such difficulties were already very clear by the mid-1970s.

Some of these difficulties had been evident for some time. Packaged precut and mixed salads had been introduced to America in the late 1950s—they first arrived in Britain courtesy of Marks and Spencer in 1986.[82] It did not take the American microbiologists long to get to work. Early studies appeared to show that the microbial population of these vegetable products was only increased by such processing techniques as antibiotic and chemical dips, that manipulation during bagging further increased and distributed the microbes, and that their numbers increased yet further with length of storage and increases in moisture.[83] Revisiting these findings in the mid-1970s, later observers were only confirmed in their views that setting microbial standards for salad products was not a practical possibility, especially given such factors as seasonal variation and many different places of production. From 1973 onward, however, the Centers for Disease Control kept a record of outbreaks of foodborne illness caused by fresh produce: a record which showed a rising trend right through to 1997 and beyond.[84]

Although Americans were solid salad eaters, they had historically subsisted by preference, as had most European populations, on a diet heavy in meat and potatoes. With postwar prosperity, this became the staple diet of the vast majority of the population. In the 1970s, however, the medical correlation of saturated fat with rising death rates from heart disease and cancer led to public information campaigns such as the National Cancer Institute's Five-a-Day for Better Health aimed at increasing the national intake of fruit and vegetables.[85] And while the healthy heart campaign succeeded in increasing the consumption of fruit and vegetables, it also changed the epidemiology of food-borne disease across the nation.

In 1994, staff of the Minnesota Department of Health's Acute Disease Epidemiology Section published a discussion of this development based on their fifteen-year-long program to "aggressively identify and investigate the occurrence of" foodborne disease in the state.[86] One result of the increased consumption of fresh fruit and vegetables, up nationally by 16 percent and 29 percent respectively between 1970 and 1990, had been the occurrence of large outbreaks of hepatitis A, shigellosis, and salmonellosis associated with salad items. Outbreaks of *Salmonella* linked to fruit and vegetables had traditionally been rare, they noted, but four such multistate outbreaks had occurred since 1990, and Minnesota itself had suffered three of these. Two—one each of *Salmonella* Chester and *Salmonella* Poona—were caused by eating cantaloupe melons, and two—one of *Salmonella* Javiana and one of *Salmonella* Montevideo—involved eating tomatoes.[87] The cantaloupe outbreaks were linked to salad bars or fruit salad, suggesting that after contamination from the rind when cut, "temperature abuse" had led to multiplication of the pathogens when the salads were left at room temperature for several hours. Noting the wider change in epidemiological pattern, the Minnesota team drew attention not only to the Five-a-Day for Better Health campaign but also to the increased consumption of food in commercial venues and to the development of large and complex networks of food distribution, which had created a pattern of low-level contamination of mass-produced or distributed food products that constituted a new challenge to public health authorities.[88] By the turn of the twentieth century, national outbreak statistics underscored this argument. Twelve fruit and vegetable outbreaks had been recorded in the 1970s, sixty in the 1980s, and 111 in the 1990s, with "salad" and vegetables contributing most to the increase.[89]

The industrialization of agriculture, which resulted in the pooling of foods from many sources of production at processing plants, the action of processing itself, and the wide distribution of such processed foods—whether bagged salad vegetables, processed cheese, or partially cooked hamburgers, to name only a few such products—by road, rail, and air across local and national boundaries often spread the effects of contamination widely. One example detailed by the Minnesota team involved mozzarella cheese, produced by a relatively small cheese plant in Wisconsin between March and May 1989, which resulted in a multistate outbreak of *Salmonella* Javiana and *Salmonella* Oranienburg infections. Contaminated cheese produced at this plant was distributed to four large processors, who shredded it, contaminating cheese from other sources in the process, before distributing their products across a number of different states.[90] Similarly, as described below, tomatoes grown by three different producers, delivered to one packing house in Ohio, packed and redistributed to restaurants in various states, and served as tomato salads, led to a multistate outbreak affecting citizens from twenty-one American states.[91]

The developments reported by the Minnesota team were not limited to the United States. In England and Wales, eighty-five outbreaks of gastroenteritis were associated with salad consumption between 1992 and 2000, including two notable *Salmonella* Typhimurium DT104 outbreaks, one of which was linked to similar outbreaks in Iceland, Scotland, Germany, and the Netherlands.[92] Across Europe, recognition of potential outbreak problems caused by increasing human travel and the removal of trade barriers within the continent led to the establishment of the Salm-Net surveillance network, operational from January 1994, and to the Foodborne Viruses in Europe network at much the same time.[93] Salm-Net consisted of microbiologists and epidemiologists responsible for national *Salmonella* surveillance in fourteen countries linked together with the aim of establishing an online database of compatible data available to all members; by September 1995, the system had already facilitated the identification and investigation of two outbreaks, one related to iceberg lettuce, the other to "an imported snack."[94]

In 1995 also, the Centers for Disease Control (CDC) implemented a new outbreak detection algorithm to analyze routine laboratory-based *Salmonella* surveillance data submitted through the Public Health Information System by state health departments. This compared the actual number of reports of each *Salmonella* serotype each week with the number expected on the basis of historical data.[95] Operational from May 1995, the algorithm immediately flagged a nationwide increase in *Salmonella* Stanley that June. A query was put through to Salm-Net, who replied that an outbreak of *Salmonella* Stanley had begun in Finland in March. Subsequent investigation revealed that in both Finland and the US, alfalfa sprouts grown from seed supplied by a Dutch shipper were responsible for the outbreaks.[96]

At the same time as microbiological communications were becoming more sophisticated, the methodology of epidemiological investigation on the ground was being refined. Into the 1970s, methods of outbreak investigation remained essentially traditional, reliant on the questioning of victims as to food consumption during the apparent period of infection, and identification of the vehicle of infection, supported by analysis of the foods concerned where and if possible. By the 1990s, new, more sharply analytical techniques were beginning to be employed in America. These included case-control studies to support the identification of the causal source of infection, followed by trace-back of the product to grower, and trace-forward investigations to determine the extent of outbreaks. In a 1996 multistate outbreak of *E. coli* O157:H7 associated with mesclun (baby leaf) lettuce affecting Illinois and Connecticut, for example, both these methods were used.[97]

E. coli O157:H7 was a relative newcomer as a food poisoner in 1996, having first been recognized in 1982 as the cause of outbreaks of bloody diarrhea in Michigan and Oregon. It has since become an important food pathogen, and one that has challenged conventional produce safety programs.[98] Its emergence has been linked to the acquisition by the usually innocuous bacterium *E. coli* of three major virulence factors: Shiga toxins, products of the pathogenicity island called the locus of enterocyte effacement, and products of the F-like plasmid pO157.[99] These events were probably selected for by changing technologies of production in the meat industries of the United States.[100] In this respect, *E. coli* O157:H7 is among the evolutionary responses, similar to antibiotic-resistant bacterial strains, to the agribusiness farm environments of large-scale containment facilities for livestock.[101] In the 1996 mesclun outbreak, the trace-back investigation focused on the retail outlets that had sold mesclun to patients. Retail invoices and bills of lading were requested, and the lettuce traced to distributors, from whom wholesale invoices were requested. The producer implicated was then visited, investigated, and the process reversed: customer information from the producer was used to identify states that had received the implicated mesclun. Analyses of *E. coli* isolates taken during the relevant period were then requested to determine if they matched the outbreak-associated type.[102]

New Testing Methods and International Cooperation

The use of these techniques rapidly developed the science of salad and understanding of the multiple routes of its contamination with pathogens. The new surveillance networks were part of a wider scientific endeavor to come to grips with this increasingly international problem. As noted by Robert V. Tauxe of the CDC, the recent dramatic changes in diet, processing practices, and commerce had changed the "familiar scenario of food-related outbreaks."[103] It was this new "outbreak scenario" that spurred the CDC to further action, although the initial spur was a hamburger-related outbreak rather than a salad one. In 1993, undercooked burgers contaminated with *E. coli* O157:H7 sold in Jack-in-the-Box fast food chain outlets in the Pacific Northwest infected more than a hundred people in several states and killed a number of children.[104] The head of the CDC's DNA-based diagnostics and subtyping laboratory found that pulsed field gel electrophoresis (PFGE), a technique for separating large DNA molecules first developed in 1984, worked well for subtyping *E. coli* strains, and successfully applied it to the Jack-in-the-Box outbreak.[105] After this the CDC set out to develop the technique as a tool for investigating foodborne infections.[106] PFGE was, a Minnesota epidemiologist observed, like the invention of the telescope: "People could always see the stars, but telescopes made one hell of a difference."[107] The procedure expanded the range of resolution for DNA fragments by some two orders of magnitude, and it rapidly became established as the gold standard in epidemiological studies of pathogens, making it easier to distinguish between different strains within families of bacteria, and so to link environmental findings with clinical infections.

The development of PFGE also made international cooperation and tracking of infections easier. While the existence of national central public health authorities such as the Centers for Disease Control in America and the Public Health Laboratory Service in Britain was undoubtedly important to the development of epidemiological techniques, disease surveillance, and observation of national trends in disease incidence, there was growing concern with the international character of many foodborne outbreaks, such as the 1995 outbreak of *Salmonella* Stanley in America and Finland, and a second outbreak of *Salmonella* Newport affecting Oregon, Canada, and Denmark.[108] The CDC had set up PulseNet (National Molecular Subtyping Network) in 1995 in collaboration with state public health laboratories and the US Department of Agriculture.[109] Using PFGE, it was possible to identify outbreak strains of such bacteria as *E. coli* O157:H7 and *Salmonella* Typhimurium, and to distribute their pattern to the laboratories via the Internet. This technology had the potential to provide a "truly national surveillance for foodborne disease."[110]

PulseNet International followed in 1996.[111] This surveillance program for food- and water-borne disease outbreaks, emerging pathogens, and acts of bioterrorism was organized through seven area networks (the United States, Canada, Latin America, Asia Pacific, Africa, the Middle East, and Europe). As of 2011, it comprised 120 laboratories in eighty countries. In America, it consisted of eighty-seven laboratories in eight regions, with headquarters at the CDC in Atlanta. The laboratory basis for PulseNet was the DNA "fingerprinting" of bacteria causing human illness, in order to define and defeat outbreaks. This organizational and scientific enterprise became possible through the development of PFGE. With pulsed field gel electrophoresis and the international coordination offered by PulseNet, the identification of complex outbreaks became much easier.

The incorporation of bioterrorism into PulseNet's remit may raise eyebrows, but in a world that has become accustomed to very much more violent terrorist atrocities, it is not widely known that one of the earliest modern terrorist plots was based on the use of *Salmonella* Typhimurium as a terrorist weapon. In September 1984, an outbreak of *Salmonella* Typhimurium associated with ten restaurants was logged in the city of The Dalles, Oregon. The outbreak was linked to the restaurants' salad bars, but they shared no common food source, and the investigators searched in vain for a common link between the food outlets. The Dalles was the county seat of Waco County, which since 1981 had been home to the Rajneeshee cult, housed on a ranch, whose relations with its neighbors had rapidly deteriorated, resulting in a threat to remove the cult's tax-free charitable status. In 1985, the cult collapsed, and its members were accused of various crimes, including sabotage of the salad bars since a phial of the same *Salmonella* as the outbreak strain was found in its laboratory. The cult's medical director had fled to Europe, and it transpired that she was a nurse practitioner who had obtained and cultivated the bacilli, and with cult members had sprinkled the liquid culture repeatedly on items in local salad bars. This was a trial run for a wholesale poisoning of the town water supply, with a view to disabling the local voters who were intending to revoke the cult's tax-free status. The fear that this example might provoke copycat events led to the suppression of the investigation report. It was only after the Sarin gas attack on the Tokyo underground in 1995 that the authors of the report on The Dalles poisoning episode were allowed to publish.[112]

In the years following the establishment of PulseNet, important salad-related outbreaks were a regular feature, and, among other issues, it became apparent that America had a large and continuing problem with tomato-related food poisoning.[113] Between 1999 and 2006, six out of eleven multistate outbreaks of salmonellosis were associated with contaminated tomatoes served in restaurants across America.[114] The final straw came in the autumn of 2006. On 3 October, the Vermont Public Health Laboratory informed American PulseNet of two *Salmonella* Typhimurium isolates with "indistinguishable" pulsed field gel pattern (CD Xba1 pattern JPXX01.0604). A few days earlier, the Minnesota Department of Health had identified four cases of *Salmonella* Typhimurium Newport isolate with a pattern matching that in Vermont. The following day, PulseNet identified a further thirty-three isolates with that pattern recorded by multiple states since 1 August—a notable leap in the record compared to the four previous years. By 19 October, the CDC was coordinating a multistate investigation, using a "dynamic iterative process of hypothesis generation."[115] To detail this process would take too long, but the outcome was the finding that all 190 known cases (which came from twenty-one US states, with one from Canada) were grouped in four district restaurant clusters and were all linked to the consumption of large, round, raw tomatoes. In a coordinated effort, state, local, and federal officials traced the source of these tomatoes to a single packing house in Ohio, supplied by three growers from twenty-five fields in three counties.[116]

This American example gives some indication of the foodborne risks that lie not only on, but in, salad vegetables. The dangers lurking in ready-prepared, bagged salads, already flagged in America in the 1960s, have been driven home by a succession of such outbreaks in Britain since 2000.[117] Yet if the herbs and baby leaves common in prepacked salads are difficult to clean because of their fragility, and salad sprouts can be contaminated from seed, both cucumbers and tomatoes can internalize bacteria with little more than minimal assistance from humans. As early as 1963, published evidence suggested that the bacterial content of cucumbers can vary from 9 percent at the core to 50 percent in the middle and 88 percent in the outer ring, including the disinfected surface peel.[118] Bacteria on warm tomatoes will pass through the skin into the tissue if the fruit is washed in cold water. Not only that, but tomatoes may be inoculated with pathogens by insects when in flower, or via the leaves and stem if irrigated with contaminated water.[119]

Tomatoes are not the only fruit to cause problems. Repeated international and interstate outbreaks of hepatitis A, norovirus, and cyclosporiasis have been caused by contaminated raspberries—a fruit which, it appears, generally goes unwashed because it is fragile, and as cane fruit is assumed to be "above" contamination.[120] In reality, infections can be carried by spray irrigation or on the hands of the pickers. There is no doubt, however, that modern surveillance systems, public health concerns, and newspaper reportage have made much of an age-old problem. In the nineteenth and even twentieth centuries most people suffered repeated bouts of diarrhea especially in summer, without much complaint, and even today only an estimated 3 percent of cases are reported.[121]

Conclusions

There is no doubt that food poisoning became more complicated in the late twentieth century as the food industries became complex and global, offering microbiological opportunity, and as modern scientific methods revealed increasing numbers of bacteria and viruses responsible for the escalating numbers of outbreaks. Until the mid-1970s, food poisoning outbreaks in America were largely associated with staphylococcal intoxication; in Europe they were mainly associated with *Salmonella*.[122] From the later 1970s, however, a lively stable of other infections began to be identified: cryptosporidium, campylobacter, listeria, *E. coli* O157:H7, and norovirus among them, all of which are blithely international in their habits. Some of these new foodborne pathogens, such as *E. coli* O157:H7, have themselves been products of the conditions of industrial agriculture. These uncomfortable discoveries came about through developments in medical science made through the twentieth century—not only in laboratory sciences like microbiology but also in epidemiology, the medical field science of detection. In both these domains, the exploration of the history of salad-related food poisoning allows us to see them developing toward the dictionary definition of science: an investigation according to rules laid down in exact science for performing observations and testing the soundness of conclusions.[123] All the surveillance systems in the world will not prevent these outbreaks, and consumers can only follow the scientific prescription to keep washing everything and trust their luck, especially when eating away from home.

Anne Hardy is an honorary professor in the London School of Hygiene and Tropical Medicine, University College, London. Her interests lie chiefly in the history of infectious disease, both social and scientific. Her most recent book, *Salmonella Infections, Networks of Knowledge and Public Health in Britain, 1880–1975* (Oxford: Oxford University Press, 2015), was awarded the Pickstone Prize of the British Society for the History of Science in 2016.

Notes

1. The British Library Catalogue lists just two volumes: Weinraub, *Salad*; Peachey, *The Book of Salads*. The term "salad science" is one I seem to have coined myself; I am not aware of having seen it used anywhere in my reading on this subject.
2. Greens, of course, were not the only source of food poisoning. *Salmonella*-contaminated raw milk posed a particular hazard, one recognized by British authorities in the early decades of the twentieth century. However, in the case of milk there were technological controls—namely, pasteurization and sanitary handling—though these took some time to be implemented nationally. Hardy, *Salmonella Infections*, 105–9.

3. Author's personal recollection.
4. Evelyn, *Acetaria*.
5. Grigson, *Food with the Famous*, 124–25.
6. Jaine, "Introduction"; Pepys, *The Diary of Samuel Pepys*; Pepys, *The Diary*, vol. 1, 139; Pepys, *The Diary*, vol. 3, 87.
7. Saunders, *Salad for the Solitary*, iii; Woodforde, *The Diary of a Country Parson*.
8. Beeton, *Mrs Beeton's Book*, 252–52, 370, 376, 387, 391.
9. De Loup, *The American Salad Book*, preface.
10. David, *Summer Cooking*, 26.
11. De Loup, *The American Salad Book*, 9. See also Shapiro, *Perfection Salad*, 71–105.
12. Evelyn, *Acetaria*, xviii–xix.
13. Beeton, *Mrs Beeton's Book*, 252.
14. Ayrton, *The Cookery*, 344–45.
15. Markets, "Covent Garden," 9. Coleworts are members of the cabbage family that do not have hearts, including kale and "greens." The name was first recorded in 1683 according to the Oxford English Dictionary.
16. Mayhew, *London Labour*, vol. 1, 145.
17. Ibid., 152–53.
18. Wynter, "The London Commissariat," 236.
19. Ibid.
20. Odling, *Memoir*, 73–74.
21. Notes of the Week, "Watercress and Typhoid," 8, 287; Ad Hoc Diseases Committee, "An Account," 224–30.
22. A Physician, "Is Watercress Wholesome?"
23. Smee, *My Garden*, 92. For Smee, see Buchanan, "Smee, Alfred (1818–1877)."
24. Smee, "Regulation," 60.
25. Notes of Observation, "Watercress"; A Physician, "Is Watercress Wholesome?"
26. Odling, *Memoir*, 72–74.
27. Smee, *My Garden*, 30–35.
28. Ibid., 93.
29. Smee, "Regulation." See London City Council, "Appendix II" of *MOAR for the Year 1904*, 2.
30. Smee, "Regulation," 60.
31. Hibberd, *Home Culture*, 4. For Hibberd see Wilkinson, "The Preternatural Gardener."
32. Hibberd, *Home Culture*, 4.
33. Hardy, *The Epidemic Streets*, Fig. 6.2, 156.
34. Hardy, *Salmonella Infections*, 43–63.
35. London City Council, *MOAR for the Year 1894*, 35.
36. London City Council, *MOAR for the Year 1900*, 10.
37. London City Council, *MOAR for the Year 1903*, 35. See also Annotation, "Enteric Fever."
38. London City Council, "Appendix II" of *MOAR for the Year 1904*.
39. De Loup, *The American Salad Book*, 9.
40. Suzanne, *A Book of Salads*, 8.
41. See for example, Mechi, "Town Sewage."
42. Ten reports were published in the British Parliamentary Papers (BPP) series, between 1901 and 1915.

43. Royal Commission on Sewage Disposal, *Methods*, 183–88.
44. See for example, Klein, "On the Behavior"; Martin, "On the Nature"; Savage, "Report."
45. Andrewes and Horder, "A Study," 708. On these debates see Amsterdamska, "Medical and Biological Constraints."
46. Hardy, "Lives, Laboratories."
47. Hardy, *Epidemic Streets*.
48. Creel, "Vegetables as a Possible Factor," 187. This article was reprinted in Britain in 1917 in *Reports of the Medical Officer*.
49. Ibid., 190–91.
50. Melick, "The Possibility."
51. Ibid., 36.
52. Ibid.
53. Geldreich and Bordner, "Fecal Contamination."
54. On the decline in typhoid death rates see Hardy, *Epidemic Streets*, 33.
55. Hardy, *Epidemic Streets*, 216–17.
56. For typhoid in America, see Hardy, "Scientific Strategy," 10–23.
57. Kratz, "The Present Status," 195; map facing 194.
58. Chapman, "The Menace."
59. Hedberg, MacDonald, and Osterholm, "Changing Epidemiology," 674.
60. Chapman, "The Menace," 4. In his words: "I am told there was absolutely no doubt that the substance was feces." Creel noted in 1912 that "even in well managed households and public eating places scrupulous care in preparing articles for the table is exceptional." Creel, "Vegetables as a Possible Factor," 193.
61. Chapman, "Menace to Life," 4.
62. Chapman, "Menace to Life," 5–6. Frederick Henry Osborn (1889–1981) became a distinguished philanthropist, and eugenicist.
63. Chapman, "Menace to Life," 6 and 5, table 1.
64. Ibid., 10, chart II.
65. Chapman, "Sewage Disposal."
66. Glaub, "Environmental Sanitation."
67. Ibid., 115, 118.
68. Winfield, *China*.
69. Scott, *Health and Agriculture*, 21–22.
70. Winfield, *China*, 117–19, 127, 129.
71. Ibid., 112.
72. Falk, "Bacterial Contamination," 1338.
73. Cruickshank, "Typhoid Fever in Devon." I am indebted to Claas Kirchhelle for sharing this source, which he identified as part of his University College Dublin Wellcome Trust University Award project, "Enslaved Viruses: Bacteriophage-Typing and Infectious Disease Control."
74. Gayler et al., "An Outbreak."
75. Keymer, "A Survey"; Keymer, "A Survey—Part II."
76. Lofton, Morrison, and Leiby, "The Enterobacteriaceae."
77. Geldreich and Bordner, "Fecal Contamination," 188.
78. Behravesh et al., "A Multistate Outbreak."
79. Tamminga, Beumer, and Kampelmacher, "The Hygienic Quality," 152.

80. Geldreich and Bordner, "Fecal Contamination," 193.
81. Ibid., 192–93.
82. See Shapiro and Holder, "Effect of Antibiotic," for American bagged salad. For British bagged salad and Marks and Spencer, see "Fact & Folklore, 1 History and Facts, BLSA fact sheet No. 1," 25 March 2016, Leafy Salad website, accessed 5 September 2020, http://www.makemoreofsalad.com/wp-content/uploads/2016/06/Salad-History-and-Etymology.pdf.
83. Shapiro and Holder, "Effect of Antibiotic," 341–45.
84. Sivapalasingam et al., "Fresh Produce."
85. Hedberg, MacDonald, and Osterholm, "Changing Epidemiology," 671.
86. Ibid., 672.
87. These are four serotypes (Chester, Poona, Javiana, and Montevideo), or different groups, within the single species *Salmonella enterica*. Typhimurium is another (mentioned below).
88. Hedberg, MacDonald, and Osterholm, "Changing Epidemiology," 672–77.
89. Sivapalasingam et al., "Fresh Produce," figure 1, 2343. The article's numbers for outbreaks due to salad were 6 (1970s), 25 (1980s), 44 (1990s); and due to vegetables, 2 (1970s), 14 (1980s), 34 (1990s).
90. Hedberg, MacDonald, and Osterholm, "Changing Epidemiology," 671–82, 675–76.
91. For the multistate tomato outbreak, see Behravesh et al., "A Multistate Outbreak."
92. Sagoo et al., "Microbiological Study," 404.
93. Fisher, "Salm-Net."
94. Ibid.
95. Martin and Bean, "Data Management Issues."
96. Mahon et al., "An International Outbreak."
97. Hilborn et al., "A Multistate Outbreak."
98. Lytton, *Outbreak*, 118–61.
99. Lim, Yoon, and Hovde, "A Brief Overview."
100. Drexler, *Emerging Epidemics*.
101. Kirchhelle, *Pyrrhic Progress*.
102. Drexler, *Emerging Epidemics*.
103. Cited in Stephenson, "New Approaches," 1337.
104. Stephenson, "New Approaches," 1339.
105. Ibid.
106. Ibid.
107. Ibid.
108. Sivapalasingam et al., "Fresh Produce," table 6, 2346.
109. For the creation of PulseNet, see Swaminathan et al., "PulseNet."
110. Hedberg, "Global Surveillance Needed," 59.
111. CDC, "About PulseNet."
112. Török et al., "A Large Community Outbreak"; See also Pendergrast, *Inside the Outbreaks*, 221–23.
113. Behravesh et al., "A Multistate Outbreak."
114. Ibid., 2054.
115. Ibid.
116. Ibid., 2057.

117. N. S. Boxall et al., "A *Salmonella* Typhimurium."
118. Samish, Etinger-Tulczynska, and Bick, "The Microflora," 261.
119. Lin and Wei, "Transfer of *Salmonella*"; Guo et al., "Survival of Salmonellae."
120. Reid and Robinson, "Frozen Raspberries"; Chalmers, Nichols, and Rooney, "Foodborne Outbreaks"; Sarvikivi et al., "Multiple Norovirus Outbreaks."
121. Behravesh et al., "A Multistate Outbreak," 2058.
122. Hedberg, MacDonald, and Osterholm, "Changing Epidemiology," 671.
123. Oxford English Dictionary.

Bibliography

A Physician. "Is Watercress Wholesome?" *Gardeners Magazine* 19 (1876): 2.

Ad Hoc Diseases Committee. "An Account of an Outbreak of Sonne Dysentery at Papworth Village Settlement." *Monthly Bulletin of the Ministry of Health* 12 (1953): 224–30.

Amsterdamska, Olga. "Medical and Biological Constraints: Early Research on Variation in Bacteriology." *Social Studies of Science* 17, no. 4 (November 1, 1987): 657–87.

Andrewes, F. W., and Thomas J. Horder. "A Study of the Streptococci Pathogenic for Man." *The Lancet* 168, no. 4333 (15 September 1906): 708–13; no. 4334 (22 September 1906): 775–83; no. 4335 (29 September 1906): 852–55.

Annotation. "Enteric Fever Spread by Water-Cress." *The Lancet* 162, no. 4189 (12 December 1903): 1671.

Ayrton, Elisabeth. *The Cookery of England, Being a Collection of Recipes for Traditional Dishes of All Kinds from the Fifteenth Century to the Present Day, with Notes on Their Social and Culinary Background.* London: A. Deutsch, 1974.

Beeton, Isabella. *Mrs Beeton's Book of Household Management.* Edited by Nicola Humble. Abridged edition. Oxford: Oxford University Press, 2000.

Behravesh, C. Barton, D. Blaney, C. Medus, S. A. Bidol, Q. Phan, S. Soliva, E. R. Daly, et al. "Multistate Outbreak of Salmonella Serotype Typhimurium Infections Associated with Consumption of Restaurant Tomatoes, USA, 2006: Hypothesis Generation through Case Exposures in Multiple Restaurant Clusters." *Epidemiology & Infection* 140, no. 11 (November 2012): 2053–61.

Boxall, N. S., G. K. Adak, E. De Pinna, and I. A. Gillespie. "A *Salmonella* Typhimurium Phage Type (PT) U320 Outbreak in England, 2008: Continuation of a Trend Involving Ready-to-Eat Products." *Epidemiology & Infection* 139, no. 12 (December 2011): 1936–44.

Buchanan, P. D. "Smee, Alfred (1818–1877), Chemist and Surgeon." In *Oxford Dictionary of National Biography*, edited by H. C. G. Matthew, 50: 992–93. Oxford: Oxford University Press, 2004.

CDC. "About PulseNet." Accessed 16 October 2019. https://www.cdc.gov/pulsenet/about/index.html.

Chalmers, R. M., G. Nichols, and R. Rooney. "Foodborne Outbreaks of Cyclosporiasis Have Arisen in North America. Is the United Kingdom at Risk?" *Communicable Disease and Public Health* 3, no. 1 (March 2000): 50–55.

Chapman, Edward N. "The Menace to Life and Health from Improper Sewage Disposal in Colorado." *Colorado Medicine* 31, no. 1 (1934): 4–10.

———. "Sewage Disposal: A Major Public Health Problem in Colorado." *Colorado Medicine* 31, no. 10 (1934): 337–43.

Creel, R. H. "Vegetables as a Possible Factor in the Dissemination of Typhoid Fever." *Public Health Reports (1896-1970)* 27, no. 6 (1912): 187-93.

Cruickshank, J. C. "Typhoid Fever in Devon: The Value of Phage-Typing in a Rural Area." *Emergency Public Health Laboratory Service Monthly Bulletin* 6 (1947): 88-96.

David, Elizabeth. *Summer Cooking.* Harmondsworth: Penguin Books, 1965.

De Loup, Maximilian. *The American Salad Book.* London: William Heinemann, 1901.

Drexler, Madeline. *Emerging Epidemics: The Menace of New Infections.* Washington, DC: Joseph Henry Press, 2002.

Evelyn, John. *Acetaria: A Discourse of Sallets (1699), with an Introduction by Tom Jaine.* Edited by Christopher P. Driver. Totnes, Devon: Prospect Books, 1996.

Falk, Lloyd L. "Bacterial Contamination of Tomatoes Grown in Polluted Soil." *American Journal of Public Health and The Nation's Health* 39, no. 10 (October 1949): 1338-42.

Fisher, I. "Salm-Net: A Network for Human Salmonella Surveillance in Europe." *Eurosurveillance,* 1 September 1995, 7-8. Accessed 5 September 2020, https://www.eurosurveillance.org/content/10.2807/esm.00.00.00194-en.

Gayler, Gilbert E., Robert A. MacCready, Joseph P. Reardon, and Bernard F. McKernan. "An Outbreak of Salmonellosis Traced to Watermelon." *Public Health Reports* 70, no. 3 (March 1955): 311-13.

Geldreich, Edwin E., and Robert H. Bordner. "Fecal Contamination of Fruits and Vegetables during Cultivation and Processing for Market: A Review." *Journal of Milk and Food Technology* 34, no. 4 (1 April 1971): 184-95.

Glaub, W. H. "Environmental Sanitation—A Colorado Major Health Problem: A Review of the Problem." *Rocky Mountain Medical Journal* 43 (1946): 99-118.

Grigson, Jane. *Food with the Famous.* London: Michael Joseph, 1979.

Guo, Xuan, Jinru Chen, Robert E. Brackett, and Larry R. Beuchat. "Survival of Salmonellae on and in Tomato Plants from the Time of Inoculation at Flowering and Early Stages of Fruit Development through Fruit Ripening." *Applied and Environmental Microbiology* 67, no. 10 (October 2001): 4760-64.

Hardy, Anne. *The Epidemic Streets: Infectious Disease and the Rise of Preventive Medicine, 1856-1900.* Oxford: Oxford University Press, 1993.

———. "Lives, Laboratories, and the Translations of War: British Medical Scientists, 1914 and Beyond." *Social History of Medicine* 30, no. 2 (1 May 2017): 346-66.

———. *Salmonella Infections, Networks of Knowledge, and Public Health in Britain, 1880-1975.* Oxford: Oxford University Press, 2015.

———. "Scientific Strategy and Ad Hoc Response: The Problem of Typhoid in America and England, c. 1910-50." *Journal of the History of Medicine and Allied Sciences* 69, no. 1 (23 December 2013): 3-37.

Hedberg, Craig W. "Global Surveillance Needed to Prevent Foodborne Disease." *California Agriculture* 54, no. 5 (1 September 2000): 54-61.

Hedberg, Craig W., Kristine L. MacDonald, and Michael T. Osterholm. "Changing Epidemiology of Food-Borne Disease: A Minnesota Perspective." *Clinical Infectious Diseases* 18, no. 5 (1994): 671-80.

Hibberd, Shirley. *Home Culture of the Water-Cress: A Practical Guide to the Cultivation of the Water-Cress in Pans, Troughs, Beds, and Forcing Frames, for the Supply of the Household in All Seasons.* London: E. W. Allen, 1878.

Hilborn, Elizabeth D., Jonathan H. Mermin, Patricia A. Mshar, James L. Hadler, Andrew Voetsch, Christine Wojtkunski, Margaret Swartz, et al. "A Multistate Outbreak of *Escherichia coli* O157:H7 Infections Associated with Consumption of Mesclun Lettuce." *Archives of Internal Medicine* 159, no. 15 (9 August 1999): 1758–64.

Jaine, Tom. "Introduction." In *Acetaria: A Discourse of Sallets (1699), with an Introduction by Tom Jaine*, edited by Christopher P. Driver. Totnes, Devon: Prospect Books, 1996.

Keymer, I. F. "A Survey and Review of the Causes of Mortality in British Birds and the Significance of Wild Birds as Disseminators of Disease." *The Veterinary Record* 70 (1958): 713–20.

———. "A Survey and Review of the Causes of Mortality in British Birds and the Significance of Wild Birds as Disseminators of Disease, Part II." *The Veterinary Record* 70 (1958): 736–40.

Kirchhelle, Claas. *Pyrrhic Progress: The History of Antibiotics in Anglo-American Food Production*. New Brunswick: Rutgers University Press, 2020.

Klein, Edward. "On the Behaviour of the Bacillus of Enteric Fever and of Koch's Vibrio in Sewage." *British Parliamentary Papers*, vol. LI (1895), Appendix B no. 2, 885–88.

Kratz, F. W. "The Present Status of Full-Time Local Health Organization." *Public Health Reports (1896–1970)* 57, no. 6 (1942): 194–96.

Lim, Ji Youn, Jangwon Yoon, and Carolyn J. Hovde. "A Brief Overview of Escherichia Coli O157:H7 and Its Plasmid O157." *Journal of Microbiology and Biotechnology* 20, no. 1 (January 2010): 5–14.

Lin, Chia-Min, and Cheng-I Wei. "Transfer of *Salmonella* Montevideo onto the Interior Surfaces of Tomatoes by Cutting." *Journal of Food Protection* 60, no. 7 (July 1997): 858–62.

Lofton, C. B., S. M. Morrison, and P. D. Leiby. "The Enterobacteriaceae of Some Colorado Small Mammals and Birds, and Their Possible Role in Gastroenteritis in Man and Domestic Animals." *Zoonoses Research* 1, no. 15 (1962): 277–93.

London County Council. "Appendix II: Report on Watercress and Watercress-Beds in the Neighbourhood of London." In *Medical Officer's Annual Report for the Year 1904*, 13th Annual Report: 1–18. London: Southwood, Smith & Co., Ltd., 1905.

———. *Medical Officer's Annual Report for the Year 1894*. 3rd Annual Report. London, 1895.

———. *Medical Officer's Annual Report for the Year 1900*. 9th Annual Report. London: Jas. Truscott & Son, Printers, 1901.

———. *Medical Officer's Annual Report for the Year 1903*. 12th Annual Report. London: Jas. Truscott & Son, Ltd., 1904.

Lytton, Timothy D. *Outbreak: Foodborne Illness and the Struggle For Food Safety*. Chicago: University of Chicago Press, 2019.

Mahon, Barbara E., Antti Pönkä, William N. Hall, Kenneth Komatsu, Stephen E. Dietrich, Anja Siitonen, Gary Cage, et al. "An International Outbreak of Salmonella Infections Caused by Alfalfa Sprouts Grown from Contaminated Seeds." *Journal of Infectious Diseases* 175, no. 4 (1997): 876–82.

Markets. "Covent Garden." *Gardeners Magazine*, 1876.

Martin, Sidney. "On the Nature of Antagonism of Soil to the Typhoid Bacillus." *British Parliamentary Papers*, vol. XXVI (1901), Appendix no. 5, 487–510.

Martin, S. M., and N. H. Bean. "Data Management Issues for Emerging Diseases and New Tools for Managing Surveillance and Laboratory Data." *Emerging Infectious Diseases* 1, no. 4 (1995): 124–28.

Mayhew, Henry. *London Labour and the London Poor.* 2 vols. New York: Dover Publications, 1968.

Mechi, J. T. "Town Sewage." *Gardeners Magazine* 19 (1876): 676.

Melick, C. O. "The Possibility of Typhoid Infection through Vegetables." *Journal of Infectious Diseases* 21, no. 1 (1917): 28–38.

Notes of Observation. "Watercress." *Gardeners Magazine* 18 (1875): 619.

Notes of the Week. "Watercress and Typhoid." *Medical Officer* 7, no. 8 (1912): 287.

Odling, Elizabeth. *Memoir of the Late Alfred Smee F.R.S. by His Daughter.* London: G. Bell and Sons, 1878.

Peachey, Stuart. *The Book of Salads 1580–1660.* Bristol: Stuart Press, 1993.

Pendergrast, Mark. *Inside the Outbreaks: The Elite Medical Detectives of the Epidemic Intelligence Service.* Boston: Houghton Mifflin Harcourt, 2010.

Pepys, Samuel. *The Diary of Samuel Pepys, A New and Complete Transcription.* Edited by Robert Latham and William Matthews. 11 vols. London: Bell (vols. 1–9); Bell and Hyman (vols. 11–12), 1970–71.

———. *The Diary of Samuel Pepys, A New and Complete Transcription.* Vol. 1, *1660.* Edited by Robert Latham and William Matthews. London: Bell, 1970.

———. *The Diary of Samuel Pepys, A New and Complete Transcription.* Vol. 3, *1662.* Edited by Robert Latham and William Matthews. London: Bell, 1971.

Reid, T. M., and H. G. Robinson. "Frozen Raspberries and Hepatitis A." *Epidemiology and Infection* 98, no. 1 (February 1987): 109–12.

Royal Commission on Sewage Disposal. *Methods of Treating and Disposing of Sewage.* 5th Report. Vol. Cmd. 4278. London: Wyman & Sons, Limited, 1908.

Sagoo, S. K., C. L. Little, L. Ward, I. A. Gillespie, and R. T. Mitchell. "Microbiological Study of Ready-to-Eat Salad Vegetables from Retail Establishments Uncovers a National Outbreak of Salmonellosis." *Journal of Food Protection* 66, no. 3 (March 2003): 403–9.

Samish, Zdenka, R. Etinger-Tulczynska, and Miriam Bick. "The Microflora within the Tissue of Fruits and Vegetables." *Journal of Food Science* 28, no. 3 (1963): 259–66.

Sarvikivi, E., M. Roivainen, L. Maunula, T. Niskanen, T. Korhonen, M. Lappalainen, and M. Kuusi. "Multiple Norovirus Outbreaks Linked to Imported Frozen Raspberries." *Epidemiology & Infection* 140, no. 2 (February 2012): 260–67.

Saunders, Frederick. *Salad for the Solitary, by an Epicure.* London: R. Bentley, 1853.

Savage, William G. "Report on the Distribution of Organisms of the Gaertner Group in the Animal Intestine." *British Parliamentary Papers*, vol. XXX (1908), Appendix B no. 7, 275–300.

Scott, James Cameron. *Health and Agriculture in China: A Fundamental Approach to Some of the Problems of World Hunger.* London: Faber and Faber, 1952.

Shapiro, Jennie E., and Ian A. Holder. "Effect of Antibiotic and Chemical Dips on the Microflora of Packaged Salad Mix." *Applied Microbiology* 8, no. 6 (November 1960): 341–45.

Shapiro, Laura. *Perfection Salad: Women and Cooking at the Turn of the Century.* New York: Farrar, Straus, and Giroux, 1986.

Sivapalasingam, Sumathi, Cindy R. Friedman, Linda Cohen, and Robert V. Tauxe. "Fresh Produce: A Growing Cause of Outbreaks of Foodborne Illness in the United States, 1973 through 1997." *Journal of Food Protection* 67, no. 10 (October 2004): 2342–53.

Smee, Alfred. *My Garden, Its Plan and Culture Together with a General Description of Its Geology, Botany, and Natural History.* London: Bell and Daldy, 1872.

———. "Regulation of Sewage Grounds: Proposed Heads of Legislation for the Regulation of Sewage Grounds." *Gardeners Magazine* 19 (1876): 23–24, 60–62, 88–90.

Stephenson, Joan. "New Approaches for Detecting and Curtailing Foodborne Microbial Infections." *JAMA* 277, no. 17 (7 May 1997): 1337–40.

Suzanne, Alfred. *A Book of Salads: The Art of Salad Dressing.* Edited by Charles Herman Senn. 3rd ed., revised and augmented by C. Herman Senn. London: Food and Cookery Publishing Agency, 1914.

Swaminathan, B., T. J. Barrett, S. B. Hunter, R. V. Tauxe, and CDC PulseNet Task Force. "PulseNet: The Molecular Subtyping Network for Foodborne Bacterial Disease Surveillance, United States." *Emerging Infectious Diseases* 7, no. 3 (2001): 382–89.

Tamminga, S. K., R. R. Beumer, and E. H. Kampelmacher. "The Hygienic Quality of Vegetables Grown in or Imported into the Netherlands: A Tentative Survey." *Journal of Hygiene* 80, no. 1 (February 1978): 143–54.

Török, Thomas J., Robert V. Tauxe, Robert P. Wise, John R. Livengood, Robert Sokolow, Steven Mauvais, Kristin A. Birkness, Michael R. Skeels, John M. Horan, and Laurence R. Foster. "A Large Community Outbreak of Salmonellosis Caused by Intentional Contamination of Restaurant Salad Bars." *JAMA* 278, no. 5 (6 August 1997): 389–95.

Weinraub, Judith. *Salad: A Global History.* London: Reaktion Books, 2016.

Wilkinson, Anne. "The Preternatural Gardener: The Life of James Shirley Hibberd (1825–90)." *Garden History* 26, no. 2 (1998): 153–75.

Winfield, Gerald F. *China: The Land and the People.* New York: W. Sloane Associates, 1948.

Woodforde, Parson James. *The Diary of a Country Parson.* Edited by John Beresford. 5 vols. Oxford: Oxford University Press, 1981.

Wynter, Andrew. "The London Commissariat." In *Curiosities of Civilization.* London: R. Hardwicke Detroit, Singing Tree Press, 1968.

CHAPTER 2

Radioactive Diet
Food, Metabolism, and the Environment, ca. 1960

Soraya de Chadarevian

In 1961, at the first open day for the press at the Radiobiological Research Unit, situated at the UK Atomic Energy Research Establishment in Harwell, one project attracted particular attention. Three scientists from the unit, including the director, had been on a month-long diet consisting of bread and milk enriched with radioactive strontium. The experiment was designed to test the uptake of strontium in the body from different types of food. To distinguish the food source for the radioactive strontium found in the body, the foodstuffs were contaminated with two different unstable isotopes of the same element—the milk with strontium-85 and the bread with strontium-90.[1] Preliminary results indicated that there was little difference in the absorption of strontium from the milk or the bread.[2] While this finding could perhaps dispel the fear that young children drinking milk were especially strongly at risk, it also disproved theories that strontium contamination in grain—the staple food of many people—was less harmful. Many other experiments—all dealing with radiation-related effects—were mounted for the open day, but the diet experiment provided the headline or leading story for most of the press reports. Major scientific and medical journals such as *Nature* and the *British Medical Journal*, as well as broadsheets such as the *Guardian* and the *Daily Telegraph* carried reports on the experiment.[3] With a few months' delay, the *Sunday Pictorial* (the forerunner of the *Sunday Mirror*, sister paper of the *Daily Mirror*) featured one of the participants of the diet experiment, Terry Carr, 37, a father of four, on its front page, celebrating his courage in volunteering as a human guinea pig—"so that YOU may live without fear."[4] Press photos showed Carr cheerfully looking up from a table featuring his daily ratio of radioactively enriched food and stoically lying under a whole body radiation counter for the ingested radiation to be measured (figures 2.1 and 2.2). Carr was portrayed as "laughing off thoughts of danger." Reportedly he summarized his position thus: "I won't say the experiment was without danger, but the danger was negligible because the quantity of radioactivity was very small. It was like driving

Figure 2.1. Photograph of Terry Carr, experimental subject in radioactive diet experiment. *Sunday Pictorial*, 14 October 1962, p. 21. Reproduced with permission.

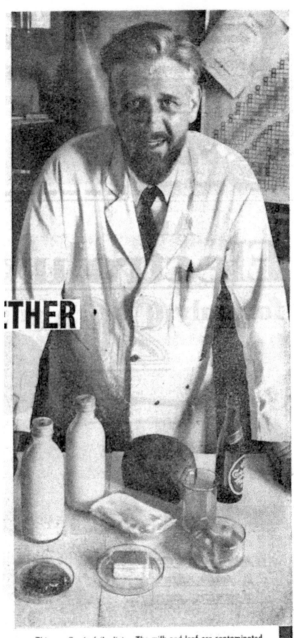

This was Carr's daily diet.   The milk and loaf are contaminated.

Figure 2.2. "This Man Is Radioactive!" Cover of *Sunday Pictorial*, 14 October 1962. Reproduced with permission.

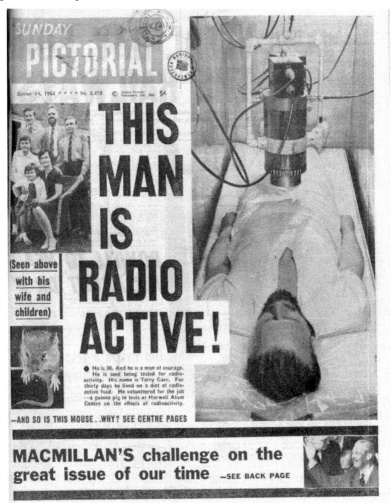

home from work—you can have an accident, but the chances are a million to one."[5] Indeed, the scientists were looking for more volunteers, especially women and children, for further investigations.

By the early 1960s, the radioactive contamination of food with elements such as strontium that accumulated in the human body was becoming a new and most alarming feature of the atomic age. The contamination of such common products of the human diet as bread and milk epitomized the daily and inescapable threat from radioactive fallout from the atomic weapon tests—a situation the diet experiments were designed to throw some light on. The experiments may well have had particular resonance in Britain where, just a few years earlier, in 1957, an accident at the new plutonium processing plant at Windscale in Cumbria, followed by a significant release of radiation into the atmosphere, had led to a six-week ban on the consumption of milk from a 200-square-mile exclusion area.

This chapter reviews the debates about the radioactive contamination of food in the 1950s and 1960s. In particular, it considers how the element strontium and milk moved into the center of attention with respect to concerns about radioactive food and how these developments affected—or did not affect—notions of food risks, the purity and safety of foodstuff, and respective legal frameworks. It explores monitoring programs; research into ecological transport cycles, food chains, and metabolism; political, legal, technical, and industrial solutions proposed to contain the perceived contamination; and the role of the press and public responses in framing the issues. The chapter focuses on events in Britain and the public management of the crisis following the Windscale fire, the world's most serious nuclear accident at the time, which compounded rising concerns on the effects of global fallout on human diet and the environment. In atomic nations like Britain, all issues around nuclear energy were wrapped up in questions of national security. Nevertheless, the emerging understanding of risks from environmental radioactivity created a template for apprehending chemical contaminants in the environment, giving debates over radioactive isotopes in milk a salience to broader debates over food safety.

Bone-Seeker

Uranium fission produces over two hundred isotopes of thirty-four elements.[6] Research on the fission products, their half-life, and their accumulation in the body, as well as their cancer-inducing effects (carcinogenicity), was pursued in animal experiments since the early 1940s under the auspices of the Manhattan Project. Strontium-90 quickly emerged as the radioisotope that raised most concerns. There were various reasons for this.

Strontium-90 has a half-life of twenty-nine years, which is long in relation to the human lifespan. Furthermore, it is an analogue of calcium and thus readily accumulates in bone tissue. Because of the slow turnover of bone tissue, once embedded in the bone, strontium, like other "bone seekers," is retained in the body for a long time and acts as an internal emitter. Animal experiments conducted at the Metallurgic Laboratory in Chicago indicated that strontium-90 was a potent carcinogenic agent, leading to high leukemia and bone cancer incidences.[7] For all these reasons, researchers agreed early on that ingested strontium-90 would represent the most serious long-term hazard from radioactive fallout.[8] There were other radioisotopes from fallout that were biomedically relevant, notably iodine-131 and cesium-137, which accumulate in the thyroid gland and in muscle, respectively. Yet iodine-131 has a half-life of only eight days, while cesium, although long-lived with a half-life of thirty-three years, is excreted rapidly from the body. These characteristics made these substances seem less of an issue at the time.[9] It was only in the aftermath of the Windscale accident that the accumulation of iodine in the food chain and the severity of the health issues involved with that element were recognized.

The results of the wartime experiments in Berkeley and Chicago reached the open literature in the first postwar years. Research in the United States continued under the auspices of the Atomic Energy Commission (AEC). From the mid-1940s, metabolic studies of strontium-90 were also pursued in Britain as part of the Medical Research Council (MRC) radiation protection program.[10] At this time, researchers were considering hypothetical scenarios of weapons deployment and attempts to determine "practical limits" for using atomic weapons. Yet events in the early 1950s—especially increasing concerns regarding the effects of fallout at the Nevada test site on US soil and the detonation of the first thermonuclear bombs, which produced global fallout—made it clear that the problem of strontium-90 was much more acute and widespread. This set in motion wide-scale monitoring efforts as a way to assess and manage the situation.

Worldwide Monitoring

In 1953, the AEC and the RAND Corporation, in collaboration with Manhattan veteran and deviser of the radiocarbon dating technique Willard F. Libby from the University of Chicago, launched a secret project code-named Sunshine (possibly an allusion to the fact that RAND was located in Santa Monica, in the sunshine state) to assess the worldwide levels of strontium-90 already produced from weapons testing. Infant bone samples from India, Japan, South Africa, and South America were collected under the pretense that they were being tested for their natural radium content. Plant and soil samples were also collected and tested for strontium-90. The measurements indicated a large uniformity in strontium-90 content in different locations around the world, highlighting the global character of the phenomenon. Project Sunshine remained secret until after global fallout became a central public concern following the infamous US Castle Bravo H-bomb test on Bikini Island in 1954.[11]

The Lucky Dragon accident and the exposure of its fishermen to, in one case, lethal doses of radiation from the fallout of the H-bomb test is well known. Perhaps less well known is that following the incident the Japanese government put in place an extensive monitoring program of the water and the organisms in the ocean, including especially plankton and fish.[12] The elevated radiation measurements found in fish in the ocean and in fishers' catches led to a collapse of the local fish market. The results made clear the importance of ingestion as a pathway for strontium into the body as well as the problem of the accumulation of radioactivity in the food chain. More generally, the events in Japan became a watershed moment for the realization that fallout put food supplies at risk.[13]

The results of the Japanese monitoring efforts and the subsequent publication of the data of the Sunshine Project prompted a wave of new monitoring programs in countries far removed from the actual test sites. They were organized on a local, national, and international level and, together, further brought home the global character of the phenomenon. Unlike the United States, Britain did not have a test site on its national territory, but being committed to an atomic energy program and contributing to global fallout through its atomic weapons tests in Australia, it nonetheless stood under strong pressure to address the problem.

In response to the situation, in 1955 Britain launched a national bone collection program to assess the amount of radioactive strontium that had already accumulated in the human skeleton. Bones from persons of all ages, both male and female, were provided by pathologists from routine autopsies around the country and sent to the Woolwich Outstation of the Atomic Energy Research Establishment, where researchers determined the strontium-90 content in the ash of the bones. To standardize procedures, pathologists provided whole or longitudinal sections of the thighbone (this was later changed to vertebrae), except for infants, for which a more varied selection of bones was sent in for analysis. Infant thighbones on their own did not provide enough material for analysis. At the same time, the strontium-90 content in various body parts of infants was found not to vary as much as in adults, so samples were still comparable. In addition to the level of strontium-90 (corrected against the naturally occurring strontium-90 in bones), a record was kept of the type of sample, the date of death, the age at death, and the county in which the individual was living at the time of death. Altogether, 3,417 specimens of human bones were tested in the national monitoring program. Surveys conducted at the Royal Hospital for Sick Children in Glasgow, at the University of Cambridge, and in West London provided additional data, leading to a total of just under 6,000 samples (2,084 in the Glasgow survey; 419 in the West London survey, and 79 samples in the Cambridgeshire survey).[14]

From 1957, the MRC, together with the Agricultural Research Council (ARC), took responsibility for these studies. Monitoring was expanded to include not just human bones and other tissues, but also testing of the air and rainwater and all products that formed part of the human diet.[15] The results were published in a Monitoring Report Series issued by the government publishing office (HMSO) and were regularly reviewed in the scientific literature. Nineteen official reports on the monitoring program were issued between 1960 and 1973.[16] The reports showed that the strontium content in the environment and in human bones rose steadily (figure 2.3). The figures peaked in 1963–64 following an increased number of atmospheric detonations of nuclear weapons carried out by the USSR and the United States in the early 1960s. With atmospheric testing ending in 1964 as an effect of the partial test ban treaty, strontium levels started to drop. Both effects showed up most dramatically in infants because of the fast turnover rates of their bones (figure 2.4).[17] The media keenly seized on the findings. Headlines in the mid-1960s ranged from the alarming: "Strontium-90 Rise Continues" and "More Strontium-90 in Babies' Bones," to the reassuring "Strontium-90 Up, but Safe."[18]

The data accumulated for monitoring provided unique opportunities for further research. Researchers at the MRC Radiobiological Research Unit at Harwell used the correlation of measurement data for strontium-90 in the human diet and the bone to investigate the uptake and turnover rates of the element (and hence of bone material more generally) in the human skeleton and the way these parameters varied with age. The investigations indicated that in children up to the age of about ten, approximately half the mineral strontium in the skeleton is replaced annually. In adolescents, the annual turnover rate is

Figure 2.3. Cumulative deposition of strontium-90 from fallout in Britain between 1954 and 1957. Antoinette Pirie, ed., *Fall Out: Radiation Hazards from Nuclear Explosions* (London: MacGibben & Kee, 1958), p. 98, figure 1. Reproduced with permission of HarperCollins Publishers Ltd. © Pirie 1958.

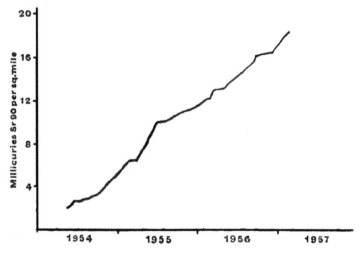

Figure 2.4. Strontium-90 accumulation in human bones in different age groups from London surveys, 1963–65. Medical Research Council, *Assay of Strontium-90 in Human Bone in the United Kingdom: Results for 1966 with Some Further Results for 1965*, MRC Monitoring Report No. 15 (London: Her Majesty's Stationery Office, 1967). Public domain.

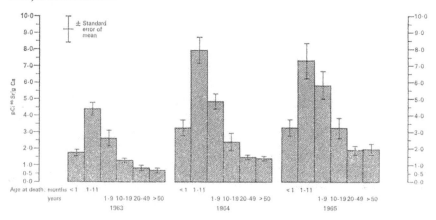

about 20 percent, dropping to an even lower figure in adults. The percentage uptake of strontium-90 in the body from diet showed a similar age dependency.[19] The diet experiments conducted at Harwell in the 1960s mentioned at the beginning of this chapter were designed to further investigate these relationships and supply additional data. In contrast to the data extracted from the "unplanned exposure" of radioactive fallout, they were set up under controlled laboratory conditions.[20]

Eventually, a mathematical formula was found that described the correlation between the different factors. It led to the abolition of the costly bone monitoring project in 1970 as accumulation rates of strontium-90 in the body could now be predicted from data on the human diet, which continued to be monitored. Data on milk, especially, continued to be collected. The same formula could also be used to calculate the radiation doses received by bone tissue from intake of strontium-90.[21] In the decades following World War II, artificially produced radioisotopes were widely used as tracer elements to investigate biochemical pathways and metabolic processes.[22] In the case of the fallout studies, the radioisotopes and their movement through soil, plants, animals, and humans functioned as their own tracer experiments.

The ultimate purpose of all the fallout studies at this stage was the assessment of its potential hazard to man.[23] The model reproduced here of the strontium transport from the environment (soil, plants, and milk) to the human body synthesized much of the contemporary thinking and knowledge about the issue (figure 2.5). It highlights the role of the human metabolism in differ-

Figure 2.5. Model for transport of strontium-90 from soils to man. Wright H. Langham, "Considerations of Biospheric Contamination by Radioactive Fallout," in *Radioactive Fallout, Soils, Plants, Foods, Man*, ed. Eric B. Fowler (Amsterdam: Elsevier, 1965), p. 11, figure 6. Reproduced with permission of Elsevier.

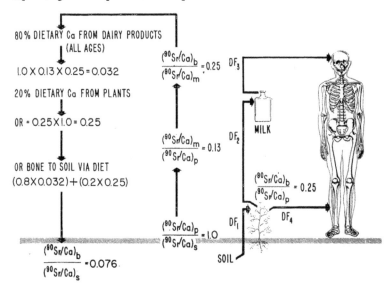

entially absorbing and retaining specific elements from the diet. It also points to the centrality of milk in the scheme. Providing 80 percent of the calcium intake for humans of all ages on a Western diet, milk and its products functioned as conduits of the contamination of the soil and plants to humans.

Milk Pathway

Milk had long been at the center of national discourses on food and food safety. If on the one hand milk was hailed as a complete and essential food—especially for young children—on the other hand it was seen as carrier of germs, especially tuberculosis bacilli. With increasing urbanization it was difficult to keep milk fresh. Indeed, "fresh milk" very soon came to mean pasteurized milk.[24]

In the 1950s, fresh milk was considered an essential part of children's diet in Western countries. In Britain, the School Milk Act of 1946 provided free milk—a third of a pint a day—in schools to all children under the age of eighteen (figure 2.6). This provision was in place until 1968, when it was gradually reduced and finally abolished; similar programs existed in the United States as part of the school lunch program. The importance of milk in national diets in Western countries, together with its importance in young children's diet

Figure 2.6. Schoolchildren in Britain drinking their free daily one-third pint of milk (1944). Photograph D 20552 from the Collections of the Imperial War Museums. Image created by Ministry of Information. Public domain.

and its association with purity, explains the crystallization of concerns around the contamination of milk in the debate around radioactive fallout in Britain and the United States, even before the pasture-to-cow-to-human transfer, or "milk pathway," and the metabolic effects of milk contaminants were fully understood (figure 2.7). The milk ban following the Windscale accident in 1957 made the consequences tangible. The accident at the plutonium production plant in Cumbria is often cited as a turning point for the realization that milk was a key conduit for radiation ingestion. It expanded concerns from strontium to iodine as a contaminant in milk. Indeed, milk, once considered as "wholesome and healthy," now came to symbolize the risks of the nuclear world.[25] More generally, the Windscale accident and its aftermath sensitized a broader public to the problem of the accumulation of contaminants in the food chain and of food as a major risk factor in the exposure to environmental

Figure 2.7. Uptake and accumulation of radiation in the cow and in cow milk. Deutsches Museum, München, Archiv, L1918-15. Reproduced with permission.

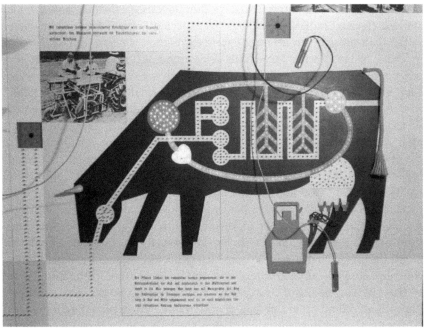

health hazards. For British consumers, fish in the Pacific was one thing; British milk, another.

Windscale Fire and Its Aftermath

The plutonium production plant at Windscale caught fire in October 1957 during a maintenance procedure. The Windscale piles were one of the cornerstones of Britain's nuclear weapons project. The fire at one of the two graphite-moderated reactors was the worst nuclear accident at the time. Plumes of radioactive material escaped from the chimney of the plant and were released into the atmosphere. Efforts to contain the fire at the plant were still underway when routine air samples in the close surroundings showed increased radioactivity levels. Operators at the plant immediately shifted their attention to the question of whether there was any health hazard for the local population through inhalation or contamination of the food chain.[26] Vans with monitoring equipment were dispatched up and down the coast to collect air samples and determine gamma radiation levels as a quick indicator of radiation levels.

Biological monitoring, including the testing of vegetation and various food-stuffs, would take longer, but the contamination of pastures and milk was a concern from the beginning. Two days after the crisis at the pile, milk samples showed elevated radiation readings. Unexpectedly, the iodine-131 readings, rather than those of strontium-90, were the biggest concern. In fact, no clear guidance existed for the permissible limit of iodine-131 exposure to the general population, as the health hazards for iodine were not yet fully recognized. Moreover, iodine, with its short half-life, was normally not present in fallout in Britain, as much of its radioactivity was depleted before fallout from the weapons tests in the Pacific reached Britain. In contrast, as would become clear in retrospect, the Windscale accident released more volatile iodine in the atmosphere, while higher proportions of other fission products, including strontium-90, were retained in the pile.[27] The International Commission on Radiological Protection (ICRP) had made recommendations for internal irradiation limits for iodine-131 but had considered only lifetime limits for radiation workers. It had not yet addressed the question of limits for emergency exposure of the general population. Furthermore, it had developed a model for a "standard man" but not for a "standard child" or "infant," the most vulnerable population group. However, there was a recent recommendation issued by the ARC for the use of civil defense authorities in case of a nuclear attack. The proposed limit for iodine-131 in milk was 0.3 microcurie per liter to keep the total radiation dose for infants below the level known to produce clinical symptoms. The figure was not directly applicable to the present case, but it was the only available one. With readings of local milk depots between 0.4 and 0.8 microcuries per liter, a milk ban seemed inevitable. Health physicists of the Atomic Energy Authority (AEA) hastily made their own calculations. Based on a series of assumptions about the size of an infant's thyroid, the daily intake of cow's milk, the retention of iodine-131 in the child's body, the concentration of iodine-131 in the thyroid, and the total radiation dose that it would deliver, they came up with a safety limit of iodine-131 of 0.1 microcurie per liter. As they commented later, "their 'standard child' . . . was conceived and delivered in one evening."[28] The same evening, on the advice of the AEA and the MRC, the Ministry of Agriculture, Fisheries and Food (MAFF) issued a ban for milk in an eighty-square-mile radius, later extended to a 200-square-mile radius. Although the iodine values prompted the ban, strontium-90 remained a concern and, more than any other element, continued to symbolize the "atomic poison" released from nuclear technologies.[29]

The ban remained in place for six weeks. Parallel to this measure, the AEA put in place an extensive monitoring program for both iodine-131 and strontium-90, in milk and other foodstuffs such as potatoes (for its importance in the daily diet) and kale (because of its high calcium content), in herbage, and in soil.[30] Food monitoring continued well beyond the actual crisis, with the MRC and ARC taking over responsibility for the program, together with the human bone monitoring program that was already under way. By 1958, Britain had, in its own estimation, a more extensive food sampling program than any other country.[31] Furthermore, an "experiment" was run in a grazing field adjacent to the Windscale plant that was most contaminated with strontium-90. The field was divided into plots, some of them remained unplowed, others where ploughed and reseeded. The grass and the dung of the animals feeding on the various plots were tested and compared. The plan was to check "under 'natural' conditions the results which have been obtained when the radioactive material is applied artificially."[32] As in other incidences, contamination was exploited to gain fundamental knowledge about food chain cycles, metabolic turnover rates, and the biological mechanisms involved.

Managing the milk ban meant striking a highly delicate balance between public health concerns and protecting the livelihood of local farmers. At a meeting called by the local National Farmers' Union branches ten days into the ban and attended by about 350 farmers as well as the Windscale general manager, other AEA representatives, officers of the MAFF, and the press, tensions were evident and emotions were running high. A report about the meeting drafted by the regional controller of the MAFF pointed to "strongly voiced resentments" expressed by some of the farmers. The uncertainty surrounding the ban and the actual dangers was one aspect. Other aspects regarded concerns about the health of the affected cows, mutations, and sterility; the possibility of moving the animals to other areas and their value on the market; compensations for the lost business; the depreciation of the value of the farm; and concerns about the safety of other foodstuffs, especially vegetables. MAFF representatives tried hard to calm the mood. They countered rumors that cows had died from the received radiation. They declared that there was no problem to buy stock from the area covered by the ban and that milk from cows moved outside the area would be "safe in every way" after twenty-four hours. Leaning on the MRC clearance of all other foodstuffs, they assured those assembled that "everything except milk could be eaten without danger." To back up their claims, they pointed to the widespread testing and sampling that was being done. Testing, the writer of the memo pondered, was undertaken "not because we expected to find danger but because it was the best way to build up confidence, enabling us to say that we had tested but there was no danger." The memo writer concluded that the meeting had been successful in restoring confidence and improving public relations.[33]

Even after the lifting of the ban, the National Farmers' Union—sensitized by the Windscale incident—remained deeply concerned about the strontium problem. The December 1958 issue of the union's own weekly publication, the *British Farmer*, dedicated several pages to the issue. The leading piece consisted of a solicited statement on the matter by the ARC, crafted with the advice of the MRC. The statement was drafted in lieu of the publication of a "Farmer's guide to radioactivity," proposed initially. Much discussion preceded the drafting of the piece, with the MRC being particularly concerned about not wanting to sound "minimally alarming" but also being wary of appearing to defend government policies, which would undermine its status as an independent scientific body.[34] To strike that delicate balance, the article focused on the monitoring and the radiation experiments that were being conducted. Facts at hand, it affirmed that the strontium-90 content in the human diet had "nowhere reached levels that would lead to concerns regarding human health." The article also conveyed that widespread monitoring and research continued, ensuring that "stringent levels of safety" for all foodstuffs, especially milk, were maintained.[35] A second essay, more journalistic in tone, provided readers with a hands-on view of the scientific work in progress. A rich series of photographs depicting "scientists at work" accompanied the text. The article assured its readers that half as much was known about the long-term effects of feeding antibiotics or spraying hormone weed killer or on the effects of diesel fumes than about the effects of radioactivity on the body. The overall message emanating from the monitoring and the laboratory investigations was that "if the present rate of nuclear-weapon testing continues, the danger to plant and animal life from radiation due to fall-out can be described as 'quite negligible.'"[36]

Milk remained a sensitive topic beyond the farmer community and not everyone agreed with the official statements on the issue. Colonel Geoffrey Taylor, a former army doctor and professor of medicine, and prospective liberal candidate for a council in Somerset, focused his election campaign on the health risks caused by the fallout from nuclear bomb tests. In June 1958, the *Observer*, an established Sunday broadsheet, published four letters by the Milk Marketing Board addressed to Taylor in response to his campaign. The letters expressed concern about the damaging effect of his speeches on milk producers. Taylor rebutted, further exposing the government's "gigantic and ghastly mistake" to misinform the public on atomic fallout.[37] He followed suit with further articles, arguing that against common supposition there was no safe threshold for strontium contamination.[38] The article in the *Observer* prompted responses by concerned readers as well as detractors. In letters addressed to the prime minister and forwarded to the MRC, one reader affirmed that she did not want to die of "strontium-in-the-bone" for drinking milk and demanded a public inquiry. Another letter writer was appalled that the government was slowly poisoning all children through the school milk program. All requested a change in policy.[39] Critics deplored Taylor's "highly tendentious misrepresentations" of sensitive matters and, citing facts and figures, accused the writer of suppressing facts, misleading readers with tendentious generalizations and reckless deductions, and of just making up things.[40] The debate caused considerable consternation in political circles. Asked for advice on the matter, the MRC did not regard it as appropriate to respond in detail to any of the letters. Instead, it kept to its official position as outlined in its recent report to Parliament that stated that at the current moment radiation from fallout was negligible in relation to other sources of radiation.[41]

Destrontification

A key question following the issuing of the milk ban in the wake of the Windscale fire was what to do with the contaminated milk. In the event, two avenues were pursued. Some of the milk was collected and thrown down a sewer leading directly to the sea.[42] Other farmers were instructed to dig trenches and bury the milk. Overall, about two million liters of milk were poured away, and the compensation costs to farmers ran to about sixty thousand British pounds (corresponding to 1.4 million pounds today).[43] Yet in view of rising concerns about the worldwide contamination of milk from fallout, efforts focused on technical ways to decontaminate milk. Researchers in the United States, Canada, and Britain were working on the project. The most promising idea was to run milk over ion-exchange columns that would selectively filter out the radioactive strontium. In 1960, a pilot plant that supposedly could extract as much as 95 percent of strontium-90 from milk without altering its calcium content was under study in the United States (figure 2.8). Some preliminary tests were also run at the Atomic Energy Research Establishment in Harwell and at the National Institute for Research in Dairying in Britain. Yet given the resources put into the development of the technique in the United States, British authorities decided to wait on the results of tests conducted on the other side of the Atlantic.[44] The technology was expected to be of possible use in atomic emergencies. Current strontium readings in milk did not seem to warrant any costly intervention on a routine commercial basis, although that option was not completely excluded either.[45]

Historian Kendra Smith-Howard has pointed out that whereas citizens had once welcomed the technological transformation of milk as a tool for its improvement, postwar consumers had become more critical of the social, environmental, and health effects of science and technology in altering milk.[46] Concerns about strontium-90 in milk from weapons testing undoubtedly contributed to this change in attitude. There was, nevertheless, also an opposing trend. In the age of space flight and the futuristic kitchen, technologically processed and synthetic (or artificial) food had become not only more acceptable but outright cool. Destrontified milk (next to pasteurized and fortified milk) could fit into this mold. At the same time, the food industry also experimented with using ionizing radiation to produce food and keep perishable food fresh.

Figure 2.8. Equipment for analyzing milk for radionuclides. J. Newell Stannard, *Radioactivity and Health: A History* (Springfield, VA: Office of Scientific and Technological Information, 1988), p. 998, figure 12.25. No copyright holder could be identified.

It might seem ironic that the same technology that led to the contamination of fresh milk was proposed as an approach to keep food fresh.[47]

Next to technological fixes, there was also the idea that other food could counteract the effect of radiation. In particular, findings suggested that some alginates derived from seaweed could reduce the absorption of strontium-90 in food by three-quarters.[48] For the British public the press translated this into the news that custard (a quintessential British dessert ingredient), next to ice lollies, both of which supposedly contained alginate, could combat fallout.[49] Stanley C. Skorna, director of the gastrointestinal research laboratories at McGill University in Montreal, identified as the active ingredient an acid polysaccharide called PE 211. His communication at the International Congress of Gastroenterology in Brussels caused considerable stir.[50] Alginate Industries, a London-based firm, was keen to move into the business and proposed to produce artificial alginates for testing. Yet the MRC that was consulted on the matter determined that more research would be necessary before the effectiveness, toxicity, and practicality of large alginate intake in children could be established.[51]

Conclusions

In its final report on the health and safety aspect of the Windscale accident, the MRC concluded that on the basis of extensive monitoring, it was safe to assume that it was "in the highest degree unlikely that any harm has been done to the health of anybody," whether workers at the plant or members of the general public. The same conclusions were repeated in the follow-up report to Parliament on *The Hazards to Man of Nuclear and Allied Radiations*. The milk ban was hailed as particularly useful to contain any hazard to health.[52] More generally, the alarm around radioactively contaminated food, especially milk, receded from public attention with the Partial Nuclear Test Ban Treaty (banning all nuclear weapons tests except those conducted underground) coming into effect in 1964 and all relevant indicators about strontium-90 contamination in soil, food, and bone going down. The signing of the Test Ban Treaty was a diplomatic decision, linked to strategic arms control as much as to concerns about food contamination. Yet with global fallout diminishing, the political decision brought a certain closure to the issue.[53]

Nevertheless, the Windscale accident in particular helped sensitize the public on the issue of food as a conduit of radiation into the body. Moreover, as a result of the Windscale accident, the MRC set in place the most extensive food monitoring system for radioactive contamination at the time. It also issued recommendations on the "maximum permissible level" of dietary contamination for the relevant isotopes after accidental release of radioactive material from a nuclear reactor. The proposed levels were based on the "fullest information available" on the radiation doses that would be delivered to different body tissues and at different ages by different isotopes and on the way they would enter the body.[54] Much of the data came from measurements and experiments performed in the wake of the Windscale accident.

The extent to which the radioactive food scare affected food regulations and standards more generally is complex and warrants further investigation. The answer may also vary from country to country. It has been argued that discussions around fallout and radioactive pollution was too tightly wrapped up with national security issues to spark a general discussion on the contamination of foodstuffs and the environment. Chemical pollution exposed by Rachel Carson was sufficiently removed from national security aspects to be able to start a wider discussion.[55] It is nevertheless significant that strontium was the first element mentioned by Carson in her highly influential exposé of the problem.[56] In Britain it seems indeed the case that the discussion about and regulation of other food contaminants, notably in the case of antibiotics that entered the food chain through new practices in husbandry in the 1950s, happened without any substantial reference to parallel concerns on the effects of fallout on the human diet.[57] In contrast, in Germany the concept of low-dose, chronic mutagenic risk to which a large part of the population is exposed—first developed in the context of risk assessment of nuclear technologies—also shaped regulatory frameworks of food additives.[58] Other authors, too, have stressed how investigations on the risk of nuclear technologies informed the philosophy as well as the tools underpinning regulatory frameworks for toxic substances more generally, in the United States and internationally. This pertains especially to the concept of low-dose exposure, the absence of any threshold for the effect of carcinogenic and mutagenic substances, and the persistence of risk for subsequent generations, as well as the emergence of the idea of a gradual poisoning of the whole planet.[59]

Unfortunately, these same debates continue to surround nuclear technologies. They resurfaced most dramatically around the Chernobyl and Fukushima nuclear disasters in the 1980s and 2010s. On both occasions, monitoring of foodstuffs by both official authorities and critical citizen groups was an essential aspect of managing and living with the crisis, as well as protesting against official declarations of safety. At the center of these efforts then and now was the Geiger counter, the instrument used to detect and measure ionizing radiation. The portable instrument makes invisible pollution audible by emitting crackling sounds as soon as a radiation source is detected. This distinguishes radiation detection from other contaminants, but unfortunately it has not helped to eliminate the risk. It is in the liminal disaster zones—in Chernobyl and Fukushima as in Windscale in 1957 during the global fallout crisis—that public attitudes to risk and regulatory frameworks are tested, contested, and remade.

Acknowledgments

Many thanks to the organizers and participants of the "Risk on the Table" workshop in Princeton in March 2017 and the follow-up book workshop in January 2020, where the paper on which this chapter is based was first presented and extensively discussed. Allison Carruth, Josh McGuffie, and Alexander von Schwerin provided additional generous feedback.

Soraya de Chadarevian is a professor in the Department of History and the Institute for Society and Genetics at the University of California, Los Angeles. She is the author and editor of numerous books, including *Designs for Life: Molecular Biology after World War II* and *Models: The Third Dimension of Science.* Her new book *Heredity under the Microscope: Chromosomes and the Study of Human Genome* was published by University of Chicago Press in 2020.

Notes

1. Strontium-90, with a half-life of 28.9 years, and strontium-85, with a half-life of 64.8 days, are the two longest-lived unstable isotopes of strontium among about thirty known ones. In addition, four stable isotopes of strontium are known.
2. "Radiobiological Research Unit, Harwell," 429. According to Stannard, milk contributes the most to the strontium intake in the body, but with milk the body distinguishes

better between strontium and calcium than with other food such as bread; see Stannard, *Radioactivity and Health*, 996.

3. See for instance "Radiobiological Research," 1562–63; "Radiobiological Research Unit, Harwell."

4. "This Man is Radioactive!" 20.

5. Ibid., 21.

6. Stannard, *Radioactivity and Health*, 299.

7. Ibid., 317–19.

8. Hacker, *Elements of Controversy*, 181–82.

9. Pirie, *Fall Out*; Stannard, *Radioactivity and Health*, 305.

10. "Metabolism of Nuclear Fission Products and the Effects of Radioactive Dusts," 16 July 1946, file FD1/468, National Archives, Richmond, UK (hereafter NA).

11. Libby started publishing results of the Sunshine Project in 1956; another series of studies performed under the same project heading by researchers at the Lamont Geological Laboratory at Columbia University were published in *Science* a year later, followed by regular updates; see Kulp, Eckelmann, and Schulert, "Strontium-90 in Man," 219–25; Eckelmann, Kulp, and Schulert, "Strontium-90 in Man, II," 266–73; Kulp, Schulert, and Hodges, "Strontium-90 in Man, III," 1249–55; Kulp, Schulert, and Hodges, "Strontium-90 in Man, IV," 448–54. On Project Sunshine, see also *Advisory Committee on Human Radiation Experiments*, 637–45.

12. Pirie, *Fall Out*, 71; Hamblin and Richards, "Beyond the Lucky Dragon," 36–56; Onaga, "Measuring the Particular," 265–304.

13. Stannard, *Radioactivity and Health*, 914. On the impact of the weapons testing on the Marshall Islanders, their food sources and health as well as their role as a test population, see Smith-Norris, *Domination and Resistance*.

14. Papworth and Vennart, "Uptake and Turnover," 1045–61.

15. Ibid., 1045. Eventually monitoring was reduced to cow's milk; ibid., 1050.

16. For the first and last report, see Medical Research Council, *Assay*, Report No.1; Medical Research Council, *Assay*, Report No.19.

17. The baby tooth survey, launched by the Greater St. Louis Citizen's Committee on Nuclear Information in the United States in 1958 and repeated in other cities was specifically aimed at determining the exposure of children to radioactive isotopes from fallout; see Fowler, "Rising Level," 53–55; Smith-Howard, *Pure and Modern Milk*, 131. In response to the results of the baby tooth survey, women across the United States organized and campaigned against the nuclear arms race and the dangers of fallout, with the safety of the nation's milk supply and children's health representing the most important rallying point; see "Women Strike," 144–48. That children were more susceptible to and faced higher risks from environmental contaminants than adults was only slowly realized in the second half of the twentieth century; see Wargo, *Our Children's Toxic Legacy*.

18. "Strontium-90 Rise Continues," *Times*, 26 October 1965; "More Strontium-90 in Babies' Bones," *Daily Worker*, 26 October 1965; "Strontium-90 Up, but Safe," *Daily Express*, 26 October 1965. All articles from press cuttings, MRC Library, Harwell.

19. Papworth and Vennart, "Retention"; Papworth and Vennart, "Uptake."

20. Papworth and Vennart, "Retention," 169.

21. Papworth, and Vennart, "Uptake."

22. Creager, *Life Atomic*.
23. Langham, "Considerations of Biospheric Contamination," 13.
24. Freidberg, *Fresh*; Smith-Howard, *Pure and Modern Milk*. On the history of milk and milk consumption, see also DuPuis, *Nature's Perfect Food*.
25. Smith-Howard, *Pure and Modern Milk*, 132. On the symbolic importance of the radioactive contamination of food, and especially milk, see also Creager, "Radiation, Cancer, and Mutation," 39.
26. For a detailed description of the chain of events at the pile and the decision on the environmental hazards leading to the milk ban, see Arnold, *Windscale*. The explosion of the reactor and a much larger disaster were only narrowly avoided. "Had it exploded," wrote British historian Peter Hennessy in his introduction to the fifty-year anniversary edition of Arnold's book, "1957 would be remembered globally for two things: the Russian 'Sputnik' (which was in space just a few days before the graphite caught alight) and the Windscale fire"; see Hennessy, "Foreword," vii. Following the accident, the Windscale piles were shut down.
27. Pirie, *Fall Out*, 119–20; Arnold, *Windscale*, 54.
28. Quoted after Arnold, *Windscale*, 57.
29. Creager, *Life Atomic*, 172–73.
30. Arnold, *Windscale*, 71–73.
31. See "The entry of strontium-90 into human diet: investigations in the United Kingdom," draft, summary point 5 [no author, no date], file FD 23/684, NA.
32. T. Williamson to L. D. C. McLees, "Atomic radiation," 27 November 1957, file MAF 164/13, NA.
33. F. M. Kearns (MAFF), [Report about meeting with farmers at Gosforth, 24 October 1957], file MAF 164/13, NA. See also T. Williamson (County Advisory Officer), "The Windscale incident: N.F.U. sponsored meeting on 24 October, 1957," same folder.
34. Dr. Lush to Dr. Cements, [MRC house internal note] 19 February 1959, file FD23/684, NA.
35. "Strontium-90—a Statement." See also further background material to the publication of the article in file FD23/684, NA.
36. "Strontium-90—Scientists at Work."
37. "Milk Board's Letters."
38. Taylor, "Hazards." For a further follow up, see Taylor, "[Letter to the Editor]" and correspondence in file "Correspondence arising from Col. Geoffrey Taylor's statement on radiostrontium in milk," FD23/682, NA.
39. Redacted letters (no author, date or place) in file "Correspondence arising from Col. Geoffrey's statement on radiostrontium in milk," FD23/682, NA.
40. Carbon copy of letter [unsigned] to the editor of *Time and Tide*, 12 May 1959, file FD23/682, NA.
41. Medical Research Council, *Hazards*.
42. On sea dumping of radioactive waste, see Hamblin, *Poison in the Well*.
43. Arnold, *Windscale*, 157.
44. On the US efforts in this respect, as well as on the role of consumers and farmers in shaping response to the destrontification of milk, see Smith-Howard, *Pure and Modern Milk*, 121–46.

45. See correspondence in file "Ion-exchange devices to remove fission products from water and milk (1960)," FD23/709, NA. A "fall-out cleaner," based on the addition of calcium phosphate advertised by Dr. Joseph Silverman of Radiation Application, Inc., in Long Island was looked upon more critically; see correspondence in same file.
46. Smith-Howard, *Pure and Modern Milk*, 122.
47. Food irradiation was proposed as an alternative to pasteurization, heat sterilization, and chemical methods of sprout control. Radiation and radioisotopes were also used for mutation breeding, increase of crop yield, and pest control in the production of food. On "nuclear agriculture" and nuclear food science, see Zachmann, "Peaceful Atoms"; Curry, *Evolution Made to Order*.
48. "Radiobiological Research Unit," 1550.
49. "Seaweed to Stem Strontium Uptake," 791; "Custard Can Combat Fall-Out." These and more press cuttings from MRC Library, Harwell.
50. "With a Pinch of Seaweed."
51. See correspondence between, among others, the MRC head office, J. Loutit at the MRC Radiobiological Research Unit at Harwell, and the ARC; file "Seaweed taken with meals as a method for absorbing radioactive strontium from contaminated food (alginates)," FD23/774, NA. In the relevant experiments with rats, alginates represented up to 20 percent of their diet.
52. Medical Research Council, *Hazards*, 137. On later, more complete assessments of the health impacts of the Windscale accident, considering revised estimates of the release of radiation, knowledge of new pathways to human beings, the abandonment of the threshold hypothesis, and the adoption of a collective dose, see Arnold, *Windscale*, 136–53.
53. On the test ban treaty, see Divine, *Blowing on the Wind*; Greene, *Eisenhower*.
54. Medical Research Council, *Hazards*, Appendix J.
55. Watkins, "Radioactive Fallout," 303.
56. Carson, *Silent Spring*, 6.
57. Kirchhelle, *Pyrrhic Progress*.
58. Schwerin, "Vom Gift im Essen." On the discussion about the regulation of food additives in Germany in the 1950s, see also Stoff, "Zur Kritik der Chemisierung."
59. Boudia and Jas, *Toxicants*; Boudia, "From Threshold to Risk."

Bibliography

Archives

National Archives, Richmond, UK

Publications

Advisory Committee on Human Radiation Experiments: Final Report, October 1995. Washington, DC: US Government Printing Office, 1995.

Arnold, Lorna. *Windscale 1957: Anatomy of a Nuclear Accident*. New York: St Martin's Press, 1992.

Boudia, Soraya. "From Threshold to Risk: Exposure to Low Doses of Radiation and Its Effects on Toxicants Regulation." In *Toxicants, Health and Regulations since 1945*, edited by Soraya Boudia and Nathalie Jas, 71–87. London: Pickering & Chatto, 2013.

Boudia, Soraya, and Natalie Jas, eds. *Toxicants, Health and Regulation since 1945*. London: Pickering & Chatto, 2013.

Carson, Rachel. *Silent Spring*. Boston: Houghton Mifflin, 1962.

Creager, Angela N. H. *Life Atomic: A History of Radioisotopes in Science and Medicine*. Chicago: University of Chicago Press, 2013.

———. "Radiation, Cancer, and Mutation in the Atomic Age." *Historical Studies in the Natural Sciences* 45 (2015): 14–48.

Curry, Helen Anne. *Evolution Made to Order: Plant Breeding and Technological Innovation in Twentieth Century America*. Chicago: University of Chicago Press, 2016.

"Custard Can Combat Fall-Out." *Birmingham Post*, 16 June 1966.

Divine, Robert A. *Blowing on the Wind: The Nuclear Test Ban Debate*. New York: Oxford University Press, 1978.

DuPuis, E. Melanie. *Nature's Perfect Food: How Milk Became America's Drink*. New York: New York University Press, 2002.

Eckelmann, Walter R., J. Laurence Kulp, and Arthur R. Schulert. "Strontium-90 in Man, II." *Science* 127 (1958): 266–73.

Fowler, John M. "The Rising Level of Fallout." In *Survival: A Study of Superbombs, Strontium 90 and Fallout*, edited by John M. Fowler, 51–67. London: MacGibbon & Kee.

Freidberg, Susanne. *Fresh: A Perishable History*. Cambridge, MA: Belknap Press, 2010.

Greene, Benjamin P. *Eisenhower, Science Advice, and the Nuclear Test-Ban Debate, 1945–1963*. Stanford: Stanford University Press, 2007.

Hacker, Barton C. *Elements of Controversy: The Atomic Energy Commission and Radiation Safety in Nuclear Weapons Testing, 1947–1974*. Berkeley: University of California Press, 1994.

Hamblin, Jacob Darwin. *Poison in the Well: Radioactive Waste in the Oceans at the Dawn of the Nuclear Age*. New Brunswick: Rutgers University Press, 2008.

Hamblin, Jacob Darwin, and Linda M. Richards. "Beyond the Lucky Dragon: Japanese Scientists and Fallout Discourse in the 1950s." *Historia Scientiarum* 25 (2015): 36–56.

Hennessy, Peter. "Foreword to the Third Edition." In *Windscale 1957: Anatomy of a Nuclear Accident*, by Lorna Arnold, vii–viii. Houndmills: Palgrave Macmillan, 2007.

Kirchhelle, Claas. *Pyrrhic Progress: The History of Antibiotics in Anglo-American Food Production*. New Brunswick: Rutgers University Press, 2020.

Kulp, J. Laurence, Walter R. Eckelmann, and Arthur R. Schulert. "Strontium-90 in Man." *Science* 125, no. 3241 (1957): 219–25.

Kulp, J. Laurence, Arthur R. Schulert, and Elizabeth J. Hodges. "Strontium-90 in Man, III." *Science* 129, no. 3358 (8 May 1959): 1249–55.

———. "Strontium-90 in Man, IV." *Science* 132, no. 3425 (19 August 1960): 448–54.

Langham, Wright H. "Considerations of Biospheric Contamination by Radioactive Fallout." In *Radioactive Fallout, Soils, Plants, Foods, Man*, edited by Eric B. Fowler, 3–18. Amsterdam: Elsevier, 1965.

Medical Research Council. *Assay of Strontium-90 in Human Bone in the United Kingdom: Further Results for 1959*. Medical Research Council Monitoring Report Series No. 1. London: Her Majesty's Stationery Office (HMSO), 1960.

————. *Assay of Strontium-90 in Human Bone in the United Kingdom: Results for 1966 with Some Further Results for 1965*. Medical Research Council Monitoring Report No. 15. London: HMSO, 1967.

————. *Assay of Strontium-90 in Human Bone in the United Kingdom: Results for 1970*. Medical Research Council Monitoring Report No. 19. London: HMSO, 1973.

————. *The Hazards to Man of Nuclear and Allied Radiations*. London: HMSO, Cmd 9780, 1956.

————. *The Hazards to Man of Nuclear and Allied Radiations: A Second Report to the Medical Research Council*. London: HMSO, Cmd 1225, 1960.

"Milk Board's Letters on Radioactivity." *The Observer*, 29 June 1958.

"More Strontium-90 in Babies' Bones." *Daily Worker*, 26 October 1965.

Onaga, Lisa. "Measuring the Particular: The Meanings of Low-Dose Radiation Experiments in Post-1954 Japan." *Positions* 26, no. 2 (2018): 265–304.

Papworth, D. G., and J. Vennart. "Retention of ^{90}Sr in Human Bone at Different Ages and the Resulting Radiation Doses." *Physics in Medicine and Biology* 18, no. 2 (1973): 169–86.

————. "The Uptake and Turnover of ^{90}Sr in the Human Skeleton." *Physics in Medicine and Biology* 29, no. 9 (1984): 1045–61.

Pirie, A., ed. *Fall Out: Radiation Hazards from Nuclear Explosion: Revised Edition Including a Report on the Windscale Disaster and an Analysis of the United States Congress Report on Radioactive Fall Out and Its Effects in Man*. London: MacGibbon & Kee, 1958.

"Radiobiological Research." *British Medical Journal*, 9 December 1961, 1562–63.

"Radiobiological Research Unit." *British Medical Journal*, 18 June 1966, 1550.

"Radiobiological Research Unit, Harwell." *Nature* 193 (3 February 1962): 428–29.

Schwerin, Alexander von. "Vom Gift im Essen zu chronischen Umweltgefahren: Lebensmittelzusatzstoffe und die risikopolitische Institutionalisierung der Toxikogenetik in der Bundesrepublik, 1955–1964." *Technikgeschichte* 81 (2014): 251–74.

"Seaweed to Stem Strontium Uptake." *New Scientist* 30 (23 June 1966): 791.

Smith-Howard, Kendra. *Pure and Modern Milk*. Oxford: Oxford University Press, 2014.

Smith-Norris, Martha. *Domination and Resistance: The United States and the Marshall Islands during the Cold War*. Honolulu: University of Hawai'i Press, 2016.

Stannard, J. Newell. *Radioactivity and Health: A History*. Springfield, VA: Office of Scientific and Technological Information, 1988.

Stoff, Heiko. "Zur Kritik der Chemisierung und Technisierung der Umwelt: Risiko- und Präventionspolitik von Lebensmittelzusatzstoffen in den 1950er Jahren." *Technikgeschichte* 8 (2014): 229–49.

"Strontium-90 Rise Continues." *Times* (London), 26 October 1965.

"Strontium-90 up, but Safe." *Daily Express*, 26 October 1965.

"Strontium-90—A Statement." *British Farmer*, 6 December 1958, 12–13.

"Strontium-90—The Scientists at Work." *British Farmer*, 6 December 1958, 14–15.

Taylor, Geoffrey. "Hazards from Nuclear Tests." *Time and Tide*, 9 May 1959, 528.

————. "[Letter to the Editor]." *Time and Tide*, 20 June 1959.

"This Man is Radioactive!" *Sunday Pictorial*, no. 2478 (14 October 1962): front page and center pages 20–21.

Wargo, John. *Our Children's Toxic Legacy: How Science and Law Fail to Protect Us from Pesticides*. New Haven: Yale University Press, 1998.

Watkins, Elizabeth S. "Radioactive Fallout and Emerging Environmentalism: Cold War Fears and Public Health Concerns, 1954–1963." In *Social Activism: A Tribute to Everett Mendelsohn*, edited by G. Allen and Roy M. MacLeod, 291–306. Dordrecht: Kluwer, 2001.

"With a Pinch of Seaweed." *New Scientist*, 6 August 1964.

"Women Strike for Peace Milk Campaign, 1961." In *Nuclear Reactions: Documenting American Encounters with Nuclear Energy*, edited by James W. Feldman, 144–48. Seattle: University of Washington Press.

Zachmann, Karin. "Peaceful Atoms in Agriculture and Food: How the Politics of the Cold War Shaped Agricultural Research Using Isotopes and Radiation in Post War Divided Germany." *Dynamis* 35 (2015): 307–31.

 CHAPTER 3

Poison and Cancer
The Politics of Food Carcinogens in 1950s West Germany

Heiko Stoff

Around 1900, Germany had both a world-leading chemical industry and an exceptionally strong natural health and lifestyle reform movement. Since the late 1860s, German firms were the main manufacturers of synthetic dyestuffs. The use of such synthetic colors of dye for industrialized food production was a fundamental aspect in the development of modern consumer societies.[1] But while at the same time there was a debate on food fraud and food adulteration in many Western countries, in Germany, the focus since the 1920s was less on the misuse of food additives than on food colorants, suspected as carcinogenic substances because of their chemical structure.

Pharmacological and toxicological research on carcinogens thereby intermingled with a critique of modern civilization, which protagonists of the German lifestyle reform movement saw as the main cause for the contamination of food with what they called "poison." The rhetoric of purity and poisoning, in this regard, resembles its use in the US "Pure Food Movement" but, through the concept of carcinogenesis, gives it a much more fundamental impact.[2] The history of food regulation in Germany, where industrial productivity merged with purist beliefs, is therefore not just an interesting case study but rather at the center of the debate about "risk on the table," especially because in the late 1940s the German lifestyle reformist discourse deeply influenced the elaboration of a new mathematical model about the dose-response relationship in carcinogenesis, which led to the conclusion of an irreversible accumulation of carcinogenic effects and the nonexistence of subthreshold doses for several carcinogens.[3] This again, during the 1950s, influenced regulatory legislatures in West Germany, Europe, and—for a short period—also the internal debate in the Joint FAO/WHO Expert Committee on Food Additives (JECFA).

During the twentieth century, the status of food additives in Germany was much more precarious than that of drugs. At least until the case of thalidomide in 1961/62, the Contergan-Skandal, there was a general public trust in pharmaceutical products. Critique was focused on their misuse, commercialization, and the special case of unethical human experimentation with drugs.[4] In contrast, synthetic food additives were suspect right from the beginning of their usage. Also, food was part of the collective identity, the idealized form of German-ness in any naturist discourse, and its contamination was regarded as an attack on the imagined community of Germans itself. Indeed, unlike in the United States, food and drugs were treated differently in a regulatory or legislative way. Until 1961, no drug law existed in Germany. In contrast, a food law had already been established in 1879 and amended in 1927, 1935, and 1958. The last of these amendments was the result of a fierce debate about food additives as poisons and carcinogenic compounds.[5]

To sum up, in Germany the anti-modern semantics of poisoning with for-eign matter informed regulatory risk perceptions of industrial food production in the twentieth century.[6] Regulatory measures of food additives were not just expert decisions but the outcome of intermingled scientific, political, social, and economic processes and negotiations. They depended on what is defined as a risk in a specific sociohistorical setting, how and why certain things be-come a problem, or why something is considered relevant and something else not.[7] The problematization of certain colorants as "cumulative poisons" had a formative influence on the processes of institutionalization, standardization, regulation, and (de)activation of food additives as precarious substances.[8] The case of food additives illustrates how binary concepts such as processed and pure food, artificial and natural agents, nutrients and adulterants, or mod-ern civilization and holistic nature shaped what was, and was not, viewed as a problem. A wide range of human actors, including pharmacologists, tox-icologists, oncologists, food chemists, health politicians, lifestyle reformists, frightened citizens, food industrialists, representatives of the pharmaceutical industry, journalists, and government officials, as well as highly influential institutions, such as the Commission for Colorants of the German Research Foundation, numerous expert associations, consumer, women's, and house-wife organizations, but also radical diet reform groups, debated legal response based on zero tolerance or calculated risk. New theories of carcinogenesis and toxic substances, such as the Druckrey-Küpfmüller equation or the concept of Acceptable Daily Intake (ADI), were linked to these actors, institutions, and discourses and validated regulatory practices. While the Druckrey-Küpfmüller equation, initiated in West Germany and Europe, but also an influence on the Delaney Clause, denied the existence of threshold doses for carcinogens, the ADI, established by JECFA, was indeed based on the idea of harmless sub-threshold doses and thereby enabled much more industry- and trade-friendly regulatory policies.

Finally, as a last introductory remark, it is important to underline the dif-ferent attitude toward toxic and carcinogenic substances. In the German dis-course since the 1920s, the political semantics of poison merged with the trope of cancer as a disease of civilization. In the early twentieth century among modern societies with an established health system, the causes of death shifted from infectious diseases to cardiovascular diseases and cancer. While numerous causes for cancer were considered, the important role of certain exogenous carcinogenic substances, researched in the Swiss and German dye industry since the 1880s, gained scientific and public acceptance. That chemicals, notably aromatic amines, could cause tumors ("aniline cancer") was a paradigm-changing moment.[9] In the debate on food additives, poison, and cancer, both toxicological and medical categories as well as nationalist and racist metaphors were linked through binary key concepts like natural and nonnatural food, increased performance and uncontrollable growth, biologically active substances (*Wirkstoffe*) and foreign matter (*Fremdstoffe*). Throughout the twentieth century, politically effective critiques of industrialized food production drew on the semantic strength of these dichotomies, while the pragmatic use of food additives in industrial food production and world trade, as I will show, was based on the fundamental differentiation between manageable toxic and uncontrollable carcinogenic substances.

Poisoned Food and Carcinogens

Poison is a key concept of both pharmacology and toxicology. According to Paracelsus's famous statement, all things are poison, and nothing is without poison, because "solely the dose determines that a thing is not a poison."[10] Until the mid-twentieth century, Paracelsus's dogma was undisputed, and it still is preserved in the threshold concept of ADI, which tries to define what dose makes a substance poisonous. But since the 1930s, poison also designates industrially produced noxious materials, which are described as carcinogenic because of their chemical structure. These chemical substances—such as certain colorants, preservatives, and pesticides—were and are discussed as ubiquitous agents of diseases of civilization, as poisons of modernity.

Poison is not only a scientific concept; it is also a political metaphor. It is something that has an ill effect on an otherwise healthy entity or situation. The substance-as-agent (poison), the practice (poisoning), and the actor (poisoner) all combine secrecy with effectiveness, thereby allowing a wide variety of significations. Poison always comes from the outside; the invisible poisoner is always the outsider who goes inside. Poison detection, therefore, is not only a toxicological, but also a highly political and criminological, practice.[11] The significance of the poison narrative in the German-speaking area manifested itself in the emergence of new and hard-to-translate terms in the first half of the twentieth century: *Genussgifte, Gewebsgifte, Zivilisationsgifte, Umweltgifte,* or *Summationsgifte*.[12] The use of *Fremdstoffe* as an overarching category for these poisons laid their political connotations bare, since before this term was used in Germany to denote certain chemical substances alien to a specific environment, it was an antisemitic phrase to degrade those who were denounced as foreign to the people or the race. *Fremdstoffe* generated a semantic field, conveying a critique of consumerism, liberalism, and individualism, along with the dramatic image of the insidious destruction of individual and collective bodies.[13]

Food additives were described as poisons since the early nineteenth century. Criticism of industrialized food production was common in all Western nations since the last third of the nineteenth century. But while this focused mainly on adulteration, in Germany the interplay of diet reform and new dietetics since the 1920s emphasized the need for healthy, natural, and pure food for the imagined racial community, the Germans (*das Volk*).[14] As Paul Rabinow states, with the emergence of a food processing industry, food itself turned into a "heterogenous commodity," which was systematically modified to meet industrial and consumer norms. During World War I and the years between the wars, "millions of people became accustomed to transformed natural products."[15] Rabinow reminds us that the cultural reaction against food classified as artificial or processed was spearheaded by a variety of lifestyle reformist groups. The calls for diet reform and fear of food poisoning are two sides of the same cultural demand, aimed at strengthening and purifying bodies for a healthy, consumerist society. German diet-reform advocates like Werner Kollath denounced processed food as denaturalized and dangerous for the fitness of the people. The sole hope for "the Germans" therefore lay in a pure natural diet.[16] This nationalist and life reformist fear of poison turned into an even more dangerous threat for the racial community, when specific azo dyes, used to color food, were identified as carcinogenic.

In the 1920s, numerous theories about carcinogenesis were discussed, including ideas of contagion and hereditary susceptibility. But there was also strong evidence that certain chemical compounds, especially polycyclic aromatic hydrocarbons, were carcinogenic agents irrespective of dose.[17] In the following decades, a "pharmacology of cancer-causing substances" began to dominate the scientific controversy about cancer in Germany. One of its main protagonists, the pharmacologist Hermann Druckrey, stated that genetic factors are not sufficient to explain the emergence of cancer. Even if an individual might be predisposed genetically, he argued, the onset of cancer is always provoked by environmental impacts and external factors. The best way to fight cancer, therefore, is the elimination of carcinogenic stimuli from the environment of humans.[18]

This scientific debate was deeply influenced by a critique of modern civilization. In 1931, the elementary teacher and journalist Curt Lenzner published a widely reviewed book titled *Gift in der Nahrung* (Poisoned food) in which he blamed the modern lifestyle for contaminating essential food substances with foreign matter. The catchphrase "poisoned food" referred to the dangers of chemical compounds used as food additives, and its invocation grabbed the attention of Germans from the Weimar Republic through the National Socialist period and right into postwar West Germany. According to Lenzner, cancer is the result of "plasmatic damages, caused by the lack of vital substances" and "the flooding of the consuming body with substances hostile to life."[19] A year later, the surgeon and eugenicist Erwin Liek, a fierce enemy of a solely scientific-technological medicine as well as the modern welfare state, claimed a close connection between urbanism and rise in cancer rates. Liek launched the semantic amalgamation of foreign matter and poison as a new concept of cancer. He proclaimed that a simple and natural lifestyle corresponded to a decrease in cancer rates of a population, and the contamination of food was the reason for the increase of cancer rates. The only solution therefore was a return to the soil and natural diet.[20] Statistics did not support this strong opinion, but Liek simply neglected such "statistical sophistry" (*statistische Klügeleien*). The crisis of cancer as a state of emergency demanded immediate measures, a reform of life itself![21] Liek's assumptions were shared by Nobel laureates, such as Gerhard Domagk, Adolf Butenandt, and Otto Warburg, as well as famous physicians, such as Karl Freudenberg and Karl-Heinrich Bauer. Lenzner and Liek set the direction, which since the 1930s was consensual in Germany: the war against cancer could be successful only through the destruction of exogenous agents. This biopolitical discourse turned lifestyle reform into a science, and turned nutrition research into an endeavor linked to lifestyle reform.[22] Some twenty-five years later, in 1956, the well-respected pharmacologist Fritz Eichholtz, director of the Pharmacological Institute at the University of Heidelberg, summarized this discourse in his warning against the "toxic condition" of modern life. His notion of a "toxic total situation" (*toxische Gesamtsituation*) informed a far-reaching debate on the boundaries of risk assessment and the dangers of chemical substances as the result of the modern lifestyle and uncontrolled industrial productivity. Modernity, civilization, and capitalism together seemed to be poisoning the Germans through chemical agents. Eichholtz stated that in modern civilization, the risk of poisoning is everywhere, and that to avoid it, one either must find regulatory measures or give up modernity altogether.[23]

The new cancer theory of exogenous agents corresponded to the narrative of the Germans being endangered by invisible foreign matter as the leitmotif of Nazi Germany.[24] While in the National Socialist era there were inner contradictions and an open dispute between propagandists of pure food and advocates of "ideal preservatives," which secured the food situation during war, Werner Kollath's distinction between natural, near-natural and nonnatural or "dead" food was widely accepted.[25] The scientific establishment was convinced that to win the total war, food additives—after all a product of the famed

German chemical industry—were essential, but, as the case of butter yellow shows, in the long run the discourse of carcinogenic foreign matter prevailed.

The Case of Butter Yellow

In the 1940s, *p*-dimethylaminoazobenzene, better known as butter yellow, an azo compound used to give butter, margarine, and cheese an attractive color, caused tremendous public interest. Butter yellow had been synthesized in the 1860s and produced and used as a colorant in Germany since the 1870s.[26] In the early 1940s, rumors spread that butter yellow, which many Germans consumed on a daily basis, was highly carcinogenic. The cancer-causing effect of azo dyes—a class of colorants based on aniline from coal tar—had been discussed since the late nineteenth century, when the surgeon Ludwig Rehn brought attention to the high appearance of bladder cancer in workers producing the aniline dye fuchsine. It took another two decades until the thesis of cancer caused by aniline derivates was reinforced in a detailed clinical study with workers in the Basel color industry.[27]

In the first decades of the twentieth century, particular polycyclic aromatic hydrocarbons were suspected of being cancer-causing substances. In 1915, Japanese pathologists Katsusaburo Yamagiwa and Koichi Ichikawa painted rabbit ears with tar and thereby produced carcinogenic effects.[28] These findings were confirmed in 1932 by the results of a working group around Ernest L. Kennaway and James W. Cook. They isolated and synthesized 3,4-benzpyrene out of coal tar and identified it as carcinogenic. Germany's leading biochemist Adolf Butenandt labeled benzpyrene as the prototype of the carcinogenic compound class of polycyclic aromatic hydrocarbon.[29] It was thereby of main importance that even a minor alteration of the constitution of these compounds could have a deep impact on their activities.[30] Another carcinogenic polycyclic aromatic hydrocarbon, methylcholanthrene, was found to be chemically related to steroids, such as the "sex" hormones. Because the newly isolated estrogens were characterized by a partially aromatized hydrocarbon framework, these biopolitical agents were suddenly the focus of cancer research.[31] As Jean-Paul Gaudillière has shown, Butenandt, who was cooperating with one of the leading producers of estrogens, the Berlin Schering AG, was absolving estrogens from the suspicion by blaming oncogenesis on genetic predispositions that could be activated only by steroids. But Butenandt also highlighted that, by contrast, butter yellow was a perfect model substance for quantitative studies on carcinogenesis.[32]

In 1934, Tomizo Yoshida demonstrated that liver tumors could be induced in rats by feeding or injecting the active component of scarlet red (o-Aminoazotoluol), which also was the source for extracting butter yellow.[33]

Between 1932 and 1937, the Japanese pathologist Riojun Kinosita orally administered either scarlet red or butter yellow to rats with the outcome that especially butter yellow was found to be a highly carcinogenic substance. Kinosita suggested that butter yellow developed carcinogenic potency because of its structure as a whole: "Of more than 50 azo and related compounds, Butter Yellow is by far the most potent in producing hepatic cancer in the rat, requiring for this only a small dose and a short period."[34] In 1940, a working group led by Druckrey confirmed Kinosita's findings and disproved that Kinosita's results were based on the genetic predisposition of the Japanese rats. For the first time, it appeared to have been proven by experiment that an exogenous substance provoked tumors in internal organs.[35] Over the following years, several scientists continued research on butter yellow. In 1943, the influential chemist Richard Kuhn declared that butter yellow was the most prominent representative among cancer-causing azo dyes.[36] A year later, the US pathologist Eugene L. Opie reported that "administration of butter yellow produces multiple foci of focal hyperplasia, cystic ducts, and cholangiofibrosis, and corresponding with these lesions, which are precursors of tumor growth, multiple tumors are formed."[37]

These research results were published even during wartime, so rumors spread that while Germans consumed a known cancer-causing food colorant, the Nazi authorities failed to respond. A ban on synthetic food additives during wartime was unrealistic, and the Nazi bureaucracy tried to cover the subject discreetly, but there was nonetheless resistance based on the assumption that the Nazis violated their own commitment to protect Germans from foreign matter. In 1941, German women's organizations, which were deeply involved in the nation's health and nutrition policies, applied pressure on Hans Reiter, president of the German Reich Health Office, to prevent the production and use of azo dyes. As historian Robert N. Proctor relates in his book on cancer research in Nazi Germany, during 1941 a member of the Göttingen branch of the NS-Frauenwerk informed Reiter that while women were certainly willing to sacrifice for the war, their accepting the presence of cancer-causing agents in food was a completely different matter.[38] It seemed that Reiter indeed urged the almighty I. G. Farben to cease the production of butter yellow. Beginning in 1942, butter and margarine were supposed to be colored with carotene, not azo dyes. In the end, it was a coalition of scientists, politicians, lifestyle reformers, and women's organizations, and also the rather cooperative chemical industry itself, that succeeded in prohibiting butter yellow. Despite their different interests, all of these actors could agree on a preventive strategy to ensure the need to secure the Germans from invasion by foreign matter.

Four years after the fall of the Nazi empire, Butenandt created a public scare and scandalized media coverage by proclaiming that, despite its proven cancer-causing effects, butter yellow was still in use.[39] In January 1950, the engineer Walter Hinnendahl of Nuremberg was one among many concerned citizens who appealed to the German federal president Theodor Heuss for elimination of potentially cancer-causing food colorants. He wrote to Heuss that he was shocked by an article titled "Deadly Butter Yellow" (*Todbringendes Buttergelb*). Hinnendahl argued that authorities who allowed the use of butter yellow as a food dye should be prosecuted for the involuntary manslaughter of millions of people.[40] While it was not even clear whether butter yellow was still industrially produced and used as a food dye in West Germany, as accusations and denials followed fast on one another, the substance nevertheless played a major role in establishing a new theory of carcinogen dose-response relations and fostering measures to ban carcinogenic food additives. The main actor of this story is the aforementioned Hermann Druckrey, whose key role in regulatory actions of the 1950s has often been overlooked.[41] Druckrey was born in 1904 and had been a member of the NSDAP and SA as early as 1931. He subsequently acquired a high rank in the SA, and was close to the SS. In the early postwar years, he was supported and whitewashed by Butenandt, who himself was a beneficiary of the Nazi regime. Though Druckrey never achieved an impressive academic career, he remained an internationally renowned expert on carcinogenic substances due to his new theory on the relationship between dose and effect, developed during 1948 together with the electrophysicist Karl Küpfmüller while residing in the Hammelburg internment camp.[42] Butter yellow came to define a new episteme by inspiring and strengthening a new theory of carcinogenesis, generating a critical discourse on modern civilization, and prompting stricter statutory measures for food additives.

The Druckrey-Küpfmüller Equation

According to Druckrey and Küpfmüller, carcinogenesis was a chronological pharmacological process following mathematical regularities.[43] Druckrey fed seven hundred rats with butter yellow, demonstrating that to produce tumors, a certain total dose was necessary, irrespective of the period of intake, in his case between 35 and 365 days. There existed a simple relation of reciprocal proportionality between the amount of the daily administration of butter yellow and the necessary duration of treatment to bring about cancer: With a daily dose of three milligrams, the latency period was 350 days, but if the dosage was ten times higher and the daily amount thirty milligrams, the latency period would be just one-tenth—that is, thirty-four days. The latency period was inversely proportional to the daily dose. Whether given three or thirty milligrams daily, the total came to one thousand milligrams of butter yellow. Druckrey concluded that the cancer-causing effect of butter yellow was, even at the smallest doses, irreversible right from the start of the experiment and throughout the entire life span of the animals. Once the critical total dose has been exceeded, tumors developed. One of the central conclusions of this study was that the ongoing consumption of even the smallest dosage of a carcinogenic substance like butter yellow could be lethal for consumers.[44] The effect could not be interpreted in terms of a "threshold dose," the amount below which exposure produced no cancer. Rather, at a certain dose low enough, the necessary induction time becomes longer than the total life span of the organism.[45] This could have led to some form of risk assessment for carcinogens as cumulative poisons, but that was not intended by Druckrey, who opted for an absolute ban on such synthetic substances.

Druckrey and Küpfmüller substantiated their thesis with an impressive mathematical methodology: in pharmacology, several substances exist whose effect is caused by the product of a constant concentration ("C") of the toxin and the time of duration ("t"). The effect of these Ct-toxins, the occupancy of cellular receptors, must be irreversible if the effects are to sum up.[46] Formalizing their thesis as the equation "dt^n = constant"—where "d" stands for the daily dose and "t" for the exposure time to effect—Druckrey and Küpfmüller determined, as Dutch toxicologist Henk Tennekes summarizes, that "irreversible receptor binding with an associated irreversible effect would lead to reinforcement of the effect by exposure time."[47] The concept of toxins irrevocably occupying specific receptors of an organism corresponded to the hit theory of the radiologist Friedrich Dessauer concerning the release of mutations through rays. Development of tumors therefore rested upon irreversible effects of physical and chemical agents, such as ionizing radiation. The common characteristic is that they affect specific receptors that are nonreplaceable functional units of the cell. According to Druckrey and Küpfmüller, this type of cumulative effect could be found in two highly important biological processes: the triggering of mutations through rays, and the inciting of cancer with dimethylaminoazobenzene.[48] Thus Druckrey also concluded in his radical theory, later called "genotoxic," that cancer-causing substances bring about irreversible alteration of inheritance.[49] His theory was not that genetic predisposition causes cancer, as Butenandt had declared for steroids, but that carcinogenic substances alter genetics.

Around 1950, the Druckrey-Küpfmüller equation played such a big role not merely because it was valid science and rarely criticized by other pharmacologists and physicians, but most of all because it fitted perfectly to lifestyle reformist assumptions. Additionally, well-respected physicians and scientists like Karl-Heinrich Bauer, Fritz Eichholtz, and Butenandt, who was a *persona grandissima* in the West German scientific community, supported it and used it to implement strict regulation policies in West Germany and Europe.

The Pharmacology of Cancer-Inducing Substances

Druckrey drew the conclusion that scientists should both seek out such cancer-causing substances in the environment and encourage elimination as far as possible of those with which humans were in daily contact, even if they occur only in trace amounts.[50] In 1949, at the first postwar congress of the German Society for Surgery, Druckrey's theory became the basis of new political demands. The surgeon Eduard Rehn, son of Ludwig Rehn, drafted a resolution with far-reaching implications, including the preparation of lists of safe and unsafe synthetic dyes.[51] Rehn based his resolution on ideas the well-respected physician Karl Heinrich Bauer had already presented to the German Society for Surgery. Bauer thereby referred to the German Research Council (Deutscher Forschungsrat), founded in 1949 by Werner Heisenberg, a short-lived but nonetheless important forerunner of the German Research Foundation (Deutsche Forschungsgemeinschaft), which had decided to convene a special commission under the direction of Butenandt and Druckrey to consider the problem of food colorants.[52] Over the following years, several commissions for colorants, foreign matter, whitening agents, and nutrition were established. When in 1951 the German Research Council merged with the Provisional Association of German Science (Notgemeinschaft) to (re-)establish the German Research Foundation, the only features of the German Research Council that survived were these commissions. Their remit was to decide on which additives could be tolerated and which could not. The commissions functioned as influential instruments of regulation and policy advice.[53]

The main cause of cancer—the "disease of the era" as the news magazine *Der Spiegel* called it—was now sought in cell-altering exogenous agents, especially artificial colorants.[54] Karl-Heinrich Bauer introduced a new oncological theory downplaying the role of genetics while emphasizing the significance of exogenous chemical and physical factors. In implicitly taking up Lenzner's and Liek's arguments, he recapitulated the thesis of a strong connection between modern civilization and an assumed rise in cancer: "Our age of chemicalization and mechanization of the environment, including nourishment, also knows cancer noxae for everyone, for all, for millions, day after day and involuntary for long, long years."[55] Bauer, born in 1890, was influenced by lifestyle reform as well as by biological and eugenicist thought. He contributed to the national socialist sterilization law without being a member of the NSDAP. In the postwar years, just like Butenandt, he turned into an influential scientific statesman. He was the first postwar head of the University of Heidelberg and, in 1964, one of the founders of the German Center for Cancer Research.[56]

It was Bauer who, referring to Ludwig Rehn, introduced the term "cancer noxa" (*Krebsnoxe*) to designate every energy resulting from radiation or any chemical reaction that can turn a healthy cell into a cancer cell.[57] Bauer was one of the leading German experts on cancer and already in the 1930s had focused on exogenous agents. In 1943, he had intervened at the Ministry of the Interior to support the ban of azo dyes.[58] According to Bauer, the increase in cancer rates in all Western nations since the 1870s was not caused by genetics or the prolongation of life, but by the rise of exogenous poisons, foreign matter, or adulterants. This also implied that cancer is not attributable to genetic fate but instead can be prevented through scientific, political, and legal measures.[59] Journalists of *Der Spiegel* precisely identified the anti-modern background that informed Bauer's well-received assumptions: "Bauer linked his daring notion with the even more daring conclusion, that grime, tar, pitch, aniline, azo dyes as well as Roentgen and radium rays are always noxae, which are 'foreign to nature' and for which modern man has developed no reactions of resilience. Human cancer is mostly linked to technology and modern civilization." And, in *Der Spiegel*'s typical ironic style, they concluded that if Bauer's assumptions are true, "then the modern man must escape from his self-made artificial, technical and chemical environment. He must give up modern semi luxury food and must scrap all inventions and achievements of the last two hundred years. To radically fight cancer therefore means a radical change in attitudes for the modern man and his world."[60]

The Politics of Carcinogens

In West Germany in the 1950s, there existed a rather heterogenous coalition advocating zero tolerance for carcinogenic food additives; participants ranged from scientific experts to journalists and rather obscure lifestyle-reform activists, to pressure groups for consumer rights. Even representatives of the pharmaceutical industry, such as Bernhard Wurzschmitt (BASF) and Ulrich Haberland (Bayer), argued in favor of approved lists for food additives, not only because they needed secure products but also because they shared the purist discourse. Catholic and Protestant women's and housewives' organizations wrote hundreds of letters to the Ministry of the Interior demanding the ban of food coloring and a much stricter food law.[61] Even more radical enemies of industrial food production, who assembled in the Society for Vital Substances, a melting pot of former Nazi nutrition experts and lifestyle reformists, tried to gain influence and indeed received intensive media coverage. Their leading members, such as the founder Hans-Adalbert Schweigart and the former protagonist of the national socialist "New German Medicine" Karl Kötschau, combined the 1930s discourse on natural food, as presented by Kollath, with the 1950s concept of a "toxic total situation," as presented by Eichholtz. Thus "bad" carcinogenic substances—foreign matter—were contrasted with solely "good" vital substances like vitamins, trace minerals, or enzymes. This concept referred less to vitalism then to the fundamental role of certain substances to strengthen the human body. In the mid-1960s, the concept of vital substances merged with politics of life protection, and the aims of the Society for Vital Substances were officially defined as "biopolitical."[62] Even if Bauer, Butenandt, and Druckrey tried to disavow these "purists," the media saw a strong resemblance between their positions.[63] In contrast, the Syndicate of Consumers' Associations (Arbeitsgemeinschaft der Verbraucherverbände), founded in 1953, was using the purist discourse and claimed the amendment of the food law in 1958 as its first victory, but, like other newly institutionalized consumer protection groups, actually focused on the nationalization of consumer policy and not the "war against cancer."[64]

Opponents to strict regulations for food additives were hard to find in the 1950s. Only a few experts, such as the toxicologist Werner Schulemann, disagreed with Druckrey and doubted the new theories of carcinogenesis altogether. The most outspoken opponents for stricter regulations, especially for preservatives, came from the food industry and their lobby groups. This concerned above all the fruit importers and the fishing industry. Their allies were in the Ministry of Economics, and they often lobbied successfully with the more radical market-oriented political parties. The amendment of the food law was successfully deferred by these activists, but it could not be prevented. In the 1950s, purist arguments were stronger than specific industrial interests.[65]

Perhaps the most important figure for the enforcement of a much stricter food law in West Germany was a state representative, the undersecretary Werner Gabel from the Department for Food Chemistry at the Ministry of the Interior, who was in constant quarrel with the Ministry of Economics. It was Gabel who supported female West German parliament delegates whose aim was to tighten the German food law. On 24 February 1956, three members of the German parliament, Christian-Democrat Hedwig Jochmus, Liberal Marie-Elisabeth Lüders, Social-Democrat Käte Strobel (later to become the first minister of health), and forty-three other female delegates presented a petition calling upon the Bundestag to request that the federal government produce a draft for a new food law. The text of the petition was drawn up by Jochmus, Strobel, and Gabel. The United Front of Female Delegates (Einheitsfront der weiblichen Abgeordneten) found strong support from the public and the media. Druckrey himself testified before the parliament as a scientific expert.[66] Finally in December 1958, after long political debates that were deeply influenced by the expertise of the commissions of the German Research Foundation, the amended German Food Law determined the general ban of "foreign matter" in food and the approval of tested food additives by way of exception.[67] In the end, it was an alliance of the female delegates of the Bundestag with the scientific experts of the commissions that enabled a new and strong food law and proved itself, for a short historical moment, more effective than the lobbyists of food industries, notably the fishing industry, and the interest groups of trade represented by the Ministry of Economics.[68]

Purifying Europe

Butenandt and Druckrey, both typical examples of "science advisers as policy-makers" and far from being isolated in the international scientific community, also fought for the new theory of carcinogenesis to become accepted as a European norm. Here, for the first time, they met strong opposition.[69] Both were involved in international organizations concerned with the rise in cancer and the role of cancer noxae, such as the Union Internationale contre le Cancer (UICC). At the UICC conference that took place in Paris in 1950, Druckrey and Butenandt demanded an "international cooperation for the protection against carcinogenic agents." The German scientists were busy establishing the Druckrey-Küpfmüller equation as a basis for international legal measures.[70] In May 1954, the Commission for Colorants of the German Research Foundation organized an international meeting for cancer prophylaxis in Bonn-Bad Godesberg, which resulted in the founding of the European Committee for the Protection of Populations against Risks of Chronic Toxicity (EUROTOX) and the announcement of the Godesberg Decrees on Food Additives. The latter played a major role in the establishment of JECFA one year later, an institution that was to dominate all means of regulations of food additives from then on. In the mid-1950s, Butenandt was absorbed by the transfer of the Max Planck Institute for Biochemistry to Munich, so Druckrey remained as the main actor to fight carcinogens in food within the frame of EUROTOX and JECFA beside the like-minded Otto Högl from the Swiss Public Health Office in Bern, Maurice Dols from the Netherlands' Ministry of Agriculture, Fisheries and Food, and René Truhaut from the Cancer Institute in Paris.[71]

According to the Godesberg Decrees, no foreign additives and no artificial colorants should be allowed in food unless they were explicitly authorized by law. To garner any official approval, synthetic additives must be demonstrably harmless for human health, there must be an actual demand for its application, the consumer must not be misled about its real value, and the amount added to food must be as low as possible.[72] "Harmless" additives were those that did not produce any toxic, carcinogenic, or germ-damaging effects in laboratory animals when given with continual dosage in the highest tolerated concentration for a lifetime.[73] The aim of the Godesberg Decrees—to establish the Druckrey-Küpfmüller equation as the basis for a European Codex—was supported by the West German government. Minister of the Interior Gerhard

Schröder (not to be confused with the later chancellor of the same name) wrote to Chancellor Konrad Adenauer that he hoped the cooperation of the World Health Organization (WHO) and the Food and Agriculture Organization (FAO) would help to strengthen the Godesberg Decrees as an international standard. Germany had to be protected from poisoned food imported from foreign countries. According to a JECFA working paper, the "Bad Godesberg meeting" was of major influence but had to be aligned with the conclusions of US senator James J. Delaney's Food Protection Committee.[74]

The year 1958 was crucial in the struggle to establish no-threshold policies concerning cancer-causing substances. Not only did Druckrey present the Druckrey-Küpfmüller equation for the first time in English at the CIBA-Symposium "Carcinogenesis," but both the amendment of the German Food Law, which was internationally recognized and seen as a model for other nations, and the implementation of the Delaney Clause in the US Food Additives Amendment took place. The latter used the formula of "generally recognized as safe" to legalize established food additives. But all new chemical substances had to be evaluated under a new clause that Senator Delaney had successfully integrated into the US food law with the help of a mobilized public and organized consumer movement: "The Food and Drug Administration shall not approve for use in food any chemical additive found to induce cancer in man, or, after tests, found to induce cancer in animals."[75]

Both the German decree and the US amendment were based on the assumption that carcinogenic substances had to be regulated differently than other toxic additives. As David A. Kessler summarizes, Delaney's anticancer clause did not permit risk-benefit analysis because there is no "known threshold below which a safe tolerance for a carcinogenic substance can be assumed." Delaney's anti-cancer clause followed the proposals made in August 1956 on a symposium held by the U.I.C.C., which again recapitulated the Druckrey-Küpfmüller equation.[76] Druckrey himself had a seminal influence on the legal acts concerning carcinogenic food additives in 1958. In the Color Additive Amendments of 1960, Delaney directly referred to Druckrey and Küpfmüller as main support for the no-threshold hypothesis.[77]

Thus, while it seemed that the "war" against carcinogens in food finally succeeded in 1958, the more sustainable debate took place at JECFA. There, on the contrary, a risk-benefit standard was established that, using the concept of an acceptable daily intake of certain additives, weighed economic benefits against health risks.[78] The chief task of JECFA's experts was to elaborate international standards for food additives to secure both health and trade.[79] This resulted in the installment of a Codex Alimentarius commission, whose aim was to establish a positive list of secure food additives.[80] The committee, which first met in December 1956 in Rome, had to set general principles for the use of food additives. But for the FAO as the representative of industry and trade, the most different socioeconomic and climatic circumstances had to be considered.[81] As one participant of the conference, Eugen Mergenthaler from the German Institute for Food Chemistry in Munich, summarized, JECFA had to secure health harmfulness, but also had to consider economic demands for stabilized food of consistent quality. Mergenthaler concluded that food additives should be banned only if they damaged the nutritional value or if they could be substituted by improved and economically reasonable new modes of production and processing.[82] Indeed, during the 1960s and especially with the help of the new concept of ADI, economic and political objections superseded Druckrey's no-threshold policy.[83]

At JECFA, Geoffrey Malcolm Badger and Bernard L. Oser were the most outspoken opponents to Druckrey. Badger, a chemist from Australia and disciple of cancer researcher James W. Cook, polemicized against exaggerating the dangers of food colorants besides azo dyes. The US biochemist Oser was against any life reform positions in general. "There will always be a certain risk," he reminded his colleagues of the industrial and chemical progress dominant in the United States since the early twentieth century: risks had to be weighed against nutritional and economic benefits.[84]

Druckrey and Dols could not prevail against the disapproval of such well-respected scientists, the leading role of FAO, and the lobbying activities of global food industry companies. During the 1950s, the Council of Europe and the European Economic Community had already been working on a European Codex Europaeus for food additives, which generally had stricter standards concerning cancer causing substances. But when in 1964, in the name of the global economy, the European Codex was subjected to the Codex Alimentarius and JECFA standards, the impact of any zero-tolerance policy finally diminished.[85] The Druckrey-Küpfmüller equation thereby was successively displaced by specific regulations based on the much more flexible and industry-friendly concept of ADI, inaugurated by Truhaut, Druckrey's former comrade in arms. This threshold concept produced zones of lesser or higher risk, a calculable risk assessment.[86] Whereas the Druckrey-Küpfmüller equation produced uncertainty and suspicion, the ADI facilitated a smooth adjustment to market and trade requirements. Since the early 1960s, thresholds of food additives rest upon the ADI value determined by JECFA experts. Butenandt and Druckrey had tried to establish the Druckrey-Küpfmüller equation and the paradigm of cancer as the starting point of risk policies, but carcinogenesis was "normalized" as an exception by JECFA. The ADI value proved to be a negotiable instrument for the modulation of health policies and economic interests, an instrument that replaced policies of risk prevention with those of risk assessment.[87]

Conclusions

Druckrey's radical conclusions, drawn from the case of butter yellow and reinforced by the longstanding narrative of "poisoned food," were legally manifested in the food amendments of 1958. But the regulations following the amendment of the German food law were far less strict and informed by the lobbying groups of the food industry, just like the Delaney Clause was gradually disempowered.[88] The essential tension between the axiom that there are no harmless substances and the pragmatic drawing up of a positive list characterized the debate on food additives in the second half of the twentieth century. Whereas the German scientists failed in establishing rigid policies for food additives on a global scale, the semantics of poisoning and carcinogenesis determined the agenda of both the consumer and environmental movement.

Since the 1960s, as manifested in Rachel Carson's *Silent Spring* from 1962— in some ways a literary remake of Fritz Eichholtz's *Toxische Gesamtsituation* from 1956—an apocalyptic discourse about the poisoning of life itself faced the pragmatic risk management policies and pro-business negotiations of national governments and transnational organizations.[89] In 1961, the German physician Bodo Manstein, not only an advocate of biopolitics and life protection but also an important actor of the West German movement for nature protection, described a "contamination of all biological systems including humans."[90] In the US debate, the semantics of food poisoning climaxed around 1970, as can be detected from publications like William Frank Longgood's *The Poisons in Your Food* from 1969 or *The Chemical Feast* published by the Ralph Nader Study Group in 1970. With an interesting twist, the ecological debate was transferred into the internal environment when Gene Marine and Judith Van Allen, in their monograph *Food Pollution*, spoke of "the violation of our inner ecology."[91] In the second half of the twentieth century, the critique of modern food production in general, and especially the use of food additives, ranges between purist assumptions about "poisoned food," often inflamed by media coverage, and the arduous research work of testing and screening, which seldom reveals knowledge and definitude, but a constant flow of hypotheses and uncertainties.[92] Still in 1973, pharmacologist and physician Peter Marquardt, in his introduction to *Gift auf dem Tisch* (Poison on the table), perpetuated the narrative of poisoned food: since the end of the nineteenth century, industrialization and mechanization had changed living conditions in a dramatic way. New methods of food preservation, new lifestyle habits, and new markets for processed food had emerged in the metropolitan areas. Food colorants and preservatives played a leading role in this scenario.[93]

The purist discourse, which characterized the critique of food additives in Germany since the early twentieth century, was based on the semantic interaction of poison and cancer and was part of a broader narrative of foreign matter contaminating the imagined community of the Germans. On the other side, the inauguration of ADI and the cooperation of WHO and FAO succeeded in establishing threshold values and regulation practices that preferred economy to health. In this setting, the strict separation of toxic and carcinogenic effects was of major importance. The first could be regulated through the ADI; the latter were marginalized as an exception, for which the Druckrey-Küpfmüller equation, in general accepted as a fact by the scientific community, could be applied. Because it was much easier to ban carcinogenic substances, for the food and the chemical industry it was highly important that food additives were classified as toxic and not as carcinogenic, whereas advocates of consumer rights had to denounce chemicals as carcinogenic rather than as toxic to provoke public reaction and legal action.[94] This scenario, with each precarious substance at stake, remains today.

Heiko Stoff is a research scholar at Hannover Medical School. His work focuses on the history of science, medicine, and gender in the nineteenth and twentieth century. He recently edited, together with Alexander von Schwerin, a special issue on food additives regulation in *Technikgeschichte* 81, no. 3 (2014). He also edited, with Alexander von Schwerin and Bettina Wahrig, *Biologics: A History of Agents Made from Living Organisms in the Twentieth Century* (2013) and wrote *Die Komamethode: Willensfreiheit, Selbstverantwortung und der Anfang vom Ende der Roten Armee Fraktion im Winter 1984/85* (2020), *Gift in der Nahrung: Zur Genese der Verbraucherpolitik in Deutschland Mitte des 20. Jahrhunderts* (2016), *Wirkstoffe. Eine Wissenschaftsgeschichte der Hormone, Vitamine und Enzyme, 1920–1970* (2012), and *Ewige Jugend: Konzepte der Verjüngung vom späten 19. Jahrhundert bis ins Dritte Reich* (2004).

Notes

1. See Stoff and Travis, "Discovering"; Hisano, *Visualizing*, 41–66; Cobbold, "Introduction of Chemical Dyes."
2. Coppin and High, *The Politics*.
3. Druckrey, "Quantitative Aspects," 60.
4. Lenhard-Schramm, *Das Land*, 71–134.
5. This is elaborated in Stoff, *Gift*; and Stoff, "Die toxische Gesamtsituation." For the politics of drug regulation see Daemmrich, *Pharmacopolitics*.
6. For the imminent role of the trope "poison" in this context see Grossarth, *Die Vergiftung*; and Stoff, *Gift*. In general, see Kirchhelle, "Toxic Tales"; Bertomeu-Sánchez and Guillem-Llobat, "Following Poisons"; and Kirchhelle, "Toxic Confusion."
7. Stoff, *Wirkstoffe*, 7–24. For the concept of problematization see Foucault, *Fearless Speech*, 171; and Callon, "Struggles."
8. Balz, von Schwerin, Stoff, and Wahrig, *Precarious Matters*. Also Wunderlich, "Zur Entstehungsgeschichte."

9. Cantor, "Introduction." See also Rosenberg, "Pathologies." I owe this remark to Xaq Frohlich, whose comments helped me to restructure this article for an American readership.

10. Deichmann et al., "What Is There."

11. Stoff, "Identity."

12. Literal translations would be "pleasure poisons," "tissue poisons," "civilization poisons," "environmental poisons," and "summation poisons."

13. Stoff, *Gift*, 47–58.

14. See Treitel, *Eating Nature*; Fritzen, *Gesünder leben*; Heyll, *Wasser*; Melzer, *Vollwerternährung*; and Merta, *Wege*.

15. Rabinow, *Essays*, 104. See also Atkins, *Liquid Materialities*.

16. Kollath, *Die Ordnung*.

17. Butenandt, "Neuere Beiträge."

18. Druckrey, "Die Pharmakologie."

19. Lenzner, *Gift*, XI, 193.

20. Liek, *Krebsverbreitung*, 179; Liek, *Der Arzt*, 66, 184.

21. Liek, *Krebsverbreitung*, 30–47, 134, 142, 147. See also Proctor, *The Nazi War*, 22–27.

22. Stoff, *Gift*, 59–60; Proctor, *The Nazi War*.

23. Eichholtz, *Die toxische Gesamtsituation*.

24. Harrington, *Reenchanted Science*, 185–88.

25. Kollath, *Die Ordnung*, 14. See also Stoff, *Wirkstoffe*, 279–353; Sperling, *Kampf*; and Moser, *Deutsche Forschungsgemeinschaft*.

26. For this chapter, see especially Stoff and Travis, "Discovering," 147–52. Also Sellers, "From Poison to Carcinogen"; and, in general, Hisano, *Visualizing Taste*.

27. See Cobbold, "Responding"; Schaad, *Chemische Stoffe*; Hien, *Chemische Industrie*; and Andersen, "Roth."

28. Yamagiwa and Ichikawa, "Experimental Study."

29. Butenandt, "Neuere Beiträge," 347–48.

30. Schürch, and Winterstein, "Über die krebserregende Wirkung," 79–91.

31. Butenandt, "Neuere Beiträge," 348.

32. Gaudillière, "Hormones"; Gaudillière, "Biochemie." See also Stoff, "Oestrogens."

33. Bauer, "Über Chemie." For scarlet red, see Schmähl, "Krebserzeugende Stoffe."

34. Kinosita, "Studies"; and Bauer, "Über Chemie," 26–28. See also Stoff and Travis, "Discovering," 152–56.

35. Brock, Druckrey, and Hamperl, "Die Erzeugung."

36. Kuhn and Beinert, "Fermentgift."

37. Opie, "The Pathogenesis."

38. Proctor, *Nazi War*, 165–70.

39. "Krebs."

40. Walter Hinnendahl to Theodor Heuss, 1 January 1950, Bundesarchiv Koblenz, B 116/419. See also Stoff and Travis, "Discovering," 150–51.

41. For example, he is not mentioned in Vogel, *The Politics*.

42. Wunderlich, "Mit Papier"; and Wunderlich, "Zur Entstehungsgeschichte."

43. Hermann Druckrey, "Begründung für die Schaffung eines speziellen Institutes für die pharmakologische Lebensmittelforschung," undated, Archiv der Max-Planck-

Gesellschaft, III. Abt., Rep. 84/1, Nr. 429. See also Wunderlich, "Zur Entstehungsgeschichte," 371.

44. Druckrey, "Versuche"; Druckrey and Küpfmüller, "Quantitative Analyse"; and Hermann Druckrey, "Begründung für die Schaffung eines speziellen Institutes für die pharmakologische Lebensmittelforschung," undated, Archiv der Max-Planck-Gesellschaft, III. Abt., Rep. 84/1, Nr. 429. For an English summary, see Druckrey, "Quantitative Aspects." See also Stoff and Travis, "Discovering," 153–54.

45. Druckrey, "Pharmacological Approach."

46. Druckrey and Küpfmüller, "Quantitative Analyse," 259.

47. Tennekes, "A Critical Appraisal."

48. Druckrey and Küpfmüller, "Dosis." See also Wunderlich, "Dosis"; and Schwerin, "Vom Gift."

49. Druckrey, "Versuche," 47; Druckrey and Küpfmüller, "Quantitative Analyse," 260. See also Schwerin, "Vom Gift," 259–62.

50. Druckrey and Küpfmüller, "Quantitative Analyse," 254, 259.

51. Druckrey, "Versuche," 56.

52. Bauer, "Über Chemie," 39.

53. Stoff, "Hexa-Sabbat."

54. "Krebs."

55. Bauer, "Über Chemie," 26, my translation. "Nun, unser *Zeitalter der Chemisierung und Technisierung unserer Umwelt* einschließlich unserer Nahrung kennt auch *Krebsnoxen für jedermann*, also für alle, für Millionen, Tag für Tag und unfreiwillig für lange, lange Jahre."

56. Wolgast, "Karl Heinrich Bauer." See also Moser, "Deutsche Forschungsgemeinschaft," 94–97; and Proctor, *The Nazi War*, 58–68.

57. Lau and Baier, "Über Versuche."

58. Proctor, *The Nazi War*, 64

59. Bauer, "Über Chemie," 33–34.

60. "Krebs," 27; and Bauer, "Über Chemie," 34. My translation.

61. Stoff, *Gift*, 103–22.

62. Ibid., 149–75.

63. "Werden wir vergiftet?"

64. Rick, *Verbraucherpolitik*.

65. Stoff, *Gift*, 144–45.

66. Second Deutscher Bundestag, 149th meeting, Bonn, 8 June 1956, Bundesarchiv Koblenz, B 142/1528, 7901.

67. Krusen, "Das neue Lebensmittelrecht."

68. Stoff, "Hexa-Sabbat."

69. Jasanoff, *The Fifth Branch*. See also Smith and Phillips, "Food Policy."

70. Stoff, *Gift*, 122–40.

71. Ibid., 126–30; and Jas, "Adapting."

72. Hamperl, "Ergebnisse."

73. Hermann Druckrey, "Ergebnisse einer Tagung westeuropäischer Wissenschaftler zur Verhütung des Krebses bei der deutschen Forschungsgemeinschaft (DFG) in Bad Godesberg am 1. Mai 1954," Archiv der Deutschen Forschungsgemeinschaft, Bonn, 6019.

74. Stoff, *Gift*, 130–32.
75. Hutt, "Regulation." See also Gaudillière, "DES"; and Marcus, *Cancer from Beef.*
76. Kessler, "Implementing," 822.
77. Merrill, "FDA's implementation"; Weisburger and Weisburger, "Food Additives."
78. Ramsingh, "The Emergence"; and Ramsingh, "The History."
79. WHO, *General Principles*, 4. See also Jas, "Adapting"; and Pestre, "Regimes."
80. Ramsingh, "The Emergence." See also Millstone and Zwanenberg, "The Evolution."
81. WHO, *General Principles*, 5–9. See also Ramsingh, "The History," 79.
82. Mergenthaler, "Tagung."
83. FAO and WHO, *Evaluation.*
84. WHO, "Conference on Food Additives/1–19, 1955," WHO Library, Geneva. See also Vogel, "From 'The Dose Makes the Poison.'"
85. For the complicated cooperation of WHO and FAO, see Ramsingh, "The History."
86. Truhaut, "The Concept"; and Truhaut, *25 Years.* See also Boudia and Jas, "Introduction"; and Pestre, "Regimes."
87. Reinhardt, "Regulierungswissen."
88. See Maffini and Vogel, this volume.
89. Carson, *Silent Spring.* See also Waddell, *And No Birds Sing.*
90. Manstein, *Im Würgegriff*, 47.
91. Longgood, *The Poisons*; Turner, *The Chemical Feast*; and Marine and Van Allen, *Food Pollution.* See also Jundt, *Greening*, 131–34.
92. See Davis, *Banned*, 221–24.
93. Marquardt, "Einleitung."
94. For the newest debate on whether the Druckrey-Küpfmüller equation might be applicable also for toxic substances, see Tennekes, "A Critical Appraisal."

Bibliography

Archives

Archiv der Deutschen Forschungsgemeinschaft, Bonn
Archiv der Max-Planck-Gesellschaft, Berlin
Bundesarchiv, Koblenz
WHO Library, Geneva

Publications

Andersen, Arne. "'Roth, blau und grün angestrichene, Schrecken erregende Gestalten.' Farbstoffindustrie und arbeitsbedingte Erkrankungen." In *Das blaue Wunder. Zur Geschichte der synthetischen Farben*, edited by Arne Andersen and Gerd Spelsberg, 162–92. Cologne: Volksblatt, 1990.
Atkins, Peter William. *Liquid Materialities: A History of Milk, Science and the Law.* Farnham: Ashgate, 2010.
Balz, Viola, Alexander von Schwerin, Heiko Stoff, and Bettina Wahrig, eds. *Precarious Matters: The History of Dangerous and Endangered Substances in the 19th and 20th Centuries.* Berlin: Preprint des Max-Planck-Institut für Wissenschaftsgeschichte, 2008.

Bauer, Karl-Heinrich. "Über Chemie und Krebs—dargestellt am 'Anilinkrebs.'" *Langenbecks Archiv für Klinische Chirurgie* 264 (1950): 21–44.

Brock, Norbert, Hermann Druckrey, and Herwig Hamperl. "Die Erzeugung von Leberkrebs durch den Farbstoff 4-Dimethylamino-azobenzol." *Zeitschrift für Krebsforschung* 50 (1940): 431–56.

Butenandt, Adolf. "Neuere Beiträge der biologischen Chemie zum Krebsproblem." *Angewandte Chemie* 53 (1940): 345–52.

Bertomeu-Sánchez, José Ramón, and Ximo Guillem-Llobat. "Following Poisons in Society and Culture (1800–2000): A Review of Current Literature." *Actes d'Història de la Ciència i de la Tècnica* 9 (2017): 9–36.

Boudia, Soraya, and Nathalie Jas. "Introduction: Risk and 'Risk Society' in Historical Perspective." *History and Technology* 23 (2007): 317–31.

Callon, Michel. "Struggles and Negotiations to Define What Is Problematic and What Is Not: The Sociologic Translation." In *The Social Process of Scientific Investigation*, edited by Karin D. Knorr, Roger Krohn, and Richard Whitley, 197–219. Dordrecht: Springer, 1981.

Cantor, David. "Introduction: Cancer Control and Prevention in the Twentieth Century." *Bulletin of the History of Medicine* 81 (2007): 1–38.

Carson, Rachel. *Silent Spring*. Boston: Houghton Mifflin, 1962.

Cobbold, Carolyn. "The Introduction of Chemical Dyes into Food in the Nineteenth Century." *Osiris* 35 (2020): 142–61.

———. "Responding to the Colourful Use of Chemicals in Nineteenth-Century Food." *Actes d'Història de La Ciència I De La Tècnica* 9 (2016): 37–54.

Coppin, Clayton A., and Jack C. High. *The Politics of Purity: Harvey Washington Wiley and the Origins of Federal Food Policy*. Ann Arbor: University of Michigan Press, 1999.

Daemmrich, Arthur A. *Pharmacopolitics: Drug Regulation in the United States and Germany*. Chapel Hill: University of North Carolina Press, 2004.

Davis, Frederick Rowe. *Banned: A History of Pesticides and the Science of Toxicology*. New Haven: Yale University Press, 2014.

Deichmann, William B., Dietrich Henschler, Bo Holmestedt, and Gert Keil. "What Is There That Is Not Poison? A Study of the Third Defense by Paracelsus." *Archives of Toxicology* 58 (1986): 207–13.

Druckrey, Hermann. "Die Pharmakologie krebserregender Substanzen." *Zeitschrift für Krebsforschung* 57 (1950): 70–85.

———. "Pharmacological Approach to Carcinogenesis." In *Ciba Foundation Symposium-Carcinogenesis: Mechanisms of Action*, edited by E. W. Wolstenholme and Maeve O'Connor, 110–30. Chichester: Wiley, 1959.

———. "Quantitative Aspects in Chemical Carcinogenesis." In *Potential Carcinogenic Hazards from Drugs: Evaluation of Risks*, edited by René Truhaut, 60–78. Berlin: Springer, 1967.

———. "Versuche zur Krebserzeugung mit Anilin oder Anilin-Derivaten." *Langenbecks Archiv für Klinische Chirurgie* 264 (1950): 45–60.

Druckrey, Hermann, and Karl Küpfmüller. "Dosis und Wirkung. Beiträge zur theoretischen Pharmakologie." *Die Pharmazie*. 8. Beiheft, 1. Ergänzungsband (1949): 514–645.

———. "Quantitative Analyse der Krebsentstehung." *Zeitschrift für Naturforschung* 3b (1948): 254–66.

Eichholtz, Fritz. *Die toxische Gesamtsituation auf dem Gebiet der menschlichen Ernährung. Umrisse einer unbekannten Wissenschaft.* Berlin: Springer, 1956.

Foucault, Michel. *Fearless Speech.* New York: Semiotext(e), 2001.

FAO and WHO, eds. *Evaluation of the Toxicity of a Number of Antimicrobials and Antioxidants. Sixth Report of the Joint FAO/WHO Expert Committee on Food Additives.* Geneva: World Health Organization, 1962.

Fritzen, Florentine. *Gesünder leben. Die Lebensreformbewegung im 20. Jahrhundert.* Stuttgart: Steiner, 2006.

Gaudillière, Jean-Paul. "Biochemie und Industrie. Der 'Arbeitskreis Butenandt-Schering' während der Zeit des Nationalsozialismus." In *Adolf Butenandt und die Kaiser-Wilhelm-Gesellschaft. Wissenschaft, Industrie und Politik im "Dritten Reich,"* edited by Wolfgang Schieder and Achim Trunk, 198–246. Göttingen: Wallstein, 2004.

———. "DES, Cancer, and Endocrine Disruptors: Ways of Regulating, Chemical Risks, and Public Expertise in the United States." In *Powerless Science? Science and Politics in a Toxic World,* edited by Soraya Boudia and Nathalie Jas, 65–94. New York: Berghahn, 2014.

———. "Hormones at Risk. Cancer and the Medical Uses of Industrially-Produced Sex Steroids in Germany, 1930–1960." In *Risks of Medical Innovation: Risk Perception and Assessment in Historical Context,* edited by Thomas Schlich and Ulrich Tröhler, 148–69. London: Routledge, 2006.

Grossarth, Jan. *Die Vergiftung der Erde: Metaphern und Symbole agrarpolitischer Diskurse.* Frankfurt am Main: Campus, 2018.

Harrington, Anne. *Reenchanted Science: Holism in German Culture from Wilhelm II to Hitler.* Princeton: Princeton University Press, 1999.

Hamperl, Herwig. "Ergebnisse einer Tagung westeuropäischer Wissenschaftler zur Prophylaxe des Krebses bei der 'Deutschen Forschungsgemeinschaft,' in Bad Godesberg am 1. Mai 1954." *Zeitschrift für Krebsforschung* 60 (1955): 616–20.

Heyll, Uwe. *Wasser, Fasten, Luft und Licht: Die Geschichte der Naturheilkunde in Deutschland.* Frankfurt am Main: Campus, 2006.

Hien, Wolfgang. *Chemische Industrie und Krebs. Zur Soziologie des wissenschaftlichen und sozialen Umgangs mit arbeitsbedingten Krebserkrankungen in Deutschland.* Bremerhaven: Wirtschaftsverlag NW, 1994.

Hisano, Ai. *Visualizing Taste: How Business Changed the Look of What You Eat.* Cambridge: Harvard University Press, 2019.

Hutt, Peter Barton. "Regulation of Food Additives in the United States." In *Food Additives,* 2nd edition, revised and expanded, edited by A. Larry Branen, P. Michael Davidson, Seppo Salminen, and John H. Thorngate, III, 198–226. New York: Dekker, 2002.

Jas, Nathalie. "Adapting to 'Reality': The Emergence of an International Expertise on Food Additives and Contaminants in the 1950s and Early 1960s." In *Toxicants, Health and Regulation since 1945,* edited by Soraya Boudia and Nathalie Jas, 47–69. London: Pickering & Chatto, 2013.

Jasanoff, Sheila. *The Fifth Branch: Science Advisers as Policymakers.* Cambridge: Harvard University Press, 1994.

Jundt, Thomas. *Greening the Red, White, and Blue: The Bomb, Big Business, and Consumer Resistance in Postwar America.* Oxford: Oxford University Press, 2014.

Kessler, David A. "Implementing the Anticancer Clauses of the Food, Drug and Cosmetic Act." *University of Chicago Law Review* 44 (1977): 817–850.

Kinosita, Riojun. "Studies on the Cancerogenic Azo and Related Compounds." *Yale Journal of Biology and Medicine* 12 (1940): 287–300.

Kirchhelle, Claas. "Toxic Confusion: The Dilemma of Antibiotic Regulation in West German Food Production (1951–1990)." *Endeavour* 40 (2016): 114–27.

———. "Toxic Tales: Recent Histories of Pollution, Poisoning, and Pesticides (ca. 1800–2010)." *NTM Zeitschrift für Geschichte der Wissenschaften, Technik und Medizin* 26 (2018): 213–29.

Kollath, Werner. *Die Ordnung unserer Nahrung. Grundlagen einer dauerhaften Ernährungslehre.* Stuttgart: Hippokrates, 1942.

"Krebs. Die Krankheit der Epoche." *Der Spiegel* 7, no. 28 (1953): 22–30.

Krusen, Felix. "Das neue Lebensmittelrecht in der Bundesrepublik." *Die Nahrung* 4 (1960): 549–62.

Kuhn, Richard, and Helmut Beinert. "Über das aus krebserregenden Azofarbstoffen entstehende Fermentgift." *Berichte der deutschen chemischen Gesellschaft* 76 (1943): 904–09.

Lau, Hans, and Paul Baier. "Über Versuche zur gelenkten Krebserzeugung durch Zusammenwirken einer spezifischen Krebsnoxe (2-Acetylaminofluoren) und einer nicht-krebsspezifischen Schädigung (Cholinmangelernährung)." *Langenbecks Archiv für Klinische Chirurgie* 278 (1954): 156–72.

Lenhard-Schramm, Niklas. *Das Land Nordrhein-Westfalen und der Contergan-Skandal: Gesundheitsaufsicht und Strafjustiz in den "langen sechziger Jahren."* Göttingen: Vandenhoeck & Ruprecht, 2016.

Lenzner, Curt. *Gift in der Nahrung. Zweite umgearbeitete und erweiterte Auflage.* Leipzig: Verlag der Dykschen Buchhandlung, 1933.

Liek, Erwin. *Der Arzt und seine Sendung. Sechste Auflage.* Munich: Lehmanns, 1927.

———. *Krebsverbreitung, Krebsbekämpfung, Krebsverhütung.* Munich: Lehmanns, 1932.

Longgood, William Frank. *The Poisons in Your Food.* New York: Simon and Schuster, 1969.

Manstein, Bodo. *Im Würgegriff des Fortschritts.* Frankfurt am Main: Europäische Verlagsanstalt, 1961.

Marine, Gene, and Judith Van Allen. *Food Pollution: The Violation of Our Inner Ecology.* New York: Holt, Rinehart and Winston, 1972.

Marcus, Alan I. *Cancer from Beef: DES, Federal Food Regulation, and Consumer Confidence.* Baltimore: Johns Hopkins University Press, 1994.

Marquardt, Peter. "Einleitung." In *Gift auf dem Tisch? Profit oder Gesundheit in unseren Lebensmitteln,* edited by Werner Gabel, Hans Glatzel, Peter Marquardt, and Konrad Pfeilsticker, 9–15. Herford: Nicolai, 1973.

Melzer, Jörg M. *Vollwerternährung: Diätetik, Naturheilkunde, Nationalsozialismus, sozialer Anspruch.* Stuttgart: Steiner, 2003.

Mergenthaler, E. "Tagung des gemeinsamen Fachausschusses für Lebensmittelzusatzstoffe der Food and Agriculture Organization (FAO) und der World Health Organization (WHO) in Rom vom 3. bis 10. Dezember 1956." *Zeitschrift für Lebensmitteluntersuchung und -Forschung* 108 (1958): 184–86.

Merrill, Richard A. "FDA's Implementation of the Delaney Clause: Repudiation of Congressional Choice or Reasoned Adaptation to Scientific Progress." *Yale Journal on Regulation* 5 (1988): 1–88.

Merta, Sabine. *Wege und Irrwege zum modernen Schlankheitskult. Diätkost und Körperkultur als Suche nach neuen Lebensstilformen 1880–1930.* Stuttgart: Steiner, 2003.

Moser, Gabriele. *Deutsche Forschungsgemeinschaft und Krebsforschung, 1920–1970.* Stuttgart: Steiner, 2011.

Millstone, Erik, and Patrick van Zwanenberg. "The Evolution of Food Safety Policy-Making Institutions in the UK, EU and Codex Alimentarius." *Social Policy & Administration* 36 (2002): 593–609.

Opie, Eugene L. "The Pathogenesis of Tumors of the Liver Produced by Butter Yellow." *Journal of Experimental Medicine* 80 (1944): 231–46.

Pestre, Dominique. "Regimes of Knowledge Production in Society: Towards a More Political and Social Reading." *Minerva* 41 (2003): 245–61.

Proctor, Robert N. *The Nazi War on Cancer.* Princeton: Princeton University Press, 1999.

Rabinow, Paul. *Essays on the Anthropology of Reason.* Princeton: Princeton University Press, 1996.

Ramsingh, Brigit. "The Emergence of International Food Safety Standards and Guidelines: Understanding the Current Landscape through a Historical Approach." *Perspectives in Public Health* 134 (2014): 206–15.

Ramsingh, Brigit. "The History of International Food Safety Standards and the Codex Alimentarius (1955–1995)." Ph.D. diss., University of Toronto, 2013.

Reinhardt, Carsten. "Regulierungswissen und Regulierungskonzepte." *Berichte zur Wissenschaftsgeschichte* 33 (2010): 351–64.

Rick, Kevin. *Verbraucherpolitik in der Bundesrepublik Deutschland. Eine Geschichte des westdeutschen Konsumtionsregimes, 1945–1975.* Baden-Baden: Nomos, 2018.

Rosenberg, Charles E. "Pathologies of Progress: The Idea of Civilization as Risk." *Bulletin of the History of Medicine* 72 (1998): 714–30.

Schaad, Nicole. *Chemische Stoffe, giftige Körper: Gesundheitsrisiken in der Basler chemischen Industrie, 1860–1930.* Zürich: Chronos, 2003.

Schmähl, Dietrich. "Krebserzeugende Stoffe." In *Gifte. Geschichte der Toxikologie*, edited by Mechthild Amberger-Lahrmann and Dietrich Schmähl, 167–96. Berlin: Springer, 1988.

Schürch, O., and A. Winterstein. "Über die krebserregende Wirkung aromatischer Kohlenwasserstoffe." *Hoppe-Seyler's Zeitschrift für physiologische Chemie* 236 (1935): 79–91.

Schwerin, Alexander von. "Vom Gift im Essen zu chronischen Umweltgefahren. Lebensmittelzusatzstoffe und die risikopolitische Institutionalisierung der Toxikogenetik in Westdeutschland, 1955–1964." *Technikgeschichte* 81 (2014): 251–74.

Sellers, Christopher C. "From Poison to Carcinogen: Towards a Global History of Concerns about Benzene." *Global Environment* 7 (2014): 38–71.

Smith, David, and Jim Phillips. "Food Policy and Regulation: A Multiplicity of Actors and Experts." In *Food, Science, Policy and Regulation in the Twentieth Century: International and Comparative Perspectives*, edited by David Smith and Jim Phillips, 1–16. Abingdon: Routledge, 2000.

Sperling, Frank. *"Kampf dem Verderb" mit allen Mitteln. Der Umgang mit ernährungsbezogenen Gesundheitsrisiken im "Dritten Reich" am Beispiel der chemischen Lebensmittelkonservierung.* Stuttgart: Deutscher Apotheker Verlag, 2011.

Stoff, Heiko. "Die toxische Gesamtsituation: Die Angst vor mutagenen und teratogenen Stoffen in den 1950er Jahren." In *Contergan: Hintergründe und Folgen eines Arzneimittel-*

Skandals, edited by Thomas Großbölting and Niklas Lenhard-Schramm, 45–70. Göttingen: Vandenhoeck & Ruprecht, 2017.

———. *Gift in der Nahrung. Zur Genese der Verbraucherpolitik in Deutschland Mitte des 20. Jahrhunderts*. Stuttgart: Steiner, 2015.

———. "'Hexa-Sabbat': Fremdstoffe und Vitalstoffe, Experten und der kritische Verbraucher in der BRD der 1950er und 1960er Jahre." *NTM Zeitschrift für Geschichte der Wissenschaften, Technik und Medizin* 17 (2009): 55–83.

———. "Identity, Precariousness, and Poison: A Brief and Political Outlook." In *Poison and Poisoning in Science, Fiction and Cinema: Precarious Identities*, edited by Heike Klippel, Bettina Wahrig, and Anke Zechner, 239–49. Cham: Palgrave Macmillan, 2017.

———. "Oestrogens and Butter Yellow: Gendered Policies of Contamination in Germany, 1940–1970." In *Gendered Drugs and Medicine: Historical and Socio-Cultural Perspectives*, edited by Teresa Ortiz-Gómez and Maria Jésus Santesmases, 23–41. Farnham: Ashgate, 2014.

———. *Wirkstoffe. Eine Wissenschaftsgeschichte der Hormone, Vitamine und Enzyme, 1920–1970*. Stuttgart: Steiner, 2012.

Stoff, Heiko, and Anthony S. Travis. "Discovering Chemical Carcinogenesis: The Case of the Aromatic Amines." In *Hazardous Chemicals*, edited by Ernst Homburg and Elisabeth Vaupel, 137–78. New York: Berghahn, 2019.

Tennekes, Henk A. "A Critical Appraisal of the Threshold of Toxicity Model for Non-Carcinogens." *Environment & Analytical Toxicology* 6 (2016): 408.

Treitel, Corinna. *Eating Nature in Modern Germany*. Cambridge: Cambridge University Press, 2017.

Truhaut, René. "The Concept of the Acceptable Daily Intake: An Historical Review." *Food Additives & Contaminants* 8 (1991): 151–62.

———. *25 Years of JECFA Achievements (1956–1981). Report Presented to the 25th Session of JECFA. March 23–April 1*. Geneva: WHO, 1981.

Turner, James S. *The Chemical Feast: Ralph Nader's Study Group on the Food and Drug Administration*. New York: Grossman, 1970.

Vogel, David. *The Politics of Precaution: Regulating Health, Safety and Environmental Risks in Europe and the United States*. Princeton: Princeton University Press, 2012.

Vogel, Sarah A. "From 'The Dose Makes the Poison' to 'The Timing Makes the Poison': Conceptualizing Risk in the Synthetic Age." *Environmental History* 13 (2008): 667–73.

Waddell, Craig, ed. *And No Birds Sing: Rhetorical Analyses of Rachel Carson's Silent Spring*. Lincoln: Southern Illinois University Press, 2000.

Weisburger, J. H., and Elizabeth K. Weisburger. "Food Additives and Chemical Carcinogens: On the Concept of Zero Tolerance." *Food and Cosmetics Toxicology* 6 (1968): 235–42.

"Werden wir vergiftet? Gebleichte, gefärbte, konservierte, 'geschönte' Lebensmittel." *Die Zeit* 24 (17 June 1954).

WHO, ed. *General Principles Governing the Use of Food Additives: First Report of the Joint FAO/WHO Committee on Food Additives*. Technical Report Series 129. Geneva: World Health Organization, 1957.

Wolgast, Eike. "Karl Heinrich Bauer. Der erste Heidelberger Nachkriegsrektor. Weltbild und Handeln 1945–1946." In *Heidelberg 1945*, edited by Jürgen C. Heß, Hartmut Lehmann, and Volker Sellin, 107–29. Stuttgart: Steiner, 1996.

Wunderlich, Volker. "Dosis und Wirkung in der Toxikologie: Die Haber'sche Regel und Ableitungen." *BIOspektrum* 25 (2019): 584–85.

———. "'Mit Papier, Bleistift und Rechenschieber.' Der Krebsforscher Hermann Druckrey im Internierungslager Hammelburg (1946–1947)." *Medizinhistorisches Journal* 43 (2008): 327–43.

———. "Zur Entstehungsgeschichte der Druckrey-Küpfmüller-Schriften (1948–1949). Dosis und Wirkung bei krebserzeugenden Stoffen." *Medizinhistorisches Journal* 40 (2005): 369–97.

Yamagiwa, Katsusaburo, and Koichi Ichikawa. "Experimental Study of the Pathogenesis of Carcinoma." *CA: A Cancer Journal for Clinicians* 27 (1977): 174–81.

CHAPTER 4

"EAT. DIE."

The Domestication of Carcinogens in the 1980s

Angela N. H. Creager

My title derives from the cover of *Science* magazine on 2 September 1983, which featured pop art by Robert Indiana from his *EAT* series.[1] The thrice-paired words "EAT" and "DIE" were cued to an article on dietary carcinogens—agents that cause cancer—in this issue (see figure 4.1). As Philip Abelson commented in an accompanying editorial, "Mutagens are present in substantial quantities in fruits and vegetables. Carcinogens are formed in cooking as a result of reactions involving proteins or fats. Dietary practices may be an important determinant of current cancer risks."[2] Twenty years earlier, both scientists and citizens had focused on the hazards of pesticides, preservatives, and colors—the chemicals added to foods, not food itself. What had happened? This new focus on cancer-causing agents inherent in food and cooking, which I have dubbed the domestication of carcinogens, had significant implications for cancer policy and chemical regulation.[3] While many scientists were persuaded of this new perspective, most people continued instead to be worried about residues of synthetic chemicals in food. In this chapter I will sketch out briefly how scientific ideas, popular perceptions, and regulatory politics in the United States interacted during these years, leading to a situation of divergent views and stalled policy changes.[4]

Concerns about the purity and safety of food, of course, predate the twentieth century. In the late-nineteenth-century United States, widespread adulteration of food prompted the development of testing methods and calls for greater oversight.[5] This is clearly captured in the wording of the 1906 Pure Food and Drug Act. Adulterants included fraudulent substances as well as dangerous ones. As observed by one commentator, "Milk became a medium through which much water was sold at a profit."[6] The industrialization of food processing, preserving, and packaging in the mid-twentieth century—what James Harvey Young called the "chemogastric revolution"—generated new issues and risks.[7] Already, scientists at the US Department of Agriculture had discovered that some food preservatives and artificial colors were hazardous.

Figure 4.1. Cover of *Science* magazine, 23 September 1983, featuring *The Green Diamond Eat and the Red Diamond Die* by Robert Indiana, 1962 (Collection Walker Art Center, Minneapolis, MN, gift of the TB Walker Foundation). Reproduced by permission of the American Association for the Advancement of Science, Artists Rights Society, and the Walker Art Center. Robert Indiana: © 2020 Morgan Art Foundation / Artists Rights Society (ARS), New York.

But because this agency promoted the food industry, these findings did not prompt regulation.[8]

In 1938, the Federal Food, Drug, and Cosmetic Act expanded government oversight over dangerous adulterants, such as highly toxic diethylene glycol, which was present in the pediatric antibiotic Elixir Sulfanilamide, leading to the deaths of over a hundred patients. The law also bolstered regulation of "'poisonous and deleterious' substances in food," including some natural toxins such as that responsible for botulism.[9] Food additives, however, were not generally suspected in connection with cancer—though, as Heiko Stoff shows in this volume, that began to change around 1940 with the identification of the "butter yellow" coloring as a carcinogen. This was part of a broader trend in the 1930s and 1940s as scientists focused attention on external agents that could cause cancer, such as various forms of radiation, tissue irritants such as coal tars, and viruses.[10]

Ordinary citizens became more aware of such external cancer-causing agents in connection with two trends after World War II.[11] First, the development of atomic energy brought about new exposures to ionizing radiation, a known carcinogen. In addition to occupational exposures, peacetime atomic weapons testing and civilian radioactive waste threatened to expose human populations to radioactivity contaminating the environment. Investigations of how radiation affects cells focused attention on the effects of low-dose and chronic exposure to carcinogens. Although the US Atomic Energy Commission offered countless reassurances that environmental exposure levels were below the threshold for health hazards, some scientists dissented from this view.[12] Disagreement over the safety of low-level radioactive contaminants became public—and political—in the late-1950s debates over fallout from atomic weapons testing. Strontium-90, a fission product that could move from pasture through grazing cattle to milk, was especially alarming.[13] In the midst of growing opposition to atomic testing, a group of geneticists advocated the idea that human cancers typically arose from mutations to somatic cells.[14] This so-called somatic mutation theory linked the genetic and carcinogenic effects of radiation exposure, which were previously treated separately. Since genetic damage showed a linear response to ionizing radiation dose, even low-level exposures might induce cancer-producing somatic mutations, albeit at low frequency. This notion challenged radiological protection standards that posited a threshold below which damage would be nonexistent. The Limited Test Ban Treaty, signed in 1963, ended above-ground testing of atomic weapons, but radioactive waste from a growing nuclear industry posed another environmental threat. The issue of how much radiation exposure was too much, in terms of safety, remained contentious through the 1970s and beyond.[15]

Second, the 1958 Delaney amendment to the US Food, Drug, and Cosmetic Act set forth a "zero tolerance" standard for carcinogenic food additives.[16] This applied to preservatives and food colors. (Pesticides, unless concentrated by food processing, were excluded because they were subject to other regulation.[17]) Food additives were not the only source of concern for regulators. In the petrochemical era of the 1950s and 1960s, tens of thousands of new chemicals entered the marketplace as drugs, pesticides, herbicides, and numerous other products. The tragic consequences of thalidomide exposure for children born to mothers treated with the drug in England and Europe, as well as the deadly effects of DDT spraying on bird populations in the US, showed that these new chemicals could cause unanticipated harm.[18]

While the Delaney clause applied only to food additives oversight by the Food and Drug Administration (FDA), US courts upheld the policy of controlling carcinogens more stringently than other toxic substances.[19] Moreover, in the wake of Rachel Carson's *Silent Spring* the federal government targeted cancer from chemicals as a key public health problem. In 1968, the National Cancer Institute (NCI) launched a "Plan for Chemical Carcinogenesis and the Prevention of Cancers."[20] During the subsequent years, the agency publicized an estimate that as much as 90 percent of human cancer was due to environmental agents.[21] But determining which substances in pollutants, pesticides, foods, and consumer products were carcinogenic was an enormous technical challenge. The NCI had developed a standard rodent carcinogenicity test, in which fifty to one hundred mice and rats were exposed from five to six weeks after birth to two years old to the chemical being tested (usually at a half the maximally tolerated dose), and were then analyzed for tumors.[22] The US government could conduct such tests on only around one hundred compounds a year, out of more than sixty thousand chemicals on the market.[23] Even the government process whereby suspect compounds were selected for testing was long and complicated.

A Test for Mutagenic Chemicals

In light of this regulatory bottleneck, researchers in the 1960s and 1970s devised ways beyond animal testing to detect potential carcinogens. The most prominent among these was a microbial screen developed by biochemist Bruce Ames at the University of California, Berkeley. As he tells the story, it was reading the back of a bag of potato chips that led him to wonder whether the many chemicals used in processing and preserving foods might cause genetic damage, including that leading to cancer.[24] The "Ames test" he developed was cheap, quick, and quantitative.[25] It built directly upon the somatic mutation theory for cancer that had gained traction in the preceding two decades. The test consisted of four mutant strains of *Salmonella* that scored specific DNA alterations as reverse mutations, or revertants.[26] On petri dishes plated with one of these customized strains and a test chemical, each bacterial colony represented a new mutation. Ames added a rat liver extract so that he could test for metabolic intermediates generated when mammalian bodies try to detoxify chemicals. Often these byproducts were more dangerous than the chemicals themselves. This innovation enabled Ames to pick up many more known carcinogens as mutagens in bacteria (see figure 4.2).

The value of Ames's *Salmonella* test relied on two powerful but vulnerable assumptions. First, as Ames put it, a carcinogen is a mutagen.[27] Most human

Figure 4.2. Pictures of Ames tests for spontaneous revertants (A) and exposure to Furylfuramide (B), Alfatoxin B1 (C), and 2-Aminofluorene (D). The mutagenic compounds in B, C, and D were applied to the six-millimeter filter disk in the center of each plate. Each petri plate contains cells of the tester strain in a thin overlay of top agar. (The strain used here is TA98, derived by adding a resistance transfer factor to a *Salmonella* tester strain, mutant *hisD3052*, that scores frameshift mutations.) Plates C and D contain, additionally, a liver microsomal activation system isolated from rats. The spontaneous or compound-induced revertants, each of which reflects a mutational event, appear in a ring as spots around the paper disk. Reproduced from Bruce N. Ames, Joyce McCann, and Edith Yamasaki, "Methods for Detecting Carcinogens and Mutagens with the *Salmonella*/Mammalian-Microsomal Mutagenicity Test," *Mutation Research* 31 (1975): 358, with permission of Elsevier.

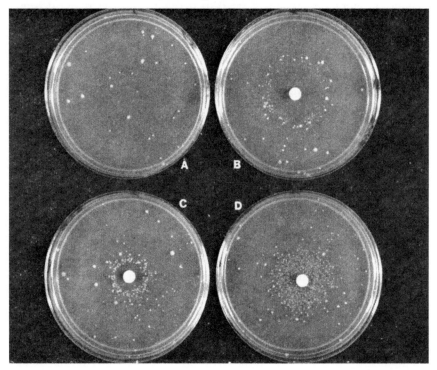

cancer, in this view, was triggered by exposure to environmental mutagens. Chemicals, like ionizing radiation, were presumed to act at the level of DNA in causing mutations and potentially cancer.[28] Second, it assumed that a microbe was a suitable model organism for detecting chemicals that were mutagenic in human cells. The test did not capture any aspects of tumorigenesis that involved factors beyond DNA damage, such as hormones, immune response, or inflammation, though the mutation theory of cancer was dominant.[29] There were compelling reasons to use the Ames test first even if the results were considered preliminary. A rodent test that followed the NCI guidelines required five hundred animals (including controls), lasted at least three and a half years, and cost at least $250,000 per substance. The Ames test took a few days and cost $300–$1000 per chemical tested.[30]

Ames's laboratory used his test to demonstrate the mutagenicity of cigarette smoke condensate, chloroacetaldehyde (derived from vinyl chloride, a common industrial chemical), and hydrogen peroxide-based hair dyes.[31] The *Salmonella* test also revealed the Japanese food preservative furylfuramide to be mutagenic; the compound was subsequently banned. Arlene Blum and Ames showed that the most commonly used flame retardant, tris(2,3-dibromopropyl) phosphate, or "Tris," was a mutagen.[32] It had only recently been introduced to products in the United States to meet the 1973 regulation for decreased flammability in children's sleepwear. Toxicologists subsequently found that Tris caused kidney cancer in rodents. In April 1977, the US government banned Tris from garments.[33] In a survey of three hundred chemicals, Ames and colleagues reported that 90 percent of bacterial mutagens also turned out to be carcinogens in animal testing.[34] Conversely a very high percentage of known carcinogens screened positive on the mutagen test. Environmentalists hailed the test for its ability to identify potentially carcinogenic chemicals, and industry was enthusiastic as well about this quick and cheap method.

By 1976, the Ames test was being used by dozens of companies. In part, this reflected Ames's openhandedness in making his strains available; he never patented them. By 1975, he had sent the testing strains out to seventy-five firms, including food and beverages companies such as Gallo, Gerber, General Foods, and Hunt-Wesson. The strains were free of charge, although as the volume of requests went up, he asked for-profit organizations to make a small donation to his laboratory to cover labor and shipping costs.[35] (Requests also came in from universities and government agencies, abroad as well as in the United States.) Many companies began using the test to screen new chemicals early in the development process. According to Ames, DuPont decided not to produce two Freon propellants because they were found to be mutagenic in *Salmonella* tests.[36] Yet the test had no official standing in federal regulation. As journalist Gina Kolata observed in 1976, "This has led to a curious situation in which industries are implicitly endorsing the tests at the same time that scientists and legislators deliberate over whether companies should be forced to use them."[37]

From the outset, Ames had been keen to get his tester strains into hands of government agency scientists. Even before he had published his first paper on the customized *Salmonella* strains, they were being "recommended by the [US] Food and Drug Administration as one of the ways for screening new drugs and food additives."[38] The screen was also being used in the US Department of Agriculture. A few years later, those advocating for more thorough regulation of chemicals through the proposed Toxic Substances Control Act (TSCA) pointed to the existence of his test and other short-term assays that could be used to pinpoint possible hazard substances.[39] Because the Ames test could be conducted swiftly and cheaply, it could be used to identify which new chemicals warranted more expensive animal testing—or which might be abandoned as commercial products. Economists argued for the cost effectiveness of this approach. As two explained, "An interesting and useful piece of biological-economic research would be to take all the chemicals introduced in a given year that might possibly be regarded as carcinogens and screen them using the Ames test. Even if all 6,000 were tested this would only cost about $3,000,000, a tiny fraction of what the researchers at the National Institute of Health spend chasing apparently nonexisting [human cancer] viruses."[40] The Environmental Mutagen Society, made up of researchers in genetic toxicology, also advocated screening of consumer products for mutagenicity.[41]

However, the version of TSCA passed by the US Congress did not require mutagenicity screening for chemicals, whether new or old.[42] The Environmental Protection Agency (EPA) had to possess compelling evidence of hazard for a chemical already on the market to require its producer to submit toxicity data. For new chemicals, a Premanufacturing Notification was required, but surprisingly few of those filed included mutagenicity data.[43] In the years after TSCA was signed into law, much of the EPA's effort went into compiling an inventory of all substances already on the market, which emphasized chemical identification (via standard nomenclature) rather than physical or toxicity data.[44]

Mutagenicity screens did not displace animal tests, which remained critical for confirmation of carcinogenicity. The Occupational Safety and Health Administration (OSHA), for example, required positive results from both an animal bioassay and a short-term test to classify a chemical as a Category I, or high-risk, toxic substance.[45] Eula Bingham, OSHA director under Carter, sought to develop a consistent no-threshold policy for carcinogens in the workplace, which encountered opposition from industry and ultimately did not withstand legal challenge.[46] For the FDA, only animal testing or human data could identify carcinogens prohibited by the Delaney Clause, and the agency targeted relatively few additives for regulation.[47] By the late 1970s, improving technologies of detection made the "zero tolerance" standard increasingly unworkable. In part due to the contested legal and political milieu for chemicals regulation, the status of new detection methods for carcinogens remained somewhat insecure. The Ames test continued to be used widely by environmental health researchers, corporate toxicologists, and government agencies, but as part of a battery of tools in regulatory science and commercial testing.

Diet as Environmental Exposure

One of the most potent carcinogens also identified as mutagenic by the Ames test was a natural substance: aflatoxin, a product of mold that grows on nuts and corn.[48] In the early 1960s, British researchers discovered that the deaths of more than one hundred thousand turkeys were due to aflatoxin-contaminated animal feed; aflatoxin was found to cause proliferation of bile duct cells.[49] Soon thereafter, aflatoxin ingestion was found to be associated with liver cancer in human populations, especially in Africa. In the case of aflatoxin and another natural carcinogen, safrole in root beer, the US government barred consumption.[50] But most scientists trying to identify carcinogens were, like Ames, focusing on industrial chemicals.

It was Japanese scientists who took the lead in considering hazards of natural compounds found in food, as well as those from cooking methods. For instance, polycyclic aromatic hydrocarbons, which are produced in the burning of organic materials, were known to be potent carcinogens.[51] The association of smoking with cancer was largely attributed to these compounds, which accumulated in the lungs from burning tobacco. Takashi Sugimura, director of the National Cancer Center in Tokyo, wondered whether grilling food might produce the same kinds of carcinogenic compounds.[52] As he recalls, one day while at home he noticed smoke in his reading room that wafted in from the kitchen, where his wife was broiling fish.[53] He and a coworker broiled fish in the laboratory, collecting the condensate of smoke on a glass-fiber filter. They dissolved the condensate in dimethyl sulfoxide and found the material to be highly mutagenic using the Ames test.[54] Other grilled proteins also produced mutagenic compounds in the smoke, as well as in the charred black material of cooked food surfaces. Along similar lines, Barry Commoner's laboratory at Washington University demonstrated that fried hamburger showed mutagenic activity in the *Salmonella* test.[55] Protein was not the only culprit. Sugimura and his coworkers went on to identify mutagens in many plants that people consume.[56]

On the one hand, this line of research enlarged the scope of materials that might be tested for mutagenicity in the name of limiting exposure. As one commentator noted, "It would appear much safer to eat a rare than a well-done steak, especially if it is cooked over charcoal!"[57] But on the other hand (and a heavy hand at that), the attention to the risks posed by *natural* agents was used by some experts to question the rationale for increased government regulation of synthetic chemicals. As the editor of *Science*, Philip Abelson, put it in an editorial, "The effort to prove a big role for industrial chemicals diverts attention from what is probably the best hope for reducing cancer incidence—careful study of foods and effects of cooking. . . . All people ingest the mutagens and carcinogens of food daily."[58] This perspective turned the ongoing concern about cancer-causing food additives on its head by pointing to the inherent carcinogenicity of many fruits and vegetables as well as of charcoal-broiled meats. By this time researchers had identified mutagenic compounds in garlic, onions, mushrooms, coffee, caramel, and a host of spices, including nutmeg, mace, ginger, cinnamon, and black pepper.[59] After two decades during which interest in so-called natural foods had expanded from hippies to urbanites, these studies raised questions about the underlying equivalence of natural and healthy.[60]

During this same period, epidemiology contributed to this emphasis on dietary carcinogens. There were longstanding observations of national differences in cancer rates, although these differences did not persist when individuals migrated.[61] For instance, residents of Japan had lower observed rates of colon cancer and breast cancer than individuals living in the United States. However, cancer rates of Japanese Americans, particularly those in the second generation, resembled those of other population groups in the United States. This suggested that environment and diet, rather than heredity, were responsible for the differences. Sugimura (among other researchers) speculated that regional foods and preparation methods might account for disparities in cancer incidence. For instance, higher rates of stomach cancer in Japan might be attributable to the nitrates and related nitroso compounds in soy products and Asian vegetables.[62] By contrast, US residents exhibited relatively low rates of stomach cancer, but (as just mentioned) higher rates of breast and colon cancer. This correlation inspired investigations into whether high levels of fat in the American diet might incite or promote these cancers.[63]

Drawing on this body of evidence, Richard Doll and Richard Peto published an influential 1981 review of cancer epidemiology. They calculated that between 10 percent and 70 percent of cancer in the United States was attributable to diet, estimating the actual figure to be around 35 percent.[64] That said, identifying the most important carcinogens in food was a formidable task. The number of naturally occurring chemical compounds present in the food supply was estimated to be over one million.[65] Government regulation focused on food additives and pesticides residues, and there existed little toxicological data on natural substances. As compared with the early 1970s, scientific and policy interest was shifting decisively to how diet and lifestyle contributed to cancer.[66] More precisely, what "environmental exposure" meant—at least for key scientists—was no longer mainly pollutants, but substances consumed: food, beverages, and cigarettes. This message reached the public, too. As Joe Jackson put it in his 1982 pop song, "Everything Gives You Cancer," "No caffeine / No protein / No booze or / Nicotine."

The controversy over saccharin in the United States illustrated the unpopularity of government action regulating substances popular with consumers. After Canadian scientists reported in 1977 that saccharin-fed rodents developed tumors, the FDA decided that the sweetener should be removed from the processed food market. Critics from industry seized on the high doses used in the animal experiments, the equivalent of eight hundred cans of soda per day, to question the validity of the tests. Congress subsequently overruled the FDA, in part because certain epidemiological studies failed to support the laboratory evidence. Given the disagreement among experts, which was itself attributable to the high degree of uncertainty in the scientific evidence, it was not difficult for manufacturers to cast doubt on the credibility of government regulation.[67] And the congressional vote overturning the FDA's ban showed the political limits of that agency's ability to safeguard the food supply.

"Nature's Pesticides"

As late as 1979, Ames was still advocating using mutagenicity testing to eliminate exposure to synthetic chemicals that might be carcinogenic.[68] That same year, he demonstrated that toxaphene, a major insecticide, was mutagenic.[69] But thereafter he changed his mind about the cancer-causing role of industrial chemicals, especially pesticide residues, in comparison to naturally occurring mutagens.[70] William Havender, a geneticist who joined Ames's lab, introduced him to economist Milton Friedman and the writing of libertarians such as Friedrich Hayek.[71] The laboratory's continuing work on DNA damage informed, and perhaps rationalized, Ames's growing distrust of stringent chemicals regulation. The mutagenicity of cigarette smoke was so high that the urine of smokers was itself mutagenic.[72] Ames and his colleagues began to argue that not all exposures were equally damaging—dose mattered, as did carcinogenic potency. One should thus, in his view, rank synthetic carcinogens alongside the many agents of DNA damage that humans were exposed to from the natural world, including food.[73] Plants, for instance, had evolved a large number of chemicals toxic to bacteria, fungi, and insects, many of them potent mutagens and carcinogens. Ames suspected that eating even natural foods contributed to the accumulation of potentially cancer-causing mutations. As he noted in 1983, "The human dietary intake of 'nature's pesticides' is likely to be several grams per day—probably at least 10,000 times higher than the dietary intake of man-made pesticides."[74] At the same time, the article in which he made this statement emphasized the benefits of a plant-rich diet, which also supplied many antioxidants, such as beta-carotene, that *protected* against DNA damage from dietary mutagens and metabolism.[75] Ames advised everyone to quit smoking, drink in moderation, eat their vegetables, and stop worrying about pesticides, whether synthetic or natural.

Cancer itself was increasingly understood by molecular biologists as a disease of aging, both from environmental exposures and the deterioration of the human body's own repair systems.[76] Normal respiration in the body generates reactive oxygen species such as hydrogen peroxide, which in turn damage DNA through single-strand breaks. Ames's group determined that this endogenous rate of oxidative DNA damage in humans was approximately ten thousand hits in each cell per day.[77] Animals with shorter lifespans exhibited even higher rates, suggesting that much of the damage resulted from ordinary metabolism (such as respiration). Mammals possess a sophisticated DNA repair system to slow the accumulated damage from ingesting and metabolizing food. These repair systems degrade over time, however, likely resulting in the higher incidence of cancer in older animals. Alvin Weinberg was among prominent scientists who argued along these lines: "Cancer is essentially a natural aging process. No matter what we eat, the huge flood of oxygen radicals produced in many metabolic processes overwhelms all but the most heavy external carcinogens, such as tobacco in heavy smokers. To be sure, anticarcinogenic substances are of benefit, but to choose a noncarcinogenic diet would probably be equivalent to starving to death."[78]

For some researchers interested in using the Ames test to screen industrially produced carcinogens, the new attention to mutagens in food was a distraction. As John Ashby put it in 1981, "While some of these 'natural' agents may, in fact, contribute to the incidence of human cancer, their confusion with commodity chemicals will not help either cause."[79] He depicted the growing concern with carcinogens in everything from chicken to coffee on a timeline, predicting that the patina on St. Paul's Cathedral and used car tires would soon join the realm of bacterial mutagens (see Figure 4.4). This prediction need not have been in jest: renovation of old buildings (particularly when covered with coal soot) was a known source of occupational hazard, and cancer epidemiology of workers in the rubber industry pointed to possible carcinogenic exposures in tires.[80] More to the point, Ashby warned that the constant alarm about carcinogens in everything had triggered public apathy.

There were also serious biological arguments against comparing natural and synthetic carcinogens. Chemical pesticides often persisted in the environment whereas the natural components of food biodegraded. In a response to Ames's 1983 article on dietary carcinogens, Samuel Epstein and Joel Swartz argued that chlorinated hydrocarbon pesticides posed a special cancer risk because of their tendency to accumulate in foods, especially fish. Their letter, which

Figure 4.3. Diagram of "issues of concern." Reproduced with permission from John Ashby, "Tests for Potential Carcinogens: Unresolved Problems," in *Short-Term Tests for Chemical Carcinogens*, ed. H. F. Stich and R. H. C. San (New York: Springer-Verlag, 1981), 481. © Springer-Verlag New York Inc. 1981.

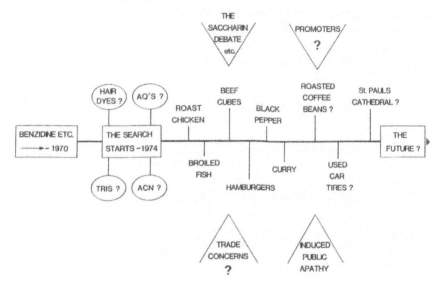

rebutted several other contentions by Ames, was cosigned by sixteen other scientists, including Eula Bingham, director of OSHA.[81] There were further evolutionary considerations. As Devra Lee Davis observed, the exposure of humans to carcinogens in plants over three million years might have selected for genetic resistance to these agents, especially since most carcinogens are also reproductive toxicants, so likely to induce spontaneous abortion. As she stated, "Through Evolution our genes may have acquired resistance to many of the naturally occurring carcinogenic compounds. We have no such genetic experience with synthetic organic materials."[82] Lactose tolerance provided one such example of nutritional adaptation through evolution. One could imagine a similar selection of resistance to natural carcinogens, or even to carcinogens produced by the pyrolysis of food through cooking. She also noted that in many cases, cooking breaks down natural toxicants.[83] Dr. I. Bernard Weinstein, who directed environmental science at Columbia Mailman School of Public Health, concurred. In discussing human exposure to synthetic halogenated hydrocarbons such as PCBs he said, "Our defense mechanisms may not be prepared to handle them, while we may be very well equipped to handle the natural pesticides."[84]

For others, the contention that many natural substances are mutagenic and/or carcinogenic necessitated updating the "zero tolerance" standard in regulation of food additives. Economists and others committed to risk assessment in regulatory decision-making gravitated to this point of view. As Lester B. Lave and Fanny Ennever put it, since "it would be difficult to have a nutritionally adequate diet if all foods with carcinogenic contaminants were banned ... Congress had no choice but to set a safety goal less stringent than the Delaney Amendment."[85] According to them, this was what led the Food and Drug Administration to join other federal agencies in establishing a level of risk regarded as negligible, namely "one cancer per million lifetimes."[86] There was also the issue of whether mutagenicity was really a valid predictor of carcinogenicity. Over the course of dozens of validation studies of the Ames test and other short-term assays, the best correlation achieved was 60 to 80 percent, lower than Ames himself had originally asserted. As scientists from the FDA and EPA observed, "Because social and regulatory concern about mutagens since the mid-1970s has been based on the published correlations between carcinogenicity and mutagenicity, the present uncertainty about that correlation is serving to undermine the whole field of environmental mutagenesis."[87]

Proposition 65

In the 1980s, Ames began questioning the value of government regulation of carcinogens, particularly consumer exposure to low-level doses. He became publicly involved on this issue in his home state of California, whose legal system allows for new legislation (in the form of "propositions") to be passed directly by voters. In 1986, David Roe of the Environmental Defense Fund and Carl Pope of the Sierra Club obtained enough signatures of their Citizens' Enforcement Law on toxic chemicals to get it on the ballot as Proposition 65. This proposed law banned businesses from discharging known carcinogens or reproductive toxicants into drinking water, and required products containing such chemicals to carry "clear and reasonable warning."[88] The measure was backed by California State Assemblyman Tom Hayden and (in part by virtue of his marriage to Jane Fonda) promoted by Hollywood celebrities including Whoopi Goldberg and Michael J. Fox.[89]

Among US states, California was the largest user of pesticides, consuming 50 percent of the national total. In addition, Silicon Valley was a major disposer of toxics, especially solvents and heavy metals.[90] The chemical refining industry, agricultural sector, and electronics firms lobbied against the proposition, as did Bruce Ames. As he stated before the California Assembly Committee on Water, Parks and Wildlife, "The main current fallacy in our approach to such pollution consists in believing that carcinogens are rare and that they are mostly man-made chemicals. Quite the contrary is the case. My estimate is that over 99.99 percent of the carcinogens Californians ingest are from natural (e.g., substances normally present in food) or traditional sources (e.g., cigarettes, alcohol, and chemicals formed by cooking food)."[91] Proposition 65 was on the same ballot as the gubernatorial candidates, so carcinogens regulation became a campaign issue. Incumbent George Deukmejian, a Republican, opposed Proposition 65, whereas his Democratic challenger, Tom Bradley, supported the measure. In the end, Deukmejian was re-elected and Proposition 65 passed as well, 63 percent to 37 percent.

The proposition required the governor to publish a list of known and potential carcinogens, specifying that the list should include those substances on the lists compiled by the International Agency for Research on Cancer (IARC), the National Toxicology Program (NTP), and OSHA. Mixtures containing 0.1 percent of a listed carcinogen were to be included as well. All told, these lists included twenty-four known carcinogens and over one hundred additional potential carcinogens. A minimum list of these chemicals, as required by the law, was estimated to be around 160.[92] The proposition specified that other chemicals could be added if "in the opinion of the state's qualified experts it has been clearly shown through scientifically valid testing according to generally accepted principles to cause cancer or reproductive toxicity."[93] In other words, the lists from other agencies were to provide the floor for regulation, not the ceiling.

In February 1987, Deukmejian published the first list of substances subject to the new law. It included only twenty-six carcinogens and three reproductive toxins.[94] What happened to all the others? As it transpired, the governor's list included only *known* human carcinogens and reproductive toxins, rather than the many more animal carcinogens considered as *likely* causes of human cancer. Moreover, the list had been vetted by a scientific advisory panel, to which Deukmejian had appointed Bruce Ames. This outraged the mainline environmentalist groups that had supported the proposition. As Carl Pope put it, "I've never seen a clearer fox-in-the-chicken-coop situation."[95] The supporters of Proposition 65 sued to expand the list, and in April 1987 the Sacramento County Supreme Court ordered the governor to add two hundred more chemicals to the published list.[96]

The law applied potentially to both natural and artificial chemicals in food, and in this sense went beyond the purview of the FDA's ban of carcinogens as additives. (The FDA exempted "generally recognized as safe" compounds, including most natural substances.[97]) Should natural chemicals be listed on warning labels? Formaldehyde exemplified this problem; it is a substance naturally occurring in many foods as well as being an industrial chemical. It is also a carcinogen, which would suggest any product containing it should, under Proposition 65, bear a warning label. But as Bruce Ames told the *San Francisco Examiner*, "Everything you look at has formaldehyde in it. Bread has formaldehyde in it, cola has formaldehyde in it. Blood has formaldehyde in it."[98] In the end, the California Department of Health and Welfare exempted chemical constituents of foods, even if they were identified as carcinogens.[99] The California Supreme Court upheld the interpretation that natural substances were exempt.

It is worth noting that, also during the 1980s, scientists, consumer advocates, and state officials were struggling to define how genetically modified foods should be regarded and regulated. Did the use of recombinant DNA techniques to create new tomatoes or strawberries render them unwholesome and unsafe? California was again at the center of such debates as biotech companies sought to field-test new products, such as "ice-minus" crops.[100] This emerging industry sought to represent genetically modified organisms (GMOs) as no less wholesome than "natural" foods, even as Ames was seeking to show that natural foods harbored their share of toxins and mutagens. These two arguments, however incompatible, reflect perturbations of the perceived boundary between "natural" and "artificial" in food, as biologists and industrialists challenged popular assumptions about the environment and health in the name of consumer choice and high-tech agriculture.

Carcinogens in a Cup of Coffee?

Debates over pesticide regulation were also occurring at the national level, which offered Bruce Ames another, larger, audience.[101] In 1988, the ABC news program *20/20* interviewed Ames, who claimed that natural carcinogens swamped out pesticide residues (and other worrisome industrial chemicals) in causing cancer. In one segment, set in a grocery store, he told TV journalist John Sulston that organic fruits and vegetables had no additional health value whatsoever, declaring that there are far more carcinogens in a cup of coffee than in the residual pesticides found on produce.[102] His message flew in the face of the environmental groups that had worked strenuously to achieve stricter regulation of industrial chemicals. As Robert Proctor has noted, in the wake of this broadcast a group of scientists and activists formed a "Bruce Ames Action Committee" to rebut his statements.[103]

By contrast, Ames's view was welcome news to industry, which sought to increase public awareness of so-called natural carcinogens. Ames and his work regularly showed up in speeches and opinion pieces originating from chemical companies. Keith C. Barrons, an agricultural technologist for Dow Chemical Company, published an editorial in *Newsweek* in 1984 titled "Not Too Much Pepper, Thank You." It featured the claim: "Some of nature's own foodstuffs are more toxic than the man-made chemicals we eat."[104] In a 1987 speech titled "The Dangerous Myth of a Risk-Free Society," Geraldine Cox, vice president and technical director of the Manufacturing Chemists Association, cited Ames extensively, particularly regarding the "misconceptions about environmental pollution" that informed Proposition 65. As she quoted Ames, "Natural carcinogens are everywhere: They are present in mushrooms, parsley, basil, celery, cola, wine, mustard, beer, peanut butter, and many more remain to be discovered."[105] Another industry speech on "Chemophobia," given at the Myrtle Beach Rotary Club that same year, also asserted that "Natural carcinogens are everywhere," listing exactly the same foods.[106] The chemical industry had already promulgated the view that synthetic chemicals were no more inherently dangerous than natural ones (particularly after the publication of *Silent Spring*). In this respect, Ames's view provided another variation on a longstanding corporate refrain.[107]

The tobacco industry was also interested in Ames. His name appears many times in the Tobacco Industry online archive, a massive collection of documents from Philip Morris, other companies, and various corporate-sponsored organizations that represented the tobacco industry. Ames was something of a conundrum for the tobacco industry: his heavy emphasis on smoking in cancer rates did not help them. At the same time, his anti-regulatory position was highly attractive. In a confidential memo "re: Bruce Ames," Arthur J. Stevens asks a colleague, "Is there any conceivable manner in which we could get this guy on our side of any of the science disputes confronting tobacco?"[108] Ames has always claimed that he never accepted industry money or testified in lawsuits, and no evidence has emerged to contradict him. Not all scientists were so scrupulous. After the influential epidemiologist Sir Richard Doll died, it emerged that "while investigating cancer risks in the industry," he had been a paid consultant for Monsanto.[109] This cast a new light on his insistence that diet, much more than exposure to industrial pollutants, contributed to human cancer rates. Countless other academic scientists of high standing, such as Philip Handler, president of the National Academy of Sciences, also consulted with industry or provided seemingly objective statements in favor of anti-regulatory positions.[110]

Despite media attention and industry publicity, Ames's assertions about omnipresent natural carcinogens did little to relieve public anxiety about the dangers of synthetic chemicals. This is well illustrated in the controversy over Alar, a plant growth regulator used by apple growers to increase storage life. The role of Alar in causing cancer in test animals had been known for a decade before a public debate erupted. Alar itself is not a carcinogen, but "one of the compound's metabolites, unsymmetrical dimethyl hydrazine (UDMH) . . . found to induce malignant tumors in young laboratory animals."[111] Alar decomposes upon heating into other compounds, including UDMH, and since heating is used in the preparation of juice and apple sauce, there was reason to be concerned about Alar spraying of apples. But this was not widely publicized until the National Resource Defense Council, a US environmental group, published a report in February 1989 called *Intolerable Risk*. It contended that spraying of produce with Alar had increased cancer risk in consumers, especially children, who consumed large quantities of apple juice and applesauce. The media repeated this conclusion widely, leading parents to stop buying apples and school boards to take apples, apple juice, and apple sauce off school lunch menus. Within a matter of days, apple sales in the United States plummeted.[112] When the deputy director of the EPA commented on 60 Minutes that he would remove Alar from the marketplace but did not have the authority, this only fueled public anger. By June, Uniroyal (which sold Alar) "voluntarily" withdrew it from circulation. It remains disputed whether the public alarm over Alar was a justifiable reaction to ineffective regulatory agencies, or hype produced by media manipulation.[113] Irrespective, the incident illustrates that Ames's argument about the relative harmlessness of chemical residues on foods did not persuade the public. If it had any effect in government circles, it was largely to bolster the use of risk assessment and cost-benefit analysis as a basis for regulatory action, a trend already in evidence.[114] But Ames's attacks on the validity of rodent testing for carcinogens drew strong criticisms from government regulators, who felt he had overstepped the bounds of his expertise, particularly in attacking a major regulatory tool.[115]

Beyond the policy realm, Ames's interest in the diverse sources of oxidative damage to DNA (including from natural compounds and metabolism) was reflected increasingly in the research literature. Many scientists, like Ames, emphasized the carcinogenicity associated with individual lifestyle—and in the byproducts of simply being alive, through metabolism.[116] A 2004 review of "Environmental and Chemical Carcinogenesis" by a group of cancer researchers focuses almost exclusively on exposure from smoking and eating (figure 4.4).[117] In the end, these researchers advocated the "design and utilization of drugs that reduce DNA damage," rather than trying to control cancer by reducing exposure to mutagenic agents.[118] Other researchers have continued to identify more carcinogens encountered in the diet, updating work begun by Sugimura. A team of Oregon State University chemists has identified ever-more-potent novel carcinogens produced by the combustion conditions that exist in vehicle exhaust or grilling meat. As the January 2014 press release for this finding stated, "These compounds were not known to exist, and raise additional concerns about the health impacts of heavily-polluted urban air or dietary exposure."[119]

Conclusions

This chapter has traced a historical arc from initial fears of food additives in the 1950s, to anxieties about pesticides in the 1960s and 1970s, and then, in the 1980s, to a belief that even natural elements of the human diet cause cancer.

Figure 4.4. Equilibrium between DNA damage and DNA repair. Reproduced from Gerald N. Wogan, Stephen S. Hecht, James S. Felton, Allan H. Conney, and Lawrence A. Loeb, "Environmental and Chemical Carcinogenesis," *Seminars in Cancer Biology* 14 (2004): 473–86, on 483, with permission of Elsevier.

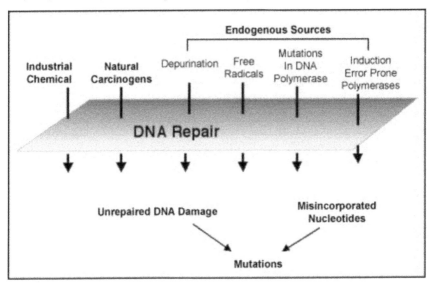

In articulating this final view, Ames was far more vocal than most scientists, but he did not fall completely outside the mainstream. A 1996 report from the US National Academy of Sciences treated the popular fear of food additives as misplaced: "Natural components of the diet may prove to be of greater concern than synthetic components with respect to cancer risk."[120] At its core, this view challenges the popular belief that nature is healthy and the artificial world dangerous. As Ames had expressed it pungently in 1983, "There are large numbers of mutagens and carcinogens in every meal, all perfectly natural and traditional. Nature is not benign."[121] Yet this domestication of risk did not convince nonscientists that natural mutagens were as dangerous as residues from synthetic chemicals on food. Consumers remained worried about poisoned apples, even as the debates left many confused and cynical about the alleged safety of foods.

By the mid-1980s, arguments over the relative roles of natural versus synthetic dietary carcinogens had little policy significance. It had become, to use an apt cliché, a tempest in a teacup. The real action in environmental regulation had already shifted to risk assessment.[122] This would not have been anticipated by the scientists who founded the Environmental Mutagen Society in 1969 or the officials at the US National Cancer Institute who called that same year for control of chemically induced cancer. In the mid-1970s, scientists and regulatory agencies remained entranced with the quest to identify mutagens as a way to control cancer risk—and companies were equally enthusiastic about this approach to increase product safety (and decrease liability). But their new laboratory methods, which seemed poised to determine which substances were safe and which were hazardous, instead gave complicated and sometimes contradictory results. One of the principal new tools, the Ames test, unexpectedly destabilized the idea that DNA damage (and hence cancer) stemmed largely from radiation and a handful of industrial chemicals. These findings contributed to a growing regulatory limbo, reflecting the weak statute for chemicals regulation passed in 1976 and the legal setbacks encountered by the FDA and OSHA in their attempts to control exposure to likely carcinogens. The rollback of expanded environmental regulation intensified under the Reagan administration, but the direction had already changed by 1980. Industry lobbying and public relations undoubtedly played a role in these developments, though companies were more effective in changing government policy than public opinion. For their part, cancer biologists turned their gaze to chasing after the key genetic targets of mutagens (especially oncogenes and tumor suppressor genes). Toxicologists and environmental researchers continued to use the Ames test to identify and classify mutagens: over ten thousand scientific papers were published using this test between 1973 and 2010.[123] And the somatic mutation hypothesis remains, in the words of two critics of it, "the prevalent theory of carcinogenesis."[124]

In describing these transitions, I have highlighted the enduring role of mutations and mutagens in orienting the science and policy of environmental carcinogens. This way of understanding cancer (and its prevention) led scientists to pinpoint some hazards and disregard others. For example, environmental health was conceived largely in terms of cancer risk to individuals, especially consumers. Hazards to other living organisms, ecosystems, and even industrial workers were sidelined. The mutational framework also relegated carcinogens that do not induce mutations to the margins. In particular, hormones and hormone-mimicking substances were invisible through the mutational spectacles that dominated tests for environmental carcinogens in the 1970s and 1980s.[125] There was the additional question of what to do with epidemiological findings that did not fit with the mutational framework. Nutritional epidemiology has generated countless studies on the role of specific foods in carcinogenesis, and the overall picture is full of contradictions. In 2013, Jonathan D. Schoenfeld and John P. A. Ionnidas published a widely cited review titled "Is Everything We Eat Associated with Cancer?"[126]

In sum, the role of dietary carcinogens remains fraught with scientific uncertainty and saturated with politics. The decision of the International Agency for Research on Cancer in October 2015 to classify processed meat as a known carcinogen and red meat as a possible carcinogen reignited public anxieties about the dangers lurking in food.[127] While this move was prompted by epidemiological rather than laboratory findings, it illustrates the growing acknowledgment of the role of dietary exposures in cancer incidence. Strikingly, the carcinogens that result from nitrates in processed meats, nitroso compounds, are encountered in foods as both preservatives and natural constituents. Another example illustrates the conundrum of vilifying food carcinogens. In the fall of 2018, the FDA revoked approval of six synthetic compounds used as flavor chemicals, all of which have been proven carcinogenic in animal tests— and all of which are also present as natural constituents of everyday foods, including grapes, basil, pineapple, mint, and hops.[128] The agency took pains to clarify that they thought these compounds were perfectly safe. If a naturally occurring compound tests positive on a rodent carcinogenicity test, should the government ban its use as an additive? Or does opening the door to carcinogenic chemical additives of any sort, whether naturally occurring or not, undermine attempts to protect consumers from a wide range of potentially hazardous chemical exposures in food? Nearly everyone agrees that exposure to environmental agents is part of the cancer puzzle. But scientists, health officials, and consumers hold widely divergent views on what counts as our carcinogenic environment, especially when it comes to exposures from cooking and eating.

Acknowledgments

An earlier version of this essay was presented as the Hans Rausing Lecture at Uppsala Universitet, Sweden and as the Fred Sanger Lecture at the Wellcome Unit for the History of Medicine, London. For their valuable comments and questions, I thank John Abraham, Richard Aspin, David Bell, Margot Canaday, Vera Candiani, Yaacob Dweck, Jean-Paul Gaudillière, Katja Guenther, Petter Hellström, Evan Hepler-Smith, Nathalie Jas, Bill Jordan, Dan Kevles, Charles Kollmer, Hannah Landecker, Ross MacFarlane, Federico Marcon, Anna-Maria Meister, Matthew Meselson, Mary Morgan, Phil Nord, Jenny Rampling, Otto Sibum, Julia Stone, and Emily Thompson, and all of the participants of the ROTT (Risk on the Table) collective. Lastly, I gratefully acknowledge Gina Surita's meticulous editing.

Angela N. H. Creager is the Thomas M. Siebel Professor in the History of Science at Princeton University, where she specializes in history of biology and biomedicine. She has published two books with University of Chicago Press, *The Life of a Virus: Tobacco Mosaic Virus as an Experimental Model, 1930–1965* (2002) and *Life Atomic: A History of Radioisotopes in Science and Medicine* (2013), which was awarded the Patrick Suppes Prize in the History of Science by the American Philosophical Society. Her current work focuses on the role of genetic tests in environmental science and regulation.

Notes

1. For photographs that include the original artwork and information on Indiana's art, see Johnson, "It Wasn't All He Needed," C31.
2. Abelson, "Dietary Carcinogens," 1249.
3. On other research aimed at understanding eating as a means of environmental exposure, see Gaudillière, "Food, Drug and Consumer Regulation"; Landecker, "Food as Exposure."
4. I am not the first historian to deal with these issues; see especially Proctor, *Cancer Wars*, 133–52; Sheila Jasanoff also comments on the "EAT. DIE." cover of *Science* magazine in Jasanoff, *Risk Management and Political Culture*, 38.
5. Goodwin, *Pure Food, Drink, and Drug Crusaders.*
6. Wharton, "Original Food and Drugs Act," 67; White, "Chemistry and Controversy," 2.
7. White, "The Chemogastric Revolution." For the attribution of the term to James Harvey Young, see White, "Chemistry and Controversy," 215.
8. White, "Chemistry and Controversy," 9.
9. Ibid., 200.
10. Creager and Gaudillière, "Experimental Arrangements"; Stoff and Travis, "Discovering Chemical Carcinogenesis."
11. Schwerin, "Vom Gift im Essen."
12. Kopp, "The Origins of the American Scientific Debate."
13. See de Chadarevian, this volume.
14. Sturtevant, "Social Implications"; Jolly, "Thresholds of Uncertainty," 485–527; Löwy, *Preventive Strikes*, 166–97; Creager, *Life Atomic*, 143–79.
15. I have simplified a long and complex debate here; see Hacker, *Dragon's Tail*; Walker, *Permissible Dose*; Boudia, "From Threshold to Risk"; Creager, "Radiation, Cancer, and Mutation."
16. Davis, *Banned*, 150; Wargo, *Our Children's Toxic Legacy*; Jas, "Adapting to 'Reality'"; von Schwerin, "Vom Gift im Essen"; Schwerin, "Cyclamates." Some of the scientific challenges posed by the Delaney Amendment are addressed in National Research Council, Food Protection Committee, and Food and Nutrition Board, *Problems in the Evaluation of Carcinogenic Hazard.*
17. Namely, the Federal Insecticide, Fungicide, and Rodenticide Act (FIFRA, 1947); see Bosso, *Pesticides and Politics.*
18. Carpenter, *Reputation and Power*; Dunlap, *DDT.*
19. Maffini and Vogel, this volume; Vogel, *Is It Safe?*, 38, 50–59; Brickman, Jasanoff, and Ilgen, *Controlling Chemicals*, 34, 38, 120–21.
20. National Cancer Institute, *Carcinogenesis Bioassay of Trichloroethylene*, ii.
21. Library of Congress, *Legislative History*, 208, 532, 539. This figure is usually attributed to Higginson, "Present Trends." Boyland went one step further and stated that the 90 percent were due to chemical components: Boyland, "The Correlation of Experimental Carcinogenesis."
22. Bucher, "The National Toxicology Program Rodent Bioassay."

23. Beyer, Beck, Lewandowski, "Historical Perspective."
24. Ames, "An Enthusiasm for Metabolism."
25. Creager, "Political Life of Mutagens."
26. Three of the strains (originally TA1531, TA1532, and TA1534) were designed to detect different kinds of frameshift mutagens. The fourth strain, TA1530, contained a base-pair change, and therefore it would detect mutations that involve base-pair substitutions. The four *Salmonella* strains were further customized with additional genetic changes that made the cells more permeable to large molecules and eliminated some kinds of DNA repair. Ames, Lee, and Durston, "An Improved Bacterial Test System."
27. Ames, "Carcinogens are Mutagens."
28. Ramel, "Advantages."
29. Sonnenschein and Soto, *Society of Cells*.
30. Maugh, "Chemical Carcinogens."
31. Kier, Yamasaki, and Ames, "Detection of Mutagenic Activity"; Ames, Kammen, and Yamasaki, "Hair Dyes are Mutagenic"; McCann et al., "Mutagenicity of Chloroacetaldehyde"; Proctor, *Cancer Wars*, 137.
32. Blum and Ames, "Flame-Retardant Additives."
33. This ban did not apply to upholstery, however, which has become a target for regulatory reform. Slater, "How Dangerous Is Your Couch?"
34. McCann, Choi, Yamasaki, and Ames, "Detection of Carcinogens"; McCann and Ames, "Detection of Carcinogens."
35. On the number of companies that received strains and made donations, see handwritten notes on letters used from January–October 1975; October–December 1975; 1–31 December 1975; and January through March 1976, Ames Papers, Box 23, Folder 5 Ames Evolution of Form Letter.
36. Ames, "Identifying Environmental Chemicals," 593n21.
37. Kolata, "Chemical Carcinogens," 1215.
38. Bruce N. Ames, Progress Report, AEC Grant AT(04-5)-34, for year beginning February 1970, with grant renewal submission and cover letter to F. B. Fry, 14 October 1970, Ames Papers, Box 29, Folder 4 AEC 1969–73.
39. Library of Congress, *Legislative History*, 413.
40. Kneese and Schulze, "Environment, Health, and Economics," 328.
41. Committee 17, Environmental Mutagen Society, "Environmental Mutagenic Hazards," 511. On the Environmental Mutagen Society, see Frickel, *Chemical Consequences*.
42. The EPA could issue a rule to require testing, but the procedural hurdle was nontrivial. Gaynor, "The Toxic Substances Control Act"; O'Reilly, "Torture by TSCA."
43. US Congress, Senate, *Hearings before Subcommittee*, 146.
44. Hepler-Smith, "Molecular Bureaucracy."
45. Brickman, Jasanoff, and Ilgen, *Controlling Chemicals*, 196–97.
46. See J. C. Hansen, "Cancer In Perspective: An Industry Viewpoint," presented at the California Manufacturers Association, Governmental Affairs Forum, 24 May 1979, DHC, Box 10, file Speeches 00464-1; Vogel, *Is It Safe?*, 72–74.
47. See Maffini and Vogel, this volume.
48. A fuller account of aflatoxin is given by Mueller, this volume.
49. The outbreak was referred to as "Turkey X," though it quickly became clear that it affected other animals and was responsible for trout hepatoma. Goldblatt, *Aflatoxin*; Lee, "Wild Toxicity"; Mueller, this volume.

50. Safrole was banned by the FDA in 1960 "on the general safety provisions of the law" rather than the Delaney Clause. Prival and Scheuplin, "Regulation of Carcinogens," 317. Interestingly, safrole did not score positive on the Ames test. Sugimura et al., "A Critical Review," 125.
51. Berenblum, *Carcinogenesis as a Biological Problem*, 8–12.
52. The carcinogenicity of "excessively heated and burnt (partly carbonized) foods," especially meats, had been recognized since the 1950s: Steyn, "Food and Beverages"; Lijinsky and Shubik, "Benzo[a]pyrene," 53–54. On the identification of benzopyrene (also called benzypyrene) as a carcinogen, see Stoff, this volume, 84.
53. Sugimura, Nagao, and Wakabayashi, "How Should We Deal," 18. See also Sugimura, "Mutagens, Carcinogens, and Tumor Promoters."
54. The Sugimura group went on to identify the presence of specific potent mutagens beyond aromatic hydrocarbons in the charred surface of muscle, particularly two mutagens in tryptophan pyrolysate. See Sugimura, "Naturally Occurring Genotoxic Carcinogens."
55. Commoner et al., "Formation of Mutagens."
56. Nagao, Sugimura, and Matsushima, "Environmental Mutagens and Carcinogens." The importance of Japanese workers in this arena is attested by an international conference held in Tokyo in 1979, "Naturally Occurring Carcinogens-Mutagens and Modulators of Carcinogenesis," which was attended by a number of American and European researchers (including Bruce Ames) as well as many Japanese scientists.
57. Fox, "*Naturally Occurring Carcinogens*," 951.
58. Abelson, "Cancer—Opportunism and Opportunity," 11.
59. Nagao, Sugimura, and Matsushima, "Environmental Mutagens and Carcinogens," 125–29.
60. Belasco, *Appetite for Change*.
61. Wynder and Gori, "Contribution of the Environment." Because the United States and Japan both had good cancer registries, the contrast between cancer rates in the two countries and the fact that Japanese migrants to the United States came to have cancer patterns more like residents there were regarded as strong evidence for dietary contributions to cancer incidence. See Wynder, "The Epidemiology of Large Bowel Cancer." The incidence of skin cancer in various regions on the globe, which correlated with average sun exposure, indicated that ultraviolet radiation was another important environmental carcinogen.
62. Nagao et al., "Mutagenic Compounds."
63. E.g., Carroll, "Experimental Evidence."
64. Doll and Peto, "The Causes of Cancer." Doll had already used epidemiological evidence to discern striking differences in incidences of human cancer in different regions of the world: Doll, *Prevention of Cancer*, especially 143. For a detailed history of the origin and reception of Doll and Peto's report see Proctor, *Cancer Wars*, 54–74.
65. National Research Council, *Carcinogens and Anticarcinogens*, 5.
66. By the end of the 1980s this shift was complete: see *Mutagens and Carcinogens in the Diet*.
67. Smith, "Latest Saccharin Tests"; Peña, *Empty Pleasures*, 141–76; Warner, *Sweet Stuff*, 181–94.
68. "Soon more sophisticated and sensitive ways of measuring DNA or other damage in people could play an essential role in complementing animal cancer tests and short-

term tests. We must identify mutagens and carcinogens with all of these methods, treat them with respect, set priorities, examine alternatives, and try to minimize human exposure. We have seen, and will continue to see, the folly of using humans as guinea pigs." Ames, "The Identification of Chemicals," 355.

69. Hooper et al., "Toxaphene."
70. On Ames's exposure to libertarian ideas and his change of heart, see Proctor, *Cancer Wars*, 143.
71. Bollier, "Leading Scientist Laughs at DDT."
72. Yamasaki and Ames, "Concentration of Mutagens from Urine"; Ames et al., "Oxidative DNA Damage."
73. The so-called HERP index that Ames and his colleagues devised as a way to rank carcinogenic hazards became the focus of much subsequent controversy. Ames, Magaw, and Gold, "Ranking Possible Carcinogenic Hazards."
74. Ames, "Dietary Carcinogens and Anticarcinogens," 1258.
75. The study of dietary anticarcinogens became an important field in its own right; see the papers in *Mutation Research*, e.g., Hayatsu, Arimoto, and Negishi, "Dietary Inhibitors of Mutagenesis."
76. Ames, Magaw, and Gold, "Ranking Possible Carcinogenic Hazards," 277.
77. Ames, "Mutagenesis and Carcinogenesis."
78. Weinberg, "Cancer and Diet," 658.
79. Ashby, "Tests for Potential Carcinogens," 481.
80. International Agency for Research on Cancer, *Overall Evaluations*, 332–33.
81. Epstein and Swartz, "Cancer and Diet."
82. Davis, "Natural Anticarcinogens," 333. See also Davis, "Paleolithic Diet," 1633.
83. Davis, "Paleolithic Diet."
84. Quoted in Marx, "Animal Carcinogen Testing Challenged," 745.
85. Lave and Ennever, "Toxic Substances Control," 79.
86. Ibid. Their claim that natural carcinogens played a role in the FDA's decisions warrants further evidence than they provide in their paper.
87. Prival and Dellarco, "Evolution of Social Concerns," 49.
88. Safe Drinking Water and Toxic Enforcement Act of 1986 (California Health and Safety Code Sections §25249.5 et seq.), originally adopted by California voters as Proposition 65, quote from §25249.6, accessed 21 August 2020, https://law.justia.com/codes/california/2007/hsc/25249.5-25249.13.html.
89. See clipped newspaper articles in Ames Papers, Box 25, Folder 21.
90. Water Education Foundation, "Layperson's Guide to Groundwater and Toxics," Draft #1, 15 July 1986, page 15, Ames Papers, Box 25, Folder 21.
91. Bruce N. Ames, "Six Common Errors Relating to Environmental Pollution," presented to the California Assembly, Committee on Water, Parks and Wildlife, 1 October 1986, Ames Papers, Box 25, Folder 21; also published, with slight modifications, as "Cancer Scares Over Trivia," *LA Times* op-ed, 15 May 1986.
92. Californians Against Toxic & Chemical Hazards, "Questions and Answers about the 1986 Initiative," 2 July 1986, copy in Ames Papers, Box 25, Folder 21.
93. Safe Drinking Water and Toxic Enforcement Act of 1986, quote from §25249.8.
94. Phipps, Allen, and Caswell, "The Political Economics."
95. Marshall, "California's Debate on Carcinogens," 1459.

96. Pease, "Identifying Chemical Hazards," 136.
97. Maffini and Vogel, this volume; Vogel, *Is It Safe?*, 36.
98. Steven A. Capps, "Environmentalists Dismayed as Prop. 65 Panel Fails to Act," *San Francisco Examiner*, 1 April 1987, copy in Ames Papers, Box 25, Folder 21.
99. Kramer and van Raveswaay, "Proposition 65."
100. Barinaga, "Field Test," 819; Rabinow, *Artificiality and Enlightenment*, 105–7.
101. In 1988, Congress passed an amendment to the Federal Insecticide, Fungicide, and Rodenticide Act requiring producers to provide more data within a ten-year time frame for continued registration of their pesticides. Wargo, *Our Children's Toxic Legacy*, 99.
102. *20/20* news segment "Much Ado About Nothing," broadcast by ABC, 18 March 1988, copy in Ames Papers. For a critique, see Bollier, "Anatomy of a Puffpiece," 15.
103. Proctor, *Cancer Wars*, 140–41.
104. Barrons, "Not Too Much Pepper," 18. He cites Ames, "Dietary Carcinogens and Anticarcinogens," n61.
105. "The Dangerous Myth of a Risk-Free Society," an Address by Geraldine V. Cox, PhD, Vice-President-Technical Director, Chemical Manufacturers Association, Presented at the 36th Annual Engineering and Science Day Assembly, Drexel University, Philadelphia, PA, 19 February 1987, 4ff., DHC, Box 1, file Speeches 00420, 4.
106. "Chemophobia: Fear, Fiction, or Fact?" Myrtle Beach Rotary Club, 9 November 1987, 9–10, DHC, Box 1, file Speeches 00420, 9.
107. See, for example, the tirade against environmentalism (and Rachel Carson) by Cleve A. I. Goring: "The Charges against Chemicals—A Campaign for Chaos," presented to the Western Agricultural Chemicals Association, San Diego, California, 5 October 1976, DHC, Box 9, file Speeches 00457.
108. A. J. Stevens, memorandum, 24 November 1993, TTID, document ID qhnh0110.
109. Boseley, "Renowned Cancer Scientist"; Becker, "Epidemiologist at Work."
110. On Handler and others, see Oreskes and Conway, *Merchants of Doubt*; Proctor, *Golden Holocaust*.
111. Wargo, *Our Children's Toxic Legacy*, 116.
112. Leiss and Chociolko, *Risk and Responsibility*, 154.
113. Marshall, "A Is for Apple."
114. National Research Council, *Risk Assessment*; Hoel, Merrill, and Perera, *Risk Quantitation and Regulatory Policy*.
115. Marx, "Animal Carcinogen Testing Challenged."
116. See, e.g., Collins et al., "Oxidative Damage to DNA."
117. Wogan et al., "Environmental and Chemical Carcinogenesis."
118. Ibid., 484. This is the last sentence of the article.
119. Simonich, "New Compounds Discovered."
120. National Academy of Sciences and National Research Council, *Carcinogens and Anticarcinogens*, 6.
121. Ames, "Dietary Carcinogens and Anticarcinogens," 1261.
122. See, e.g., Oftedal and Brøgger, *Risk and Reason*; Demortain, *Science of Bureaucracy*.
123. Claxton, Umbuzeiro, and DeMartin, "The *Salmonella* Mutagenicity Assay."
124. Sonneschein and Soto, *A Society of Cells*, xi.
125. See Vogel, *Is It Safe?*; Gaudillière and Jas, *La santé environnementale au-delà du risque?*

126. Schoenfeld and Ionnidas, "Is Everything We Eat Associated with Cancer?"
127. See International Agency for Research on Cancer, "Monographs Evaluate Consumption."
128. Berenstein, "How Activists Forced FDA."

Bibliography

Archives

Bruce N. Ames Papers (Ames Papers), circa 1952–99, BANC MSS 2000/44 c, Bancroft Library, University of California, Berkeley
Dow Historical Collection (DHC), Science History Institute Archives, Philadelphia, PA
Truth Tobacco Industry Documents (TTID), hosted by UCSF Library and Center for Knowledge Management

Publications

Abelson, Philip H. "Cancer—Opportunism and Opportunity." *Science* 206, no. 4414 (1979): 11.
———. "Dietary Carcinogens." *Science* 221, no. 4617 (1983): 1249.
Ames, Bruce N. "An Enthusiasm for Metabolism." *Journal of Biological Chemistry* 278, no. 7 (2003): 4369–80.
———. "Cancer Scares over Trivia: Natural Carcinogens in Food Outweigh Traces in Our Water." *Los Angeles Times.* 15 May 1986, sec. Op-ed.
———. "Carcinogens Are Mutagens: Their Detection and Classification." *Environmental Health Perspectives* 6 (1973): 115–18.
———. "Dietary Carcinogens and Anticarcinogens." *Science* 221, no. 4617 (1983): 1256–64.
———. "Identifying Environmental Chemicals Causing Mutations and Cancer." *Science* 204, no. 4393 (1979): 587–93.
———. "Mutagenesis and Carcinogenesis—Endogenous and Exogenous Factors." *Environmental and Molecular Mutagenesis* 14 (1989): 66–77.
———. "The Identification of Chemicals in the Environment Causing Mutations and Cancer." In *Naturally Occurring Carcinogens-Mutagens and Modulators of Carcinogenesis: Proceedings of the 9th International Symposium of the Princess Takamatsu Cancer Research Fund Tokyo, 1979*, edited by Elizabeth C. Miller, James A. Miller, Iwao Hirono, Takashi Sugimura, and Shozo Takayama, 345–58. Tokyo: Japan Scientific Societies Press, 1979.
Ames, Bruce N., Harold O. Kammen, and Edith Yamasaki. "Hair Dyes Are Mutagenic: Identification of a Variety of Mutagenic Ingredients." *Proceedings of the National Academy of Sciences of the United States of America* 72, no. 6 (1975): 2423–27.
Ames, Bruce N., Frank D. Lee, and William E. Durston. "An Improved Bacterial Test System for the Detection and Classification of Mutagens and Carcinogens." *Proceedings of the National Academy of Sciences of the United States of America* 70, no. 3 (1973): 782–86.
Ames, Bruce N., Renae Magaw, and Lois Swirsky Gold. "Ranking Possible Carcinogenic Hazards." *Science* 236, no. 4799 (1987): 271–80.
Ames, Bruce N., Robert L. Saul, Elizabeth Schwiers, Robert Adelman, and Richard Cathcart. "Oxidative DNA Damage as Related to Cancer and Aging: The Assay of Thymine

Glycol, Thymidine Glycol, and Hydroxy-Methyluracil in Human and Rat Urine." In *Molecular Biology of Aging: Gene Stability and Gene Expression*, edited by Rajindar S. Sohal, Linda S. Birnbaum, and Richard G. Cutler, 137–44. New York: Raven Press, 1985.

Ashby, John. "Tests for Potential Carcinogens: Unresolved Problems." In *Short-Term Tests for Chemical Carcinogens*, edited by H. F. Stich and R. H. C. San, 474–82. New York: Springer-Verlag, 1981.

Barinaga, Marcia. "Field Test of Ice-Minus Bacteria Goes Ahead Despite Vandals." *Nature* 326 (1987): 819.

Barrons, Keith C. "Not Too Much Pepper, Thank You." *Newsweek*, 9 April 1984.

Becker, Chris. "An Epidemiologist at Work: The Personal Papers of Sir Richard Doll." *Medical History* 43 (2002): 403–21.

Belasco, Warren J. *Appetite for Change: How the Counterculture Took on the Food Industry, 1966–1988*. New York: Pantheon, 1989.

Berenblum, Isaac. *Carcinogenesis as a Biological Problem*. Frontiers of Biology (Amsterdam), v. 34. New York: American Elsevier Pub. Co., 1974.

Berenstein, Nadia. "How Activists Forced FDA to Blacklist 'Carcinogenic' Flavor Chemicals the Agency Says Are Safe." The Counter website. Accessed 21 August 2020, https://thecounter.org/fda-carcinogenic-flavor-chemical-food-additive-lacroix-lawsuit/.

Beyer, Leslie A., Barbara D. Beck, and Thomas A. Lewandowski. "Historical Perspective on the Use of Animal Bioassays to Predict Carcinogenicity: Evolution in Design and Recognition of Utility." *Critical Reviews in Toxicology* 41, no. 4 (2011): 321–38.

Blum, Arlene, and Bruce N. Ames. "Flame-Retardant Additives as Possible Cancer Hazards." *Science* 195, no. 4273 (1977): 17–23.

Bollier, David. "Anatomy of a Puffpiece." *The Public Citizen*, October 1988, 15.

———. "Leading Scientist Laughs at DDT, Worries about Peanut Butter, Believe It or Not!" *The Public Citizen*, October 1988, 12–15, 19–20.

Boseley, Sarah. "Renowned Cancer Scientist Was Paid by Chemical Firm for 20 Years." *The Guardian*, 8 December 2006, sec. Science. Accessed 21 August 2020, https://www.theguardian.com/science/2006/dec/08/smoking.frontpagenews.

Bosso, Christopher J. *Pesticides and Politics: The Life Cycle of a Public Issue*. Pittsburgh: University of Pittsburgh Press, 1987.

Boudia, Soraya. "From Threshold to Risk: Exposure to Low Doses of Radiation and Its Effects on Toxicants Regulation." In *Toxicants, Health and Regulation since 1945*, edited by Soraya Boudia and Nathalie Jas, 71–87. London: Pickering & Chatto, 2013.

Boyland, Eric. "The Correlation of Experimental Carcinogenesis and Cancer in Man." *Progress in Experimental Tumor Research* 11 (1967): 222–34.

Bucher, John R. "The National Toxicology Program Rodent Bioassay: Designs, Interpretations, and Scientific Contributions." *Annals of the New York Academy of Sciences* 982 (2002): 198–207.

Carpenter, Daniel P. *Reputation and Power: Organizational Image and Pharmaceutical Regulation at the FDA*. Princeton: Princeton University Press, 2010.

Carroll, Kenneth K. "Experimental Evidence of Dietary Factors and Hormone-Dependent Cancers." *Cancer Research* 35, no. 11 part 2 (1975): 3374–83.

Claxton, Larry D., Gisela de A. Umbuzeiro, and David M. DeMarini. "The Salmonella Mutagenicity Assay: The Stethoscope of Genetic Toxicology for the 21st Century." *Environmental Health Perspectives* 118, no. 11 (2010): 1515–22.

Collins, Andrew R., Mária Dusinská, Catherine M. Gedik, and Rudolph Štětina. "Oxidative Damage to DNA: Do We Have a Reliable Biomarker?" *Environmental Health Perspectives* 104, Suppl. 3 (1996): 465–69.

Committee 17, Environmental Mutagen Society. "Environmental Mutagenic Hazards." *Science* 187, no. 4176 (1975): 503–14.

Commoner, Barry, Antony J. Vithayathil, Piero Dolara, Subhadra Nair, Prema Madyastha, and Gregory C. Cuca. "Formation of Mutagens in Beef and Beef Extract during Cooking." *Science* 201, no. 4359 (1978): 913–16.

Creager, Angela N. H. *Life Atomic: A History of Radioisotopes in Science and Medicine.* Chicago: University of Chicago Press, 2013.

——. "The Political Life of Mutagens: A History of the Ames Test." In *Powerless Science? Science and Politics in a Toxic World*, edited by Soraya Boudia and Nathalie Jas, 46–64. New York: Berghahn Books, 2014.

——. "Radiation, Cancer, and Mutation in the Atomic Age." *Historical Studies in the Natural Sciences* 45, no. 1 (2015): 14–48.

Creager, Angela N. H., and Jean-Paul Gaudillière. "Experimental Arrangements and Technologies of Visualization: Cancer as a Viral Epidemic (1930–1960)." In *Heredity and Infection: The History of Disease Transmission*, edited by Jean-Paul Gaudillière and Ilana Löwy, 203–41. London: Routledge, 2001.

Davis, Devra Lee. "Natural Anticarcinogens, Carcinogens, and Changing Patterns in Cancer: Some Speculation." *Environmental Research* 50, no. 2 (1989): 322–40.

——. "Paleolithic Diet, Evolution, and Carcinogens." *Science* 238, no. 4834 (1987): 1633–34.

Davis, Frederick Rowe. *Banned: A History of Pesticides and the Science of Toxicology.* New Haven: Yale University Press, 2014.

Demortain, David. *The Science of Bureaucracy: Risk Decision-Making and the US Environmental Protection Agency.* Cambridge: MIT Press, 2020.

Doll, Richard. *Prevention of Cancer: Pointers from Epidemiology.* London: Nuffield Provincial Hospitals Trust, 1967.

Doll, Richard, and Richard Peto. "The Causes of Cancer: Quantitative Estimates of Avoidable Risks of Cancer in the United States Today." *Journal of the National Cancer Institute* 66, no. 6 (1981): 1192–308.

Dunlap, Thomas R. *DDT: Scientists, Citizens, and Public Policy.* Princeton: Princeton University Press, 1981.

Epstein, Samuel S., and Joel B. Swartz. "Cancer and Diet." *Science* 224, no. 4650 (1984): 660–66.

Fox, M. "[Review of] *Naturally Occurring Carcinogens, Mutagens and Modulators of Carcinogenesis*." *British Journal of Cancer* 42, no. 6 (1980): 950–51.

Frickel, Scott. *Chemical Consequences: Environmental Mutagens, Scientist Activism, and the Rise of Genetic Toxicology.* New Brunswick: Rutgers University Press, 2004.

Gaudillière, Jean Paul. "Food, Drug and Consumer Regulation: The 'Meat, DES and Cancer' Debates in the United States." In *Meat, Medicine, and Human Health in the Twentieth Century*, edited by David Cantor, Christian Bonah, and Matthias Dorries, 179–202. London: Pickering & Chatto, 2010.

Gaudillière, Jean-Paul, and Nathalie Jas, eds. "La santé environnementale au-delà du risque: Perturbateurs endocriniens, expertise et régulation en France et en Amérique du Nord." *Sciences sociales et santé* 43, no. 3 (2016): 5–130.

Gaynor, Kevin. "The Toxic Substances Control Act: A Regulatory Morass." *Vanderbilt Law Review* 30 (1977): 1149–96.

Goldblatt, Leo A., ed. *Aflatoxin: Scientific Background, Control, and Implications*. New York: Academic Press, 1969.

Goodwin, Lorine Swainston. *The Pure Food, Drink, and Drug Crusaders, 1879–1914*. Jefferson: McFarland, 1999.

Hacker, Barton C. *The Dragon's Tail: Radiation Safety in the Manhattan Project, 1942–1946*. Berkeley: University of California Press, 1987.

Hayatsu, Hikoya, Sakae Arimoto, and Tomoe Negishi. "Dietary Inhibitors of Mutagenesis and Carcinogenesis." *Mutation Research* 202, no. 2 (1988): 429–46.

Hepler-Smith, Evan. "Molecular Bureaucracy: Toxicological Information and Environmental Protection." *Environmental History* 24 (2019): 534–50.

Higginson, John. "Present Trends in Cancer Epidemiology." *Proceedings of the Canadian Cancer Conference* 8 (1969): 40–75.

Hoel, David G., Richard A. Merrill, and Frederica P. Perera, eds. *Risk Quantitation and Regulatory Policy, Banbury Report 19*. Cold Spring Harbor: Cold Spring Harbor Laboratory, 1985.

Hooper, N. Kim, Bruce N. Ames, Marwah Adley Saleh, and John E. Casida. "Toxaphene, a Complex Mixture of Polychloroterpenes and a Major Insecticide, Is Mutagenic." *Science* 205, no. 4406: 591–93.

International Agency for Research on Cancer (IARC). "IARC Monographs Evaluate Consumption of Red Meat and Processed Meat." *IARC*, 26 October 2015. Accessed 21 August 2020, https://www.iarc.fr/wp-content/uploads/2018/07/pr240_E.pdf.

———. *Overall Evaluations of Carcinogenicity: An Updating of IARC Monographs Volumes 1–42*. Suppl. 7. Lyon: World Health Organization International Agency for Research on Cancer, 1987.

Jas, Nathalie. "Adapting to 'Reality': The Emergence of International Expertise on Food Additives and Contaminants in the 1950s and Early 1960s." In *Toxicants, Health and Regulation since 1945*, 47–70. Studies for the Society for the Social History of Medicine no. 9. London: Pickering & Chatto, 2013.

Jasanoff, Sheila. *Risk Management and Political Culture: A Comparative Study of Science in the Policy Context*. New York: Russell Sage Foundation, 1986.

Jasanoff, Sheila, Thomas Ilgen, and Ronald Brickman. *Controlling Chemicals: The Politics of Regulation in Europe and the United States*. Ithaca: Cornell University Press, 1985.

Johnson, Ken. "It Wasn't All He Needed, or All He Did." *New York Times*, 26 September 2013, sec. Arts.

Jolly, J. Christopher. "Thresholds of Uncertainty: Radiation and Responsibility in the Fallout Controversy." PhD diss., Oregon State University, 2003.

Kier, Larry D., Edith Yamasaki, and Bruce N. Ames. "Detection of Mutagenic Activity in Cigarette Smoke Condensates." *Proceedings of the National Academy of Sciences of the United States of America* 71, no. 10 (1974): 4159–63.

Kneese, Allen V., and William D. Schulze. "Environment, Health, and Economics: The Case of Cancer." *American Economic Review* 67, no. 1 (1977): 326–32.

Kolata, Gina Bari. "Chemical Carcinogens: Industry Adopts Controversial 'Quick' Tests." *Science* 192, no. 4245 (1976): 1215–17.

Kopp, Carolyn. "The Origins of the American Scientific Debate over Fallout Hazards." *Social Studies of Science* 9, no. 4 (1979): 403–22.

Kramer, Carol S., and Eileen O. van Ravenswaay. "Proposition 65 and the Economics of Food Safety." *American Journal of Agricultural Economics* 71, no. 5 (1989): 1293–99.

Landecker, Hannah. "Food as Exposure: Nutritional Epigenetics and the New Metabolism." *BioSocieties* 6, no. 2 (2011): 167–94.

Lave, Lester B., and Fanny K. Ennever. "Toxic Substances Control in the 1990s: Are We Poisoning Ourselves with Low-Level Exposures?" *Annual Review of Public Health* 11, no. 1 (1990): 69–87.

Lee, Victoria. "Wild Toxicity, Cultivated Safety: Aflatoxin and Kōji Classification as Knowledge Infrastructure." *History and Technology* 35 (2019): 405–24.

Leiss, William. *Risk and Responsibility*. Montréal: McGill-Queen's University Press, 1994.

Library of Congress. *Legislative History of the Toxic Substances Control Act*. Washington, DC: US Government Printing Office, 1976.

Lijinsky, William, and Philippe Shubik. "Benzo(a)Pyrene and Other Polynuclear Hydrocarbons in Charcoal-Broiled Meat." *Science* 145, no. 3627 (1964): 53–55.

Löwy, Ilana. *Preventive Strikes: Women, Precancer, and Prophylactic Surgery*. Baltimore: Johns Hopkins University Press, 2010.

Marshall, Eliot. "A Is for Apple, Alar, and . . . Alarmist?" *Science* 254, no. 5028 (1991): 20.

———. "California's Debate on Carcinogens." *Science* 235, no. 4795 (1987): 1459–59.

Marx, Jean. "Animal Carcinogen Testing Challenged." *Science* 250, no. 4982 (1990): 743–45.

Maugh, Thomas H. "Chemical Carcinogens: The Scientific Basis for Regulation." *Science* 201, no. 4362 (1978): 1200–5.

McCann, Joyce, and Bruce N. Ames, "Detection of Carcinogens as Mutagens in the *Salmonella*/Microsome Test: Assay of 300 Chemicals: Discussion." *Proceedings of the National Academy of Sciences of the United States of America* 73, no. 3 (1976): 950–54.

McCann, Joyce, Edmund Choi, Edith Yamasaki, and Bruce N. Ames. "Detection of Carcinogens as Mutagens in the *Salmonella*/Microsome Test: Assay of 300 Chemicals." *Proceedings of the National Academy of Sciences of the United States of America* 72, no. 12 (1975): 5135–39.

McCann, Joyce, Vincent Simmon, David Streitwieser, and Bruce N. Ames. "Mutagenicity of Chloroacetaldehyde, a Possible Metabolic Product of 1,2-Dichloroethane (Ethylene Dichloride), Chloroethanol (Ethylene Chlorohydrin), Vinyl Chloride, and Cyclophosphamide." *Proceedings of the National Academy of Sciences of the United States of America* 72, no. 8 (1975): 3190–93.

Miller, Elizabeth C., James A. Miller, Iwao Hirono, Takashi Sugimura, and Shozo Takayama, eds. *Naturally Occurring Carcinogens-Mutagens and Modulators of Carcinogenesis: Proceedings of the 9th International Symposium of the Princess Takamatsu Cancer Research Fund Tokyo, 1979*. Tokyo: Japan Scientific Societies Press, 1979.

Nagao, Minako, Kazuo Wakabayashi, Yuki Fujita, Tomoko Tahira, Masako Ochiai, and Takashi Sugimura. "Mutagenic Compounds in Soy Sauce, Chinese Cabbage, Coffee, and Herbal Teas." In *Genetic Toxicology of the Diet: Proceedings of a Satellite Symposium of the Fourth International Conference on Environmental Mutagens, Held in Copenhagen, Denmark, June 19–22, 1985*, edited by Ib Knudsen, 55–62. New York: A. R. Liss, 1986.

Nagao, Minako, Takashi Sugimura, and Taijiro Matsushima. "Environmental Mutagens and Carcinogens." *Annual Review of Genetics* 12, no. 1 (1978): 117–59.

National Cancer Institute. *Carcinogenesis Bioassay Trichloroethylene.* CAS No. 79-01-6, NCI-CG-TR-2, DHEW Publication No. (NIH) 76-802. Washington, D.C.: U.S. Department of Health, Education, and Welfare, Public Health Service, National Institutes of Health, 1976.

National Research Council, Food Protection Committee, and Food and Nutrition Board. *Problems in the Evaluation of Carcinogenic Hazard from Use of Food Additives.* Washington, DC: National Academies Press, 1959.

National Research Council. *Carcinogens and Anticarcinogens in the Human Diet: A Comparison of Naturally Occurring and Synthetic Substances.* Washington, DC: National Academy Press, 1996.

———. *Risk Assessment in the Federal Government: Managing the Process.* Washington, DC: National Academy Press, 1983.

Oftedal, Per, and Anton Brøgger, eds. *Risk and Reason: Risk Assessment in Relation to Environmental Mutagens and Carcinogens: Proceedings of a Satellite Symposium to the Fourth International Conference on Environmental Mutagens, Held in Olso, Norway, June 21–22, 1985.* New York: A. R. Liss, 1986.

O'Reilly, James T. "Torture by TSCA: Retrospectives of a Failed Statute." *Natural Resources & Environment* 25, no. 1 (2010): 43–44, 47.

Oreskes, Naomi. *Merchants of Doubt: How a Handful of Scientists Obscured the Truth on Issues from Tobacco Smoke to Global Warming.* New York: Bloomsbury Press, 2010.

Pease, William. "Identifying Chemical Hazards for Regulation: The Scientific Basis and Regulatory Scope of California's Proposition 65 List of Carcinogens and Reproductive Toxicants." *RISK: Health, Safety & Environment* 3, no. 2 (1992).

Peña, Carolyn Thomas de la. *Empty Pleasures: The Story of Artificial Sweeteners from Saccharin to Splenda.* Chapel Hill: University of North Carolina Press, 2010.

Phipps, Tim T., Kristen Allen, and Julie Caswell. "The Political Economics of California's Proposition 65." *American Journal of Agricultural Economics* 71, no. 5 (1989): 1286.

Prival, Michael J., and Vicki L. Dellarco. "Evolution of Social Concerns and Environmental Policies for Chemical Mutagens." *Environmental and Molecular Mutagenesis* 14, no. 16 (1989): 46–50.

Prival, Michael J., and Robert J. Scheuplin. "Regulation of Carcinogens and Mutagens in Foods in the States." In *Mutagens and Carcinogens in the Diet: Proceedings of a Satellite Symposium of the Fifth International Conference on Environmental Mutagens, Held in Madison, Wisconsin, July 5–7, 1989,* edited by Michael W. Pariza, Hans-Ulrich Aeschbacher, James S. Felton, and Shigeaki Sato, 307–21. New York: Wiley-Liss, 1990.

Proctor, Robert. *Cancer Wars: How Politics Shapes What We Know and Don't Know about Cancer.* New York: BasicBooks, 1995.

———. *Golden Holocaust: Origins of the Cigarette Catastrophe and the Case for Abolition.* Berkeley: University of California Press, 2011.

Rabinow, Paul. "Artificiality and Enlightenment: From Sociobiology to Biosociality." In *Essays on the Anthropology of Reason,* 91–111. Princeton: Princeton University Press, 1996.

Ramel, Claes. "Advantages of and Problems with Short-Term Mutagenicity Tests for the Assessment of Mutagenic and Carcinogenic Risk." *Environmental Health Perspectives* 47 (1983): 153–59.

Schoenfeld, Jonathan D., and John P. A. Ioannidis. "Is Everything We Eat Associated with Cancer? A Systematic Cookbook Review." *American Journal of Clinical Nutrition* 97, no. 1 (2013): 127–34.

Schwerin, Alexander von. "Cyclamates: A Tale of Uncertain Knowledge (1930s–1980s)." In *Hazardous Chemicals: Agents of Risk and Change, 1800–2000*, edited by Ernst Homburg and Elisabeth Vaupel, 179–210. New York: Berghahn Books, 2019.

———. "Vom Gift im Essen zu chronischen Umweltgefahren: Lebensmittelzusatzstoffe und die risikopolitische Institutionalisierung der Toxikogenetik in der Bundesrepublik, 1955–1964." *Technikgeschichte* 81, no. 3 (2014): 251–74.

Simonich, Staci. "New Compounds Discovered That Are Hundreds of Times More Mutagenic." *Oregon State University Newsroom*, 3 January 2014. Accessed 21 August 2020, https://today.oregonstate.edu/archives/2014/jan/new-compounds-discovered-are-hundreds-times-more-mutagenic.

Slater, Dashka. "How Dangerous Is Your Couch?" *New York Times*, 6 September 2012, sec. Magazine.

Smith, R. Jeffrey. "Latest Saccharin Tests Kill FDA Proposal." *Science* 208, no. 4440 (1980): 154–56.

Sonnenschein, Carlos, and Anna M. Soto. *A Society of Cells: Cancer and Control of Cell Proliferation*. Oxford: Bios, 1999.

Steyn, Douw G. "Food and Beverages in Relation to Cancer." In *The Symposium on Potential Cancer Hazards from Chemical Additives to Foodstuffs, held by the International Union Against Cancer*, 342–56. Rome, Italy, 1956.

Stoff, Heiko, and Anthony S. Travis. "Discovering Chemical Carcinogenesis: The Case of Aromatic Amines." In *Hazardous Chemicals: Agents of Risk and Change, 1800–2000*, edited by Ernst Homburg and Elisabeth Vaupel, 137–78. New York: Berghahn Books, 2019.

Sturtevant, A. H. "Social Implications of the Genetics of Man." *Science* 120, no. 3115 (1954): 405–7.

Sugimura, Takashi. "Mutagens, Carcinogens, and Tumor Promoters in Our Daily Food." *Cancer* 49, no. 10 (1982): 1970–84.

Sugimura, Takashi, Takashi Kawachi, Taijiro Matsushima, Minako Nagao, Shigeaki Sato, and Takie Yahagi. "A Critical Review of Submammalian Systems for Mutagen Detection." In *Proceedings of the Second International Conference on Environmental Mutagens, Edinburgh, July 11–15, 1977*, edited by David Scott, Bryn A. Bridges, and Frederick H. Sobels, 125–40. Amsterdam; New York: Elsevier, 1977.

Sugimura, Takashi, Minako Nagao, and Keiji Wakabayashi. "How We Should Deal with Unavoidable Exposure of Man to Environmental Mutagens: Cooked Food Mutagen Discovery, Facts and Lessons for Cancer Prevention." *Mutation Research* 447, no. 1 (2000): 15–25.

US Congress, Senate, Committee on Environment and Public Works, *Hearings before the Subcommittee on Toxic Substances and Environmental Oversight*, 98th Congress, 1st Session, 27, 29 July and 1 August 1983. Washington, DC: US Government Printing Office, 1984.

Vogel, Sarah A. *Is It Safe?: BPA and the Struggle to Define the Safety of Chemicals*. Berkeley: University of California Press, 2013.

Walker, J. Samuel. *Permissible Dose: A History of Radiation Protection in the Twentieth Century*. Washington, DC: Nuclear Regulatory Commission, 2000.

Wargo, John. *Our Children's Toxic Legacy: How Science and Law Fail to Protect Us from Pesticides*. New Haven: Yale University Press, 1996.

Warner, Deborah Jean. *Sweet Stuff: An American History of Sweeteners from Sugar to Sucralose*. Washington, DC: Smithsonian Institution Scholarly Press in cooperation with Rowman & Littlefield Publishers, 2011.

Weinberg, Alvin M. "Cancer and Diet." *Science* 224, no. 4650 (1984): 658–58.

Wharton, W. R. M. "The Original Food and Drugs Act of June 30, 1906: Its Inspection Evolution." In *Historic Meeting to Commemorate the Fortieth Anniversary of the Original Federal Food and Drugs Act*, 65–76. New York: Commerce Clearing House, 1946.

White, Suzanne Rebecca. "Chemistry and Controversy: Regulating the Use of Chemicals in Foods, 1883–1959." PhD diss., Emory University, 1994.

———. "The Chemogastric Revolution and the Regulation of Food Chemicals." In *Chemical Sciences in the Modern World*, edited by Seymour H. Mauskopf, 322–55. Philadelphia: University of Pennsylvania Press, 1993.

Wogan, Gerald N., Stephen S. Hecht, James S. Felton, Allan H. Conney, and Lawrence A. Loeb. "Environmental and Chemical Carcinogenesis." *Seminars in Cancer Biology* 14, no. 6 (2004): 473–86.

Wynder, Ernst L. "The Epidemiology of Large Bowel Cancer." *Cancer Research* 35, no. 11 part 2 (1975): 3388–94.

Wynder, Ernst L., and Gio Batta Gori. "Contribution of Environment to Cancer Incidence—Epidemiologic Exercise." *Journal of the National Cancer Institute* 58, no. 4 (1977): 825–32.

Yamasaki, Edith, and Bruce N. Ames. "Concentration of Mutagens from Urine by Adsorption with the Nonpolar Resin XAD-2: Cigarette Smokers Have Mutagenic Urine." *Proceedings of the National Academy of Sciences of the United States of America* 74, no. 8 (1977): 3555–59.

 CHAPTER 5

Risk on the Negotiating Table

Malnutrition, Mold Toxicity, and Postcolonial Development

Lucas M. Mueller

In September 1977, delegates from forty countries and sixteen international organizations gathered in Nairobi to tackle an urgent environmental problem of the increasingly interconnected world. Raj Malik, an official of the Food and Agriculture Organization (FAO), observed that "the world has become too interdependent to entertain a notion that any country or its produce are immune to the ill effects of mycotoxins."[1] These toxic substances, which were excreted by molds that grew on peanuts, maize, and other crops, could cause acute poisoning, cancer, and death. James Charles Nakhwanga Osogo, Kenya's minister of health, explained in his opening address that the conference took place "in one of the areas of the world where mycotoxins can represent a real hazard to the quality and quantity of human food and feed," suggesting that the substances did not pose the same hazard to every country.[2] Moreover, mycotoxins were not only a threat to food safety and security in Kenya and other tropical countries. Osogo added that the mycotoxin problem was "another specific deterrent to international trade in the produce of developing countries, since exports containing mycotoxins are sometimes found and this results in the destruction of specific shipments and often loss of entire markets and valuable foreign earnings."[3] Mycotoxins jeopardized developing countries' capacities to feed their populations and capabilities to export produce during a period when these countries burst onto the global scene, demanding a New International Economic Order (NIEO). This proposal challenged the hegemony of the North Atlantic powers and sought to reshape global trade in favor of the Global South.[4] The problem of mycotoxins was a new factor in the postcolonial struggle between industrialized and developing countries, putting health and incipient environmental politics on the table.

Scientists, food and feed manufacturers, and administrators at national and international agencies had struggled over the best solution to the problem of mycotoxins since 1960, when veterinarians isolated a previously unknown substance that caused the death of hundreds of thousands of turkeys on British farms. British and US scientists discovered that *Aspergillus* molds on peanuts in the birds' feed produced this substance and named it aflatoxin. Scientists around the globe started to study aflatoxin's effects on humans and animals and to develop detection methods for aflatoxin in crops, making aflatoxin the best studied mycotoxin. They found aflatoxin not only in peanuts but also in corn, cottonseed, and many other crops that were important foodstuffs, animal feed, and export commodities in the recently decolonized nations in Africa and Asia, creating the very problems that Osogo described at the Nairobi conference. At that meeting, Peter Thacher, a representative of the United Nations Environmental Program (UNEP), proposed a strategy to resolve the mycotoxin crisis: "UNEP wishes to assess the risks in more precise terms so that governments can have a solid basis on which to make decisions within their own values as to the level of risk they wish to accept. We believe that by helping to improve assessments of the risks we will reduce risks, and can reduce interferences to international trade and consequent economic hardship. We are very anxious to assist the exporting developing countries."[5] UNEP concentrated on risk assessments to solve the aflatoxin crisis. The UNEP proposal limited UNEP's role in conducting these assessments, while leaving decisions about the supposedly objective level of risks to be taken to national governments. UNEP thus proposed a separation of risk assessment, delegated to the international level, from value-laden political decisions about implementation left in the realm of the nation state. UNEP assumed that its work of improving risk assessments would on its own reduce health hazards and economic hardship. UNEP drew on the work of scientists and regulators who had advanced risk as the primary way to understand and solve the aflatoxin problem since 1960. However, the proposed remedy, risk assessment, was itself laden with moral assumptions.

This chapter examines how scientists and officials came to see aflatoxin through the lens of risk, from the discovery of aflatoxin in 1960 to the aftermath of the Nairobi conference, describing changing assumptions and limitations of the risk framework. I show how experiments with laboratory animals made aflatoxin into a perceptible hazard and how colonial and Cold War preoccupations with hunger and malnutrition influenced risk calculus. The risk framework espoused a zero-sum logic that justified the continued exposure of vulnerable populations in Africa to aflatoxin, sidelined developing countries' grievances about inequitable trade relations, and failed to protect vulnerable populations in Kenya. Relations between Global North and South played a central role in the calculus of aflatoxin risk. I argue that risk was decisively shaped by the international politics of decolonization during the 1960s and 1970s, when the "Third World" challenged the existing global trade system and the industrialized countries experienced "the shock of the global."[6] Researchers and officials at international organizations advanced risk as an answer to this challenge, ultimately re-entrenching global inequalities and furthering economic hardship for the developing countries.

This chapter thus challenges the narrative that risk played a role only in affluent industrial societies.[7] Sociologist Ulrich Beck argued in *Risk Society*, published in 1986, that risk became the defining feature of the new era of reflexive modernity, which began in the second half of the twentieth century.[8] In Beck's narrative, modernity, dominated by a logic of wealth production through techno-economic "progress," was succeeded by reflexive modernity, when skepticism about industrialization's scientific and technological foundations abounded, and the focus of politics shifted to risks that these very foundations generated.[9] Beck writes, "The driving force in the class society can be summarized in the phrase: *I am hungry!* The movement set in motion by risk society, on the other hand, is expressed in the statement: *I am afraid!*"[10] However, aflatoxin has contaminated foodstuffs essential for the hungry person's survival and for economic development through exports, and aflatoxin gave rise to fears about food contamination. This history of aflatoxin collapses the distinction between the "hungry world" and risk society.[11]

Aflatoxin became a problem for both worlds, intertwining them in new ways that facilitated the erasure of developing countries' economic demands. Risk combined health and wealth in one framework, which enabled scientists to move between the two domains and reshuffle them. Political demands for market access and stable commodity prices were reframed as questions of standards and acceptable risk levels, thus transforming political disputes about wealth and power into technical questions about seemingly objective levels of risk to health. These technical questions were to be answered by scientists and experts. They formed "epistemic communities" that detected and assessed risks to make what were ultimately political decisions, becoming a "fifth branch" of government. However, epistemic communities did not necessarily give unanimous advice.[12] Drawing on Sheila Jasanoff's insights into the political context of risk assessment and the displacement of political struggles into the realm of technical advisory boards, the chapter describes the process of reshuffling scientific questions and political demands on the level of international institutions in the wake of decolonization.[13]

Risk facilitated a division of responsibilities between international organizations and nation states, which differed vastly in their capacities to deal with toxicity in food and the environment.[14] For example, international organizations conducted risk assessments of aflatoxin, which left the implementation to nation-states and shifted the burden of global trade regulations onto the developing countries by excluding questions of economic justice. By considering aflatoxin risk across affluent risk societies and the hungry world, this chapter shows how the risk framework was also a process in scale-making that shaped political possibilities on the levels of nation-state and international organizations.[15] In addition to the FAO and UNEP, all major agencies of the United Nations (UN)—the World Health Organization (WHO) and the United Nations

International Children's Emergency Fund (UNICEF)—as well as the General Agreement on Tariffs and Trade (GATT) and its successor, the World Trade Organization (WTO), have been concerned with aflatoxin at some point. The international institutionalization of risk resembles the conceptual structure of international trade law, such as the GATT, which enshrined markets and movement of capital, protecting capital from demands for redistribution that national mass democratic movements and anti-colonial nationalists might voice.[16] Similarly, the international legal framework of human rights excluded questions of economic justice.[17] The lens of risk through which scientists and international agencies viewed aflatoxin circumscribed the possibilities of toxicity control in narrow ways, supporting a global system that paved the way for the expansion of markets for industrial chemicals within Europe and from the Global North, driving proliferation and harmonization of international legal regimes to regulate toxic risks in food and the environment since the 1950s.[18]

This chapter advances its argument about the centrality of North-South relations in shaping the theory and practice of aflatoxin risk in three main parts. Each part considers a different commodity that aflatoxin contaminated: peanuts as remedies for malnutrition, peanuts as export goods, and maize as staple food. The first part shows how malnutrition became a political concern that the UN Protein Advisory Group (PAG) sought to solve with peanuts as an additional source of protein. The PAG responded to the discovery of aflatoxin by determining a contamination level that allowed the continued use of peanuts, developing a calculus of aflatoxin risk. Second, I discuss how industrialized countries introduced restrictions on the import of goods potentially contaminated with aflatoxin. Developing countries challenged these measures at the UN Conference on Mycotoxins in 1977 in Nairobi, bringing together two pressing developments in international politics during the 1970s: the NIEO and environmentalism. Finally, I show how the risk-based regulatory approach failed to protect populations most vulnerable to aflatoxin, when an outbreak of acute aflatoxin poisoning in the staple food maize killed twelve people in 1981. It would not be the last outbreak of acute aflatoxin poisoning in Kenya. Together, these histories suggest that the risk-based approach ultimately evaded questions of economic justice and accountability, failing to protect the health and livelihoods of the most vulnerable people in Africa and Asia.

Sentinels for Toxicity, ca. 1960

In May 1960, farmers in southern England noticed that their turkeys suffered from a strange illness. A feed company sent veterinarian William Blount, who observed that "the birds move slowly, do not feed or drink, sink to the ground, fall quietly over on to one side, stretch out their limbs and die. It is as though the birds were in a gas chamber, and dying from the effects of some poisonous, lethal gas!"[19] Blount compared the sites of the outbreak to find common factors and thereby identify the cause of "Turkey X" disease. He discovered that the disease was caused by feed that had been produced in a specific feed mill from specific peanut batches imported on specific vessels from Brazil. Peanuts—which were imported not only from Brazil, but also from other subtropical and tropical Asian, African, and American countries—constituted the protein-rich part of the feed. These protein-rich feeds made birds gain more weight and gain it faster, enabling the rapid expansion of poultry farming in postwar Great Britain. Veterinarians, chemists, and mycologists studied the contaminated peanut batches, ultimately learning that *Aspergillus* molds excreted a poisonous molecule, which they named aflatoxin.

The researchers discovered aflatoxin in the highly artificial environment of the industrialized poultry farm. The peanuts of the contaminated feed had been stored and transported across the Atlantic before being processed into homogenous poultry feed. This process spread the toxic substance, which might have been on a few single peanuts, across large batches of feed, in a dose sufficiently high to kill a turkey, which, as a species, happened to be highly susceptible to aflatoxin. The birds became an ideal "sentinel" for a toxic substance so concentrated and distributed as to cause a visible hazard.[20] If a handful of birds had ingested a few toxic peanuts, farmers might have dismissed the dead birds as a common occurrence in poultry-keeping, in which some mortality was expected. In 1960, feed manufacturers accidentally spread aflatoxin across feed batches and industrialized farms in a way that exposed an entire avian population to levels of toxicity so high as to make their mass mortality an abnormal phenomenon, which farmers noticed and which scientists investigated with epidemiological methods. It was not only scientific methods but also industrialized agriculture and the global commodity trade that led to the discovery of aflatoxin.

The Shock of the Malnourished World

The discovery of a toxic substance in peanuts alarmed UNICEF, which was promoting peanut flour as a remedy for infant malnutrition. Aflatoxin threatened the use of this remedy in the intensely political fight against malnutrition. Malnutrition was the latest manifestation of hunger, which had transformed from an inevitable natural condition into an object of statecraft in early nineteenth-century Britain.[21] In the late nineteenth century, nutrition science began to recast hunger in the language of molecules, calories, and other calculable entities; the discourse of nutrition was embraced by states, colonial rulers, and anti-colonial activists alike, offering a new scientific language to advance and criticize government programs.[22] Nutrition became an object of interwar colonial and international statecraft for its potential to unsettle colonial rule and international relations amidst mounting fears about unfettered population growth.[23] In the interwar years, British nutrition scientists "discovered" malnutrition, the lack of specific nutrients, in East Africa.[24] British colonial rulers, the Rockefeller Foundation, and other organizations embraced the concept of malnutrition, shifting the focus from outwardly political questions of poverty and undernourishment to educational and technical interventions to train the population of the malnourished world in proper dietetics.

Hunger, food, and malnutrition did not disappear with World War II. In the context of the Cold War, the US government, supported by the Rockefeller and Ford Foundations, promoted a program of agricultural reform in Asia—later called the Green Revolution—to increase food production, fill the stomachs of hungry peasants, win their hearts and minds, and close their ears to the siren songs of communism.[25] In the late 1940s, the recently founded United Nations and its specialized agencies began to focus on malnutrition, especially kwashiorkor, in Africa. Kwashiorkor had been first observed by physician Cicely Williams in West Africa in 1933. She found that milk could cure the disease of infant malnutrition, leading her to speculate that lack of protein played a role in causing kwashiorkor.[26] Kwashiorkor remained a topic of discussion at meetings of a joint committee of nutrition experts of the World Health Organization and Food and Agriculture Organization.

In 1955, the World Health Organization established a new expert group, the Protein Advisory Group (PAG), to advise the organization on its protein-rich foods program and on new sources of protein in the fight against protein malnutrition.[27] Many prominent food and nutrition scientists from around the globe, including Nevin Scrimshaw (US), Coluchur Gopalan (India), and Benjamin S. Platt (UK), joined the PAG.[28] The PAG received funding from different US philanthropic organizations, including the Rockefeller Foundation, which funded a subgroup in protein malnutrition.[29] The PAG met regularly, developed recommendations, and issued bulletins, attempting to make the case to the international organizations that the problem of protein malnutrition was "a primary factor in susceptibility to infection and impaired growth and development of children in many developing countries" and "a primary deterrent to national social and economic development."[30] The PAG sought to fill the "protein gap," the perceived scarcity of protein, by tapping unconventional protein sources.

Peanuts featured prominently, and controversially, in this plan. Peanuts had been important cash crops in British and French African colonies since the late nineteenth century, including the infamous Groundnut Scheme in Tanganyika (present-day Tanzania), which the British government and Unilever advanced in the 1950s.[31] The PAG estimated that peanuts could provide 10 percent or more of the optimum human protein requirement (40 to 70 grams per day) in West Africa.[32] Hunger and malnutrition were thus subject to a calculus of food. While the use of groundnuts was not without critics—for example, Gopalan considered peanuts inferior to milk and vegetable proteins that UNICEF had used in nutrition programs in postwar Europe—the group continued to focus on peanuts for filling the protein gap, fighting protein malnutrition, and thus removing what they believed to be the primary obstacle to development.[33] The causes and remedies of the hungry world themselves were up for grabs.

Saving the Peanut

In fall 1961, the PAG learned that there was "a fungus among us."[34] The fungus excreted a poisonous substance and contaminated peanuts, which the PAG was promoting as a remedy for malnutrition. At the PAG meeting in 1962, members deliberated about the consequences of this discovery.[35] They set out to study aflatoxin's effects on human bodies, to develop detection methods, and to devise ways to eliminate aflatoxin contamination. The PAG wanted to be in a position to advise "the agencies as to desirable procedures and precautions to be observed in utilizing foods derived from materials subject to this kind of fungal attack."[36] The PAG would primarily use published and unpublished reports to determine aflatoxin hazards and strategies to limit exposure.

Moreover, the PAG sought to determine the hazard that aflatoxin caused in animals. They found that "insufficient information was available to allow setting of a tolerance level."[37] Instead, the PAG recommended that only peanuts that were free of toxicity should be used for human nutrition. They recommended using a duckling test that British veterinarians had first developed. A researcher would inject a sample in a duckling's esophagus and observe the reaction. If the duckling died immediately, the sample was strongly toxic. If the duckling did not die, the researcher would kill the bird after several days and examine the liver for lesions, whose presence indicate a weaker toxic potency. This test made aflatoxin visible as a direct physical harm to an animal, a duckling of the Khaki-Campbell breed. If the sample harmed a duckling, the batch contained aflatoxin and was unfit for human consumption. This test assumed that it could either detect the smallest possible amount or that ducklings were at least as sensitive as humans, permitting the continued use of peanuts in malnutrition programs.

There were alternatives to the PAG approach of justifying the continued use of peanuts. In fact, two companies that produced peanut-based nutrition supplements followed a different strategy: eliminating peanuts from food. The British pharmaceutical company Glaxo, which collaborated with the PAG, withdrew the peanut-based nutritional supplement "Amana" in late 1961. Glaxo gave the supplement's disappointing sales as the official reason. In fact, Glaxo found that most of the peanut supply was contaminated by aflatoxin and admitted privately to nutrition experts that uncertainty about the safety of the raw materials was the real reason that they withdrew the product. They said that the company "as a responsible manufacturer [was] not prepared to continue its manufacture and supply."[38] Similarly, Nestlé changed the formula of its high protein food in 1962, replacing peanuts with soy to avoid the presence of aflatoxin.[39] While industry pursued an alternative strategy, suggesting that peanuts were not an economic necessity, the PAG remained wedded to peanuts, developing a risk-based justification to continue this strategy.

In 1963, the PAG reaffirmed its recommendation that peanuts that passed the biological test could be used in nutrition programs. British researchers developed a physiochemical test for aflatoxin, using the recently introduced thin-layer chromatography (TLC). TLC was an analytical method to determine the molecular composition of a solution. Molecules were separated on a strip of adsorbent material according to physical and chemical properties. Each type of molecule produced a distinct spot, which was compared to a reference spot produced by the known substance.[40] However, the PAG was concerned that the test was not specific enough, and any batch of peanuts that tested positive on the TLC had to be subjected to the duckling test.[41] The decisive test was still whether a sample produced visible harm or lesions in an animal.

Scientists' animal experiments raised questions about how to extrapolate these findings to humans. Researchers at the Unilever Research Laboratory observed in 1961 that rats fed with contaminated peanut meal did not develop acute poisoning but liver tumors.[42] This observation prompted researchers to consider whether there was a link between aflatoxin exposure and the high liver cancer rates that scientists had noticed in Africa since the 1920s.[43] While cancer researchers generally assumed that carcinogens in even the smallest amounts could cause cancer, the available records show no discussions about this question at PAG meetings. In 1963, the PAG maintained that "there is no proof that these compounds have any harmful effect on human beings."[44] Michael Latham, an American nutritionist who did research in Tanganyika in 1963, criticized this lack of knowledge. He lamented that aflatoxin research focused on animal feed, imported products, and peanut trade—economically important for European countries—but failed to study possible effects on people in developing countries.[45]

The PAG researchers relied on animal studies to make their recommendations. They quantified aflatoxin's effects by correlating amounts of aflatoxin to specific bodily effects in ducklings and monkeys. For example, they observed that no effects were observed in ducklings for amounts smaller than 0.005 parts per million (ppm) in the diet, and results tended to be negative for aflatoxin concentration below 0.1 ppm. In monkeys, there were no "gross effects" at 0.1 ppm, 0.6 ppm caused some deaths, and higher levels would kill all monkeys.[46] These studies revealed some of the limits of such animal tests. The duckling test could not detect amounts of aflatoxin that were smaller than 0.03 to 0.05 ppm.[47] Moreover, the scientists observed that test results depended on many factors, such as the choice and preparation of a sample, or the breed, species, and age of the animal.[48] Using a seemingly simple test with live ducklings to eliminate contaminated peanut batches appeared increasingly questionable.

In the mid-1960s, the PAG members realized that aflatoxin contamination was much more widespread than initially assumed. There were hardly any sources of uncontaminated peanuts. Convinced of peanuts' nutritional value to many people, the PAG concluded that "therefore, some level of aflatoxin in the protein-rich foods and food mixtures with which the Group is concerned must be accepted if these supplements are to continue to play a role in improving human nutrition."[49] The PAG was determined to set a level of acceptable aflatoxin contamination because the importance of peanuts outweighed the hazard of aflatoxin.

The PAG proposed a level of 0.03 ppm of aflatoxin in human diets. This number corresponded exactly to the detection limit of the biological duckling test. The PAG provided a biological justification for this figure based on experiments with monkeys, which Glaxo conducted in its laboratories. These experiments revealed a sharp difference between the amount of aflatoxin that produced a change in a monkey's liver and the amount that produced no observable changes:

> The "no effect" level in monkeys is 0.3 ppm, which is the equivalent of a daily intake of 0.015 mg aflatoxin/kg body weight. Applying a safety factor of 50, a dose of 0.0003 mg/kg body weight would be an ineffective daily dose for people. This calculation means that for an infant weighing 10 kg, 0.003 mg aflatoxin per day would be a dose likely to be without effect on the liver. From this it can be calculated that a level of aflatoxin in the protein supplements of which 100 g (or 0.1 kg) is eaten should not exceed 0.03 mg/kg or 0.03 ppm.[50]

The researchers emphasized that "it is not possible to draw any conclusions from these experiments about the susceptibility of man to the toxic or carcinogenic action of aflatoxin," but they did exactly that by choosing a safety factor that confirmed the initial PAG recommendation of not using any food that tested positive in the duckling test.[51]

The PAG had maneuvered itself into such a position where it faced the stark choice between solving the problem of protein malnutrition through contaminated supplements or not having enough protein for malnourished children. The PAG justified its choice of the former:

> Although the Group would prefer to impose a lower level of aflatoxin in the foods and food mixtures concerned in order to provide a wider margin of safety, it believed that there was an even more urgent need to provide extra protein in some parts of the world so as to prevent malnutrition and starvation. These considerations outweighed the desirability of introducing measures for reducing a hypothetical health hazard by limitations which were difficult to enforce under current agricultural practices and techniques of food processing.[52]

However, the reaction by other actors, such as Glaxo or Nestlé, which substituted groundnut with other sources of protein, suggested that this was a false dilemma. There were other sources of protein. Nonetheless, the PAG stood by its recommendation, reconfirming it in 1967 and again in 1968, 1969, and 1972.[53]

In 1973, the PAG declared its approach a success in hindsight, because the "pioneering evaluation of the toxicological hazards of aflatoxin to young children in relation to their urgent need for dietary protein apparently became the basic philosophy for much regulatory action in this area subsequently by a number of governments."[54] Judging from the available records, the PAG did not discuss actual amounts of peanuts needed, but assumed that stricter limits would reduce this amount so much as to trigger a protein crisis, a crisis of malnutrition that was always already looming. Historians and anthropologists have shown how colonial rule, settlement schemes, agricultural programs, and postcolonial policies have shaped food supply and crop choice in Kenya.[55] The protein crisis perhaps justified exposing vulnerable populations to aflatoxin through a calculus of risk.

The risk assessment of aflatoxin in nutritional supplements was grounded as much in the material culture of laboratory research and animal experiments as in the late colonial and early postcolonial projects of nutritional studies and aid. The calculus of risk assessment justified the exposure of populations in the developing countries to toxic substances for their supposed benefit. The PAG, however, eschewed taking responsibility for its recommendations, stating that "the ultimate responsibility for the use of these products in human nutrition programs must rest with the governments of the territories concerned."[56] The PAG provided a justification for giving contaminated food to children and infants who were already suffering from malnutrition. This strategy set a precedent for UNEP and other international agencies that would provide risk assessment, ridden with assumptions of what counted as hunger, malnutrition, and food resources. If countries with unequal resources and expertise were left to draw their own conclusions from these assessments, they would reproduce and even exacerbate existing inequalities.

Aflatoxin and Risk Assessment in the US Federal Government

This kind of zero-sum reasoning was not limited to the PAG and the context of peanuts as nutritional supplement for the global poor. The US Food and Drug Administration (FDA) in 1978 conducted its first formal risk assessment on aflatoxin. This risk assessment justified not lowering the threshold level of acceptable aflatoxin contamination, because there would be little effect on health, and lowering the threshold would result in high economic losses. The FDA employed a *de minimis* approach for aflatoxin and other contaminants that it considered to be unavoidable food additives. This approach permitted contamination up to a certain threshold even by substances known to be carcinogenic in the tiniest amounts.[57] In 1965, the FDA set an action level of 30 parts per billion (ppb) total aflatoxin, based on the sensitivity and reliability of the available detection methods.[58] In 1969, this action level was lowered to 20 ppb. In 1974, the FDA proposed a tolerance of 15 ppb considering "the consequences of possible levels above zero."[59] The FDA took into consideration higher prices, unavailability of peanuts, which were "generally considered a highly nutritious and useful food," manufacturers' capability, and lack of "direct evidence that aflatoxin causes cancer in man or of what may be the level of no effect."[60] The FDA assumed without quantitative analyses that manufacturing costs were higher and that peanuts were an almost irreplaceable food.

In 1978, the FDA employed a quantitative risk assessment to justify the tolerance level of 15 ppb. Aflatoxin was one of the first toxic substances in food to be subjected to such a formal exercise.[61] This assessment determined the theoretical effects of lowering the threshold by 5 ppb on rates of human liver cancer. This risk assessment relied on the correlation between aflatoxin exposure and liver cancer occurrence that British researchers had established in a study in Kenya in the late 1960s.[62] Ultimately, the FDA withdrew its proposal for a tolerance level, a regulatory action that required a formal administrative hearing, and has maintained the action level—a less rigorous, informal regulation—of 20 ppb to the present day.[63]

For the PAG and the FDA, risk assessments hinged on classifying aflatoxin as an unavoidable additive. For the PAG, this classification legitimized the continued use of peanuts contaminated at low but possibly harmful dosages. For the FDA, this classification enabled a regulatory intervention under the current law. While the FDA considered manufacturers and consumers, the

PAG imagined the nutritional needs of people in the developing world. Risk assessments were not limited to places without material scarcity.

Mycotoxins, Environments, and the Third World Moment

Many national and supranational agencies set aflatoxin regulations in the late 1960s and 1970s with consequences for global trade, especially for developing countries' ability to export peanuts to Europe. For example, the International Agency for Research on Cancer (IARC), an agency of the WHO, found in 1972 that "increased frequency of liver cancer has been recorded in populations consuming diets contaminated by aflatoxins and possibly other mycotoxins, but no causal relationship has been established," giving credence to the thesis that aflatoxin was a possible human carcinogen.[64] In 1973, the European Economic Community (EEC), a major peanut importer, introduced a directive "on the fixing of maximum permitted levels for undesirable substances and products in feedingstuffs."[65] Such regulations restricted trade between Global North and South during a period in which these relations were intensely politicized.

Peanuts were one of the most important export commodities for countries such as Senegal, The Gambia, and Nigeria. Peanut commerce constituted 60 percent of the national incomes of The Gambia and Niger, 20 percent for Senegal and Sudan, and 5 percent for Mali and Nigeria. In Senegal, more than 70 percent of the population worked in the groundnut industry. The countries formed the African Groundnut Council (ACG) in 1964 in order to, among other things, "ensure reasonable prices for their products," and stabilize "prices of groundnut in the world market at remunerative level."[66] Peanut exports were threatened not only by aflatoxin but also by the competition from other oilseeds, such as soy. In April 1971, Senegal's president Léopold Senghor gave a speech at an EEC-African summit on the peanut problem, arguing that "peanut oil . . . is constantly meeting stiffer competition from various kinds of vegetable oil powerfully supported by different types of national guarantee and support systems. The production in poorer countries cannot and for a long time will not be able to have the same advantage."[67] Senghor viewed the economic trade situation not just as one of peanuts but of all oil seeds, some of which were produced and promoted by other countries, such as soy from the United States. In France, groundnut feed was being replaced by more expensive soybean cakes, which were primarily produced in the United States. Aflatoxin contamination had played a role in this shift.[68] The strict EEC directive of 1973 caused problems for many oilseed exporting countries.

A coalition of these decolonized nations had been developing a proposal for a New International Economic Order (NIEO) since the mid-1960s.[69] The

NIEO sought to overhaul the global liberal economic order by giving decolonized states sovereignty over their natural resources, control over raw material prices and commodity exports, access to markets in developed countries, and sufficient food supply.[70] These radical reforms would enable the postcolonial countries to gain real independence. The end of the Bretton Woods system in 1971 and the oil embargo by the Organization of Petroleum Exporting Countries (OPEC) in 1973 empowered the countries of the "Third World" to make a decisive push for the NIEO at the sixth special session of the UN General Assembly, which adopted the proposal. The program of action included a call to "promote exports of food products of developing countries through just and equitable arrangements, inter alia, by the progressive elimination of such protective and other measures as constitute unfair competition."[71] These measures included tariffs as well as nontariff barriers, which, for example, restricted import for health reasons.

Concern about invoking environmental pollution as a pretext to restrict market access had already been raised at the UN Stockholm Conference on the Human Environment in 1972. The conference passed a resolution that "the burdens of the environmental policies of the industrialized countries should not be transferred, either directly or indirectly, to the developing countries."[72] In the event that stricter environmental standards would result in restricting trade or negatively affecting exports, compensations should be worked out "within the framework of existing contractual and institutional arrangements."[73] Moreover, the conference suggested "internationally coordinated programmes of research and monitoring of food contamination by chemical and biological agents" be established by the FAO and WHO. The results from these programs should be used to provide early information about rising contamination levels, especially those "that may be considered undesirable or may lead to unsafe human intake."[74] The conference established the UN Environment Programme, which would pursue this goal. Further, it was recommended that the Codex Alimentarius Commission, an international body for harmonizing food regulations, be supported to develop international standards for pollutants in food and a code of ethics for international food trade.[75] FAO and UN should assist developing countries in the field of food control.

The problem of mycotoxins emerged at the intersection of the NIEO and environmental governance, both key areas of international politics of the 1970s.[76] The UN Conference on Mycotoxins in 1977 in Nairobi was the central forum for discussing this problem. Aflatoxin regulations by the EEC, Japan, and other importers had ramifications for all participants in the supply chain, including primary producers, middlemen, national exchequers, and consumers. Internationally, Osogo said, "potential problems would be the increased difficulty involved in arriving at sound international agricultural production adjustments and commodity and food security agreements."[77] The aim of the conference was to "establish an effective system of internationally agreed monitoring and control that would protect the consumer from exposure and that would safeguard the producer from unexpected, and sometimes unjustified, deprivation of the fruits of his labor and his investments."[78] While its concrete recommendations were mostly technical in nature, the conference made explicit the relationship between health and wealth that the risk calculations sought to obscure. Ultimately, the conference failed in both regards: consumers in Africa were unprotected, and producers in Africa could not sell their produce. The responsibility for the losses was ascribed to the exporter countries: "It was suggested that the risk of sending contaminated shipments was the exporter's and that it was a matter of contract and agreement between trading partners which analysis and sampling procedures were acceptable."[79] Risk was privatized.

Moreover, the conference called for regulatory harmonization through the Codex Alimentarius Commission because developing countries often lacked effective food control systems. It was concluded that "in the long run therefore, exporting countries must improve their food control systems, and surveillance and monitoring programmes to reduce contaminations." Still, the solution of the problems of balance of payments and the protection of public health with regard to mycotoxins would need to seem appropriate within "the context of the New International Economic Order and to ensure a fuller, safer, more wholesome food supply for mankind [*sic*]."[80] This approach, however, was never found or promoted. The final recommendations of the conference focused exclusively on technical assistance that would increase the "marketability" of food commodities and products through improved storage, monitoring and control, better training, and further research. Ultimately, the NIEO was not translated into actual changes on the level of policy, such as the formulation and control of risk. By 1981, the NIEO, which was already dissipating, received its final blow from US president Ronald Reagan, who refused to discuss any further changes to the global economic structure.[81] Aflatoxin remained a problem for trade, leaving the burden of regulation on the producing countries of the Global South.

Risk on the Ground

While peanuts were used as nutritional supplements and as export commodities, maize was the central staple in East Africa.[82] Researchers found that aflatoxin could also contaminate maize. On the level of policy implementation, the UN agencies provided technical assistance to improve the system of food control in Kenya and other countries. However, this system neither reduced contamination in export goods nor protected a population, which relied primarily on subsistence farming and local markets. Setting threshold levels and enforcing them depended on systematic testing, which was difficult in an agricultural system without central grain collection. In 1981, the long rain season began on 17 March in Machakos, just east of Nairobi, Kenya. "The rest of the month," the crop officer wrote, "remained very wet with low temperatures."[83] Some areas of Machakos, mostly populated by the Akamba people, had been badly hit by food shortages. Soon, doves began to die, a family's dog perished, and the fish in the Athi River also died.[84] A few months later, in early June, a family was admitted to Makueni District Hospital. The family's eight members had fallen ill at the same time with fever, vomiting, loss of appetite, and weakness. After a week, they developed jaundice. All of them, except two unweaned infants, died.[85] Kenyan researchers visited the family's homestead, where they collected foodstuff, including corn, cowpeas, sorghum, millet, and vegetables. The grains had been stored by the "traditional Akamba method" in gourds and clay pots inside a granary that was raised above the ground.[86] The samples were brought to National Public Health Laboratory Service in Nairobi, where the researchers detected 3,200 ppb of aflatoxin in the corn samples. The researchers noted that "the previous year was extremely dry and, as a result, the harvest was poor, necessitating the storage of grains that might normally have been discarded as spoiled."[87]

After the Stockholm Conference, the FAO and UNEP had organized a program to address the problem of food contamination in Kenya, especially relating to aflatoxin. They trained local laboratory staff, prepared analytical manuals, and established a library. Yet, one WHO consultant warned early on that "whereas economic interest in food production, especially food for export, will have dictated the necessary preventive measures, subsistence farmers likely to be at highest risk due to their primitive food storage conditions, are badly in need of protection against the hazards arising from aflatoxin intake."[88] The Kenyan government's focus had been on exports, so many of the efforts of aflatoxin control and sampling were centered on the ports.[89] Programs to monitor toxic substances in food and workplaces were often spotty, understaffed, and underfunded. The director of medical services appealed to his regional officers to "please help the *Wananchi* [citizens] with preventive aspects by taking regular grain samples in your Provinces—this should include also maize meal from the shops and from the maize mills themselves—and send these to the National Public Laboratory Services for analysis."[90] This laboratory, however, lacked essential chemicals. For example, in 1986, the Mycotoxin Laboratory of the National Public Health Laboratory Service analyzed a total of 109 food and feed samples, but "due to lack of essential reagents and chemicals which are required for aflatoxin analysis work has presently come to standstill."[91] As such the laboratory could illuminate the "edges of exposure" but not conduct a comprehensive surveillance program of food.[92] After the NIEO fizzled out in the early 1980s, international donor organizations imposed structural adjustment programs on African countries, resulting in the further defunding of health and other public services.[93] The highly abstract system of threshold levels, which were determined by risk assessment, required a complex laboratory infrastructure to take hold in the world. Without such infrastructure, the system reached its limits in the poor countries before it ever really began. For international regulators, the markets and consumers in the rich countries seemed to have mattered more than the decolonized nations and their farmers.

Conclusions: Risk under the Table

The history of aflatoxin regulation across rich and poor countries in the 1960s and 1970s reveals a complex landscape of expertise, international relations, and commodity trade. Aflatoxin with its twofold toxicity of acute poisoning and carcinogenicity became visible to scientific research in the artificial environment of the industrialized farm. The carcinogenic and toxic effects of aflatoxin on humans remained an object of protracted epidemiological studies. The nutrition researchers of the PAG employed an approach of quantitative risk assessment to justify the continued use of potentially contaminated goods in food and nutrition programs in the face of a perceived scarcity of protein. This approach had ramifications for the global trade of potentially contaminated commodities that were primarily exported by developing countries.

This chapter shows that toxicity and food contamination were not limited to risk societies in affluent industrialized countries. Beck wrote that "the struggle [of the 'Third World'] against hunger and for autonomy forms the protective shield behind which the hazards, imperceptible in any case, are suppressed, minimized and, by virtue of that, amplified, diffused and eventually returned to the wealthy industrial countries via the food chain."[94] This chapter reveals a more complex story, in which hazards have not merely been suppressed and minimized in the "Third World," but the process of making visible, assessing, and quantifying risk crisscrossed Global North and South. For Beck, risks know no social or national differences. Wealth and health hazards cannot be contained within a nation-state in an "interdependent" or "globalized" world. Ultimately, Beck believes that only the UN would be in a position to tackle risks. However, this empirical study of the efforts of the UN and its agencies to tackle aflatoxin risk dampen this hope. Risk-based global regulation required a level of abstraction that removed questions of inequality of trade relations and that was impossible to translate into public health programs in resource-poor settings. The risk approach failed to increase exports and to protect the health of people in Kenya and other places, where grains were produced and consumed locally. The history of aflatoxin shows how the concept of risk has tended to mask the global distribution of wealth and health, raising the question whether risk itself, as a quantitative category, should remain on the table at all.

Acknowledgments

I would like to thank Austin Cooper, Kit Heintzman, David S. Jones, and Harriet Ritvo for feedback on earlier versions of this text. I am grateful to Angela Creager and Jean-Paul Gaudillière and the other participants of the Risk on the Table workshop for helpful comments and suggestions. The support from Mary Mwiandi and David Masika at the University of Nairobi and the funding from the National Science Foundation (Award No. 1555448), the Science History Institute, and the Mellon/ACLS Dissertation Completion Fellowship made the research for this chapter possible.

Lucas M. Mueller is a postdoctoral researcher at the University of Geneva. He holds a PhD in History, Anthropology, and Science, Technology, and Society from the Massachusetts Institute of Technology.

Notes

1. Food and Agriculture Organization of the United Nations, *Report of the Joint FAO/ WHO/UNEP Conference on Mycotoxins*, 3.
2. Ibid., 48.
3. Ibid.
4. Gilman, "New International Economic Order."
5. Food and Agriculture Organization of the United Nations, *Report of the Joint FAO/ WHO/UNEP Conference on Mycotoxins*, 49.
6. Ferguson et al., *Shock of the Global.*
7. Building on Gabrielle Hecht's work on the market for nuclear materials and postcolonial power relations, this chapter shows how risk assessment in the United States and beyond was shaped not only by Cold War nuclear weapon planning but also by postcolonial North-South relations. See Hecht, *Being Nuclear;* Nash, "From Safety to Risk."
8. Beck, *Risikogesellschaft.*
9. Historians have called into question the sharp transition from class to risk society, adducing examples of earlier concerns about environmental destruction. See Bonneuil and Fressoz, *Shock of the Anthropocene.*
10. Beck, *Risk Society,* 49.
11. Cullather, *The Hungry World.*
12. The literature on the roles of scientific experts in politics is far too expansive to do justice to it here. The works of political scientist Peter Haas and STS scholar Sheila

Jasanoff are starting points of this literature. Haas introduces the concept of "epistemic communities" to describe how experts shape governments' multilateral coordination and foreign policy decisions that appear opposed to state interests at first sight. Jasanoff introduces a nuanced analysis that shows that the scientists forming epistemic communities are neither as disinterested, nor acting necessarily as coherently, as Haas suggests. See Haas, "Introduction"; Jasanoff, *The Fifth Branch*; Jasanoff, "Science and Norms." For the role of lay knowledge in risk society, see Wynne, "May the Sheep Safely Graze?"

13. See, for example, Jasanoff, "Science, Politics, and the Renegotiation"; Jasanoff, "Songlines of Risk."

14. Vogel, *The Politics of Precaution*; Tousignant, *Edges of Exposure*.

15. Hecht, "Interscalar Vehicles."

16. Slobodian, *Globalists*.

17. Moyn, *Not Enough*.

18. The literature on standards, risk, and economic interests has taken different approaches to understand these interrelations. Most authors agree that risk-based standards implicitly configure a moral economy of relations. Proctor, Oreskes, and other show that industry actively produces ignorance by hiring "faux experts," who distort facts and cast doubt on scientific studies. Others, such as Nathalie Jas and Michelle Murphy, argue that measurements and standards have always already been shaped by specific regimes of perceptibility. For example, Jas describes how European scientists developed the measure of the "acceptable daily intake" (ADI) in the 1950s to take into consideration economic development in setting health standards. Graham Burnett has emphasized the role of epistemic and disciplinary allegiances in the regulation of whaling. This chapter draws on the last two approaches to understand aflatoxin regulation and standard setting as a reconfiguration of relations between countries and people. See Busch, "Moral Economy"; Murphy, *Sick Building Syndrome*; Boudia and Jas, "Introduction"; Proctor and Schiebinger, *Agnotology*; Oreskes and Conway, *Merchants of Doubt*; Burnett, *Sounding of the Whale*; Jas, "Adapting To 'Reality'"; Jas, "Gouverner les substances chimiques dangereuses"; Stoff, *Gift in der Nahrung*.

19. William Blount, "A New Disease of Turkeys," June 1960, Disease of Turkey Poults and Groundnut Toxicity (Aflotoxin) [*sic*], Ministry of Agriculture and Fisheries, The National Archives, Kew, UK: MAF 287/41.

20. Lakoff and Keck, "Preface."

21. Vernon, *Hunger*.

22. Cullather, "Foreign Policy of the Calorie"; Amrith, "Food and Welfare in India."

23. Amrith, *Decolonizing International Health*; Amrith and Clavin, "Feeding the World"; Bashford, *Global Population*.

24. Worboys, "The 'Discovery' of Colonial Malnutrition."

25. While the Cold War shaped US foreign policy, including funding for food and nutrition projects, the Cold War was not the primary lens through which experts in postcolonial Great Britain and African countries viewed agricultural and scientific problems. These experts were much more concerned with the problems that the division between Global North and South posed. For the history of US food policy in the Cold War, see Cullather, *The Hungry World*; Connelly, "Taking Off the Cold War Lens."

26. Tappan, *Riddle of Malnutrition*.

27. Ruxin, "United Nations Protein Advisory Group."
28. "Current and Past Members of PAG," PAG Bulletin, No. 7, April 1967, FAO Archives, Rome: 12-ESN-536.
29. At the same time, the Rockefeller Foundation promoted a molecular approach to the study of life that "aimed to map the pathways in the molecular labyrinth of the human soma and psyche in order to control biological destiny." Scientists studied the biological structure and function of proteins as part of these efforts. See Kay, *The Molecular Vision of Life*, 48.
30. FAO/WHO/UNICEF Protein Advisory Group, "The PAG: Its History, Function and Work Programme," 28 September 1971, Accession No. 18378, FAO Library.
31. See chapter 1 in Mueller, "Toxic Relationships."
32. Max Milner, "Peanut as Protein Resource in International Feeding Programs," 31 January 1962, Accession No. 060838, FAO Library.
33. C. Gopalan to C. G. King, 2 February 1960, in WHO Protein Advisory Group, NU 13/4, FAO Archives: 12-ESN-495; "Progress Report on Peanut Flour," PAG Meeting, June 1961, Accession No. 0608009, FAO Library.
34. Milner, "There's a Fungus among Us."
35. PAG. "Memorandum. Proposed Agenda for the next PAG Meeting," 15 September 1961, NU 13/4 –Meetings Vol. II, FAO Archives: 12-ESN-498.
36. Milner, "There's a Fungus among Us."
37. "Minutes of Meeting, Protein Advisory Group," 26–29 March 1962, NU 13/4—Meetings, Vol. III, FAO Archives: 12-ESN-498.
38. Minutes of Marketing Policy Committee, 7 December 1961, 2/6/13/2, GSK Heritage Archives, Brentford, Middlesex, UK.
39. Zbinden, *History of Nestlé's Infant and Dietetic Preparations*, n.d., 313, Nestlé Archives, Vevey, Switzerland.
40. Wintermeyer, *Die Wurzeln der Chromatographie*.
41. "Aflatoxin Recommendation," 3 August 1963, NU 13/4 "Meeting 1963," FAO Archives: 12- ESN-498.
42. Lancaster, Jenkins, and Philp, "Toxicity Associated with Certain Samples."
43. Le Breton, Frayssinet, and Boy, "Sur l'apparition d'hépatomes 'spontanés' chez le Rat Wistar"; Pirie, "Hepatic Carcinoma in Natives"; Mueller, "Cancer in the Tropics."
44. "PAG Minutes. (1963 Meeting). Progress Report," NU 13/4 "Meeting 1963," FAO Archive: 12- ESN-498.
45. Latham, "Hazards of Groundnuts."
46. L. J. Teply, "Memo for the Record," 26 August 1963, NU 13/4 "Meeting 1963," FAO Archive: 12- ESB-498.
47. "Minutes. Meeting—WHO/FAO/UNICEF Protein Advisory Group," 5–7 July 1965, FAO Archives: 12-ESN-498.
48. E. M. DeMaeyer to V. N. Patwardhan et al., 17 September 1963, FAO Archives: 12-ESN-498.
49. "Recommendation on Aflatoxin, Meeting of the WHO/FAO/UNICEF Protein Advisory Group, 17–19 August 1966," 30 September 1966, NU-13/4 PAG (Meeting 1966), FAO Archives: 12-ESN-498.
50. Ibid.

51. W. F. J. Cuthbertson, "Effect of Groundnut Meal Containing Aflatoxin on Cynomologus Monkeys," October 1967, Accession No. 105383, FAO Library.

52. "Recommendation on Aflatoxin, Meeting of the WHO/FAO/UNICEF Protein Advisory Group, 17–19 August 1966," 30 September 1966, NU-13/4 PAG (Meeting 1966), FAO Archives: 12-ESN-498.

53. "Report on the FAO/WHO/UNICEF PAG Meeting," 1967, Accession No. 105402, FAO Library.

54. Protein Advisory Group, "PAG Activities: Retrospect and Prospect," 15 May 1973, FAO Archives: 12-ESN-536.

55. For Kenya, see Mackenzie, *Land, Ecology, and Resistance in Kenya*; McCann, *Maize and Grace*; Moskowitz, *Seeing Like a Citizen*. For an example of cash crops in Africa, see Cooper, "'Ray of Sunshine on French Tables.'"

56. "PAG New Bulletin No. 3," February 1964, NU 13/4 "Meeting 1963," FAO Archives: 12- ESB-498.

57. Vogel, *Is It Safe?*

58. Daniel Banes, "Aflatoxin in Peanut Products," 29 January 1965, Folder 428 Peanut Aflatoxin, Box 3694, Records of the Food and Drug Administration, Record Group 88, General Service Files 1938—1974, National Archives at College Park, MD.

59. Schmidt, "Aflatoxin in Shelled Peanuts and Peanut Products Used as Human Foods," 42750.

60. Ibid.

61. Bureau of Foods, FDA, *Assessment of Estimated Risk*.

62. Mueller, "Cancer in the Tropics."

63. The US Supreme Court upheld the FDA's approach of using action levels instead of tolerance levels for aflatoxin regulation when challenged in 1986. See Hutt, Merrill, and Grossman, *Food and Drug Law*, 507–21.

64. IARC, *IARC Monographs on the Evaluation of Carcinogenic Risks to Humans*, 1:153.

65. "Council Directive of 17 December 1973 on the Fixing of Maximum Permitted Levels for Undesirable Substances and Products in Feedingstuff," *Official Journal of the European Communities*, no. L 38/31 (February 11, 1974).

66. "Convention Creating the African Groundnut Council," 18 June 1964, No. 12123, 846 United Nations Treaty Series (UNTS), 215, Hein Online World Treaty Library.

67. "Senghor, Diouf Address Peanut Conference," in *Le Soleil* (Dakar) 23 May 1971, 1, in Translations on Africa, No. 1021, 15 April 1971, Joint Publication Research Service.

68. Food and Agriculture Organization of the United Nations, *Perspective on Mycotoxins*, 159.

69. Murphy, *The Emergence of the NIEO Ideology*; Prashad, *The Poorer Nations*.

70. Ogle, "State Rights against Private Capital"; Gilman, "The New International Economic Order"; Moyn, *Not Enough*.

71. UN General Assembly, Resolution 3202, Programme of Action on the Establishment of a New International Economic Order, A/RES/S-6/3202 (1 May 1974), http://www.un.org/en/ga/search/view_doc.asp?symbol=A/RES/3202(S-VI).

72. United Nations, *Report of the United Nations Conference on the Human Environment*.

73. Ibid.

74. Ibid., 21.

75. Winickoff and Bushey, "Science and Power in Global Food Regulation."
76. The link between the NIEO and global environmental governance has received little attention. See Conca, *An Unfinished Foundation.*
77. Food and Agriculture Organization of the United Nations, *Perspective on Mycotoxins,* 145.
78. Food and Agriculture Organization of the United Nations, *Report of the Joint FAO/ WHO/UNEP Conference on Mycotoxins, Held in Nairobi, 19–27 September 1977,* 2.
79. Ibid., 8.
80. Food and Agriculture Organization of the United Nations, *Perspective on Mycotoxins,* 166.
81. Gilman, "The New International Economic Order."
82. McCann, *Maize and Grace.*
83. "Crop and Food Situation Report for the Month of March 1981," 22 April 1981, Kenya National Archives, Nairobi (KNA): KD 4/1.
84. W. Koinage to Provincial Medical Officers, 6 July 1981, KNA: BY 4/255.
85. "Jaundice Due to Aflatoxin Outbreak in Machakos District, Kenya, 1981," KNA: BY 4/255.
86. Ngindu et al., "Outbreak of Acute Hepatitis Caused by Aflatoxin Poisoning in Kenya."
87. Ibid.
88. Director to Regional Director, AFRO, "Food Safety Programme," 3 February 1977, KNA: BY/25/18, KNA.
89. "Food and Drugs Control," 1979—1985, KNA: BY/14/72.
90. W. Koinage to Provincial Medical Officers, 6 July 1981, KNA: BY 4/255.
91. "Ministry of Health, National Public Health Laboratory Services, Annual Report, 1986," Library of the National Public Health Laboratory Services, Nairobi, Kenya.
92. Tousignant, *Edges of Exposure;* Mutongi, *Matatu.*
93. Geissler, "Introduction."
94. Beck, *Risk Society,* 42.

Bibliography

Archives

Food and Agriculture Organization (FAO) Archives, Rome, Italy
Food and Agriculture Organization (FAO) Library, Rome, Italy
GlaxoSmithKline (GSK) Heritage Archives, Brentford, Middlesex, United Kingdom
Kenya National Archives (KNA), Nairobi, Kenya
The National Archives, Kew, United Kingdom
Nestlé Archives, Vevey, Switzerland

Publications

Amrith, Sunil S. *Decolonizing International Health: India and Southeast Asia, 1930–65.* Basingstoke, UK: Palgrave Macmillan, 2006.
———. "Food and Welfare in India, c. 1900–1950." *Comparative Studies in Society and History* 50, no. 4 (2008): 1010–35.

Amrith, Sunil S., and Patricia Clavin. "Feeding the World: Connecting Europe and Asia, 1930–1945." *Past & Present* 218, suppl. 8 (2013): 29–50.

Bashford, Alison. *Global Population: History, Geopolitics, and Life on Earth*. New York: Columbia University Press, 2014.

Beck, Ulrich. *Risikogesellschaft. Auf Dem Weg in Eine Andere Moderne*. Frankfurt am Main: Suhrkamp, 1986.

———. *Risk Society: Towards a New Modernity*. London: Sage, 1992.

Bonneuil, Christophe, and Jean-Baptiste Fressoz. *The Shock of the Anthropocene: The Earth, History and Us*. London: Verso, 2016.

Boudia, Soraya, and Nathalie Jas. "Introduction: Risk and 'Risk Society' in Historical Perspective." *History and Technology* 23, no. 4 (2007): 317–31.

Bureau of Foods, FDA. *Assessment of Estimated Risk Resulting from Aflatoxins in Consumer Peanut Products and Other Food Commodities*. Washington, DC: FDA, 19 January 1978. Obtained under the Freedom of Information Act from the US FDA, requested and received in July 2017.

Burnett, D. Graham. *The Sounding of the Whale: Science and Cetaceans in the Twentieth Century*. Chicago: University of Chicago Press, 2012.

Busch, Lawrence. "The Moral Economy of Grades and Standards." *Journal of Rural Studies* 16, no. 3 (2000): 273–83.

Conca, Ken. *An Unfinished Foundation: The United Nations and Global Environmental Governance*. Oxford: Oxford University Press, 2015.

Connelly, Matthew. "Taking Off the Cold War Lens: Visions of North-South Conflict during the Algerian War for Independence." *American Historical Review* 105, no. 3 (2000): 739–69.

Cooper, Austin R. "'A Ray of Sunshine on French Tables': Citrus Fruit, Colonial Agronomy, and French Rule in Algeria (1930–1962)." *Historical Studies in the Natural Sciences* 49, no. 3 (2019): 241–72.

Cullather, Nick. "The Foreign Policy of the Calorie." *American Historical Review* 112, no. 2 (2007): 336–64.

———. *The Hungry World: America's Cold War Battle against Poverty in Asia*. Cambridge: Harvard University Press, 2010.

Ferguson, Niall, Charles S. Maier, Erez Manela, and Daniel J. Sargent, eds. *The Shock of the Global: The 1970s in Perspective*. Cambridge: Harvard University Press, 2011.

Food and Agriculture Organization of the United Nations. *Perspective on Mycotoxins: Selected Documents of the Joint FAO/WHO/UNEP Conference on Mycotoxins, Held in Nairobi, 19–27 September 1977*. FAO Food and Nutrition Paper, no. 13. Rome: FAO, 1979.

———. *Report of the Joint FAO/WHO/UNEP Conference on Mycotoxins, Held in Nairobi, 19–27 September 1977*. FAO Food and Nutrition Paper, no. 2. Rome: FAO, 1977.

Geissler, P. Wenzel. "Introduction: A Life Science in Its African Para-State." In *Para-States and Medical Science*, edited by P. Wenzel Geissler, 1–44. Durham: Duke University Press, 2015.

Gilman, Nils. "The New International Economic Order: A Reintroduction." *Humanity* 6, no. 1 (2015): 1–16.

Haas, Peter M. "Introduction: Epistemic Communities and International Policy Coordination." *International Organization* 46, no. 1 (1992): 1–35.

Hecht, Gabrielle. *Being Nuclear: Africans and the Global Uranium Trade.* Cambridge: MIT Press, 2012.

———. "Interscalar Vehicles for an African Anthropocene: On Waste, Temporality, and Violence." *Cultural Anthropology* 33, no. 1 (2018): 109–41.

Hutt, Peter Barton, Richard A. Merrill, and Lewis A. Grossman. *Food and Drug Law: Cases and Materials.* St. Paul: Foundation Press, 2014.

IARC. *IARC Monographs on the Evaluation of Carcinogenic Risks to Humans.* Vol. 1. Lyon: International Agency for Research on Cancer, 1972.

Jas, Nathalie. "Adapting To 'Reality': The Emergence of an International Expertise on Food Additives and Contaminants in the 1950s and Early 1960s." In *Toxicants, Health and Regulation since 1945*, edited by Soraya Boudia and Nathalie Jas, 25–46. London: Pickering & Chatto, 2013.

———. "Gouverner les substances chimiques dangereuses dans les espaces internationaux." In *Le gouvernement des technosciences: gouverner le progrès et ses dégâts depuis 1945*, edited by Dominique Pestre, 31–63. Paris: La Découverte, 2014.

Jasanoff, Sheila. *The Fifth Branch: Science Advisers as Policymakers.* Cambridge: Harvard University Press, 1990.

———. "Science and Norms in Global Environmental Regimes." In *Earthly Goods: Environmental Change and Social Justice*, edited by Fen Osler Hampson and Judith Reppy, 173–97. Ithaca: Cornell University Press, 1996.

———. "Science, Politics, and the Renegotiation of Expertise at EPA." *Osiris* 7 (1992): 194–217.

———. "The Songlines of Risk." *Environmental Values* 8, no. 2 (1999): 135–52.

Kay, Lily E. *The Molecular Vision of Life: Caltech, the Rockefeller Foundation, and the Rise of the New Biology.* Oxford: Oxford University Press, 1993.

Lakoff, Andrew, and Frédéric Keck. "Preface: Sentinel Devices." *Limn*, no. 3 (2013). Accessed 1 July 2015, http://limn.it/preface-sentinel-devices-2/.

Lancaster, M. C., F. P. Jenkins, and J. McL. Philp. "Toxicity Associated with Certain Samples of Groundnuts." *Nature* 192, no. 4807 (1961): 1095–96.

Latham, M. C. "Hazards of Groundnuts." *British Medical Journal* 2, no. 5412 (1964): 819–20.

Le Breton, Eliane, Charles Frayssinet, and Jacques Boy. "Sur l'apparition d'hépatomes 'spontanés' chez le Rat Wistar. Rôle de la toxine de l'Aspergillus Flavus. Intérêt en pathologie humaine et cancérologie expérimentale." *Comptes Rendus de l'Académie Des Sciences (Paris)* 225 (23 July 1962): 784–86.

Mackenzie, Fiona. *Land, Ecology, and Resistance in Kenya, 1880–1952.* Portsmouth, NH: Heinemann, 1998.

McCann, James. *Maize and Grace: Africa's Encounter with a New World Crop, 1500–2000.* Cambridge: Harvard University Press, 2005.

Milner, M. "There's a Fungus among Us." In *The PAG Compendium: The Collected Papers Issued by the Protein-Calorie Advisory Group of the United Nations System, 1956–1973*, D: D263–75. New York: Worldmark Press, 1975.

Moskowitz, Kara. *Seeing Like a Citizen: Decolonization, Development, and the Making of Kenya, 1945–1980.* Athens: Ohio University Press, 2019.

Moyn, Samuel. *Not Enough: Human Rights in an Unequal World.* Cambridge: Harvard University Press, 2018.

Mueller, Lucas M. "Cancer in the Tropics: Geographical Pathology and the Formation of Cancer Epidemiology." *BioSocieties* 14, no. 4 (2019): 512–28.

———. "Toxic Relationships: Health and the Politics of Science and Trade in the Postcolonial World." PhD diss., Massachusetts Institute of Technology, 2019.

Murphy, Craig. *The Emergence of the NIEO Ideology.* Boulder, CO: Westview, 1984.

Murphy, Michelle. *Sick Building Syndrome and the Problem of Uncertainty: Environmental Politics, Technoscience, and Women Workers.* Durham: Duke University Press, 2006.

Mutongi, Kenda. *Matatu: A History of Popular Transportation in Nairobi.* Chicago: Chicago University Press, 2017.

Nash, Linda. "From Safety to Risk: The Cold War Contexts of American Environmental Policy." *Journal of Policy History* 29, no. 1 (2016): 1–33.

Ngindu, A., B. K. Johnson, P. R. Kenya, J. A. Ngira, D. M. Ocheng, H. Nandwa, T. N. Omondi, et al. "Outbreak of Acute Hepatitis Caused by Aflatoxin Poisoning in Kenya." *Lancet* 1, no. 8285 (1982): 1346–48.

Ogle, Vanessa. "State Rights against Private Capital: The 'New International Economic Order' and the Struggle over Aid, Trade, and Foreign Investment, 1962–1981." *Humanity* 5, no. 2 (2014): 211–34.

Oreskes, Naomi, and Erik M. Conway. *Merchants of Doubt: How a Handful of Scientists Obscured the Truth on Issues from Tobacco Smoke to Global Warming.* New York: Bloomsbury, 2010.

Pirie, J. H. Harvey. "Hepatic Carcinoma in Natives and Its Frequent Association with Schistosomiasis." *South African Medical Record* 20, no. 1 (1922): 2–8.

Prashad, Vijay. *The Poorer Nations: A Possible History of the Global South.* London: Verso, 2012.

Proctor, Robert, and Londa L. Schiebinger, eds. *Agnotology: The Making and Unmaking of Ignorance.* Stanford: Stanford University Press, 2008.

Ruxin, Joshua N. "The United Nations Protein Advisory Group." In *Food, Science, Policy, and Regulation in the Twentieth Century: International and Comparative Perspectives,* edited by David F. Smith and Jim Phillips, 151–66. London: Routledge, 2000.

Schmidt, A. M. "Aflatoxin in Shelled Peanuts and Peanut Products Used as Human Foods." *Federal Register* 39, no. 236 (6 December 1974): 42748–52.

Slobodian, Quinn. *Globalists: The End of Empire and the Birth of Neoliberalism.* Cambridge: Harvard University Press, 2018.

Stoff, Heiko. *Gift in der Nahrung: Zur Genese der Verbraucherpolitik Mitte des 20. Jahrhunderts.* Wiesbaden: Franz Steiner Verlag, 2015.

Tappan, Jennifer. *The Riddle of Malnutrition: The Long Arc of Biomedical and Public Health Interventions in Uganda.* Athens: Ohio University Press, 2017.

Tousignant, Noémi. *Edges of Exposure: Toxicology and the Problem of Capacity in Postcolonial Senegal.* Durham: Duke University Press, 2018.

UN General Assembly. "Resolutions Adopted by the General Assembly during Its Sixth Special Session," 9 May 1974.

United Nations. *Report of the United Nations Conference on the Human Environment. Stockholm, 5–16 June 1972.* New York: United Nations, 1973.

Vernon, James. *Hunger: A Modern History.* Cambridge, MA: Harvard University Press, 2006.

Vogel, David. *The Politics of Precaution: Regulating Health, Safety, and Environmental Risks in Europe and the United States*. Princeton: Princeton University Press, 2012.

Vogel, Sarah A. *Is It Safe? BPA and the Struggle to Define the Safety of Chemicals*. Berkeley: University of California Press, 2012.

Winickoff, David E., and Douglas M. Bushey. "Science and Power in Global Food Regulation: The Rise of the Codex Alimentarius." *Science, Technology & Human Values* 35, no. 3 (2010): 356–81.

Wintermeyer, Ursula. *Die Wurzeln der Chromatographie: historischer Abriss von den Anfängen bis zur Dünnschicht-Chromatographie*. Darmstadt: GIT Verlag, 1989.

Worboys, Michael. "The 'Discovery' of Colonial Malnutrition." *Society for the Social History of Medicine Bulletin* 40 (1987): 23–26.

Wynne, Brian. "May the Sheep Safely Graze? A Reflexive View of the Expert-Knowledge Divide." In *Risk, Environment and Modernity: Towards a New Ecology*, edited by Scott Lash, Bronislaw Szerszynski, and Brian Wynne, 44–83. London: Sage, 1996.

CHAPTER 6

Contaminated Foods, Global Environmental Health, and the Political Recalcitrance of a Pollution Problem

PCBs from 1966 to the Present Day

Aurélien Féron

PCBs (polychlorinated biphenyls) are among the very few chemicals whose uses have been progressively banned at the global scale for environmental and sanitary reasons after many decades of industrial production and ubiquitous usage.[1] Used mainly in capacitors and electrical transformers, but also in hydraulic systems, cooling systems, heat transfer systems, in sealants and coatings, inks, paints, adhesives, plasticizers, and many other products and devices, this family of substances has become present in all living and working spaces, inside and outside factories, in technoindustrial equipment and infrastructures, as well as in offices and households.

At the end of the 1960s, these substances began to be considered—by many scientists as well as many political authorities—as major environmental pollutants and potential hazards for public health. They became the subject of several waves of recommendations and decisions promulgated by international and supranational organizations, as well as states, from the 1970s. During this decade, regulatory foundations were laid for transnational political initiatives to combat the problem of PCB pollution through restriction of their uses and managing disposal of many products and articles containing them. The global level of PCB production began to decrease at the end of the same decade. However, environmental pollution by these substances and a large range of problems associated with them, far from disappearing from political agendas, have kept reemerging over the last few decades and remain unresolved.

The purpose of this chapter is to highlight to what extent food risk issues have impacted the trajectory of the PCB problem—on various scales—from the late 1960s to the present day. I will expose how studying the place of food issues in the history of PCBs illuminates what I call the *political recalcitrance* of this problem. This expression highlights the conundrum encountered when trying to control certain persistent, toxic contaminants: even though, from the late 1960s to the present day, PCBs have been the subject of a succession of increasingly radical regulatory measures, the problems associated with these substances have not disappeared, with the consequence that public affairs and controversies continue to crop up.[2]

Many countries enacted regulatory measures to limit the environmental dispersal of PCBs in the early and mid-1970s, just a few years after scientists began documenting the extent, behavior, and dangers represented by this pollution. Since then, other regulatory measures have been promulgated, with the same objective of lowering the level of environmental pollution and contamination of living organisms. More broadly, since the beginning of the 1970s, various national and international public policies have gradually restricted their use, organized their disposal, and imposed ways of managing the waste, the environmental media, and the food contaminated by these substances.[3] PCBs are thus listed among the first twelve pollutants covered by the Stockholm Convention, an international agreement that was adopted in 2001 and ratified by 152 countries before it came into force in 2004. The Convention has been amended several times since, with the same aim of reducing and eventually eliminating releases of "persistent organic pollutants" into the environment.[4] Today however—about fifty years after the first transnational actions were initiated—PCBs are still used in some countries, and the elimination goal has not yet been achieved.[5] The Stockholm Convention requires the phase-out of PCB use by 2025 and "the environmentally sound management of PCB waste" by 2028.[6]

The *recalcitrance of the PCB problem* refers not only to the present but is also a critical historical fact. From the early 1970s to the present, many problems linked to the production and circulation of PCBs have emerged. As the following examples show, these problems have often appeared as food issues. Urs K. Wagner (ETI Environmental Technology Ltd.), whom the secretariat of the Stockholm Convention presented as an international expert on PCBs, pointed in 2010 to the continuing contamination problems in Europe:

> In Switzerland, the consumption of certain fishes from specific rivers was forbidden in the spring of 2010 due to PCB concentrations far above the allowed maximum levels in Europe. Recently, it has been reported that 90 percent of German sheep livers have concentrations of PCBs above accepted levels. High PCB concentrations originating from a transformer treatment plant have recently made vegetables inedible in a big German city.[7]

In France, for similar reasons of regulatory threshold values being exceeded, authorities banned fishing or consumption of river fish in many areas of the country between 2005 and 2013 (see below for more details), and in 2011, farmers located in the vicinity of a PCB disposal facility saw their contaminated herds slaughtered.[8]

Looking at how food risk issues have impacted the political trajectory of the PCB problem offers three analytical advantages. First, it sheds critical light on the ways in which PCB problems have been put on political agendas at different times and at different scales. Second, it helps to clarify how food—and more specifically actors who have addressed food risk issues—have contributed to making environmental and health-related hazards (more) visible. And third, because food risks have provided the language for managing the environmental and health risks the persistence of PCBs creates, looking at "PCBs on the table" confirms the analytic perspective offered by Soraya Boudia and Nathalie Jas on the various "modes of government" of dangerous substances and their deleterious effects deployed since 1945.

This chapter makes use of different disciplines and approaches, including science and technology studies, microhistory, environmental history, history of health, and the sociology of public problems. Thus, I speak of a "problem" only when actors have defined a situation as problematic, and I speak of a "public problem" only when actors have succeeded in giving a certain publicity to the problematization they propose.[9] From this perspective, it is also important to explore how the actors involved in a given situation understand the behavior and effects of the contaminants, and how this plays into the ways in which they understand the problematic character of the situation. In other words, I am interested in the events, actors, actions, arguments, and social dynamics that result in global and national definitions of PCB problems, but also in specific, local definitions of PCB problems.[10]

Such perspective immediately brings in the question of materiality.[11] Looking at the way actors consider the materiality of the pollution not only is important to understanding how they define the problems, but also appears crucial for the social scientist who tries to understand the peculiar recalcitrance of the PCB problem. Indeed, PCBs have been used because of their high physical and chemical "stability." They have also been described since the late 1960s as "persistent" pollutants, in the sense of *biochemical* persistence. As the science advisor and specialist on pollutants Mitchell D. Erickson has written: "[PCBs] are highly chemically stable and resist microbial, photochemical, chemical, and thermal degradation. They are physically stable with very low vapor pressures and water solubility. Thus, PCBs do not readily degrade in the environment and are lipophilic. As a result, they persist and tend to bioaccumulate."[12] This knowledge about the material tendency of PCBs to persist and bioaccumulate has been scientific consensus for several decades. Their ability to persist in the environment and in the tissues of living organisms played a central role in the consensus that led states and international and supranational organizations to promulgate the regulatory measures mentioned above.[13] Biochemical persistence is, of course, a factor contributing to recalcitrance. However, as the second and third parts of this chapter will show, biochemical persistence is only one element among others explaining the *political* recalcitrance of the PCB problem.

Based on existing literature as well as archival documents, the first part of this chapter describes how the status of PCBs has changed in the 1960s from miracle product to global environmental health problem. Using the same kind of sources, the second part specifies how our knowledge about the effects of PCBs, and the regulations concerning them, evolved over the next two decades. In the third part, drawing on scientists', associations', and political authorities' archives and interviews I conducted, the chapter explores two affairs that occurred in France between the mid-1980s and today, focusing on the role that food risks played in the re-emergence of the PCB problems.

From Miracle Product to Global Environmental Health Problem

The Industrial Success of PCBs

"Polychlorinated biphenyls" or "PCBs" are a family of synthetic chemicals theoretically comprised of 209 molecules. What are commonly called "PCBs" are mixtures of various compounds of the PCB family, which look like fluids of a more or less oily or resinous appearance. These mixtures are also known by various trade names, including Arochlor, which is probably the most known at the international level.[14]

Industrial production of PCBs began in 1929 in Anniston (Alabama, USA).[15] These molecules were mass-produced in the most industrialized countries primarily between the 1930s and 1980s, and in a particularly massive way after the mid-1950s. As summarized in a recent report of the International Agency for Research on Cancer (IARC): "Production peaked in the 1960s and 1970s, and had ceased in most countries by the end of the 1970s or early 1980s."[16] According to data collected by the scientific community and various international organizations, the countries and companies that produced them most (far ahead of the others) were the United States, Germany, the USSR, and France, by, respectively, the firms Monsanto, Bayer, Orgsteklo, and Prodelec (see table 6.1). "Estimates of the total cumulative worldwide production of PCBs indicate that 1 to 1.5 million tonnes (or more) of commercial PCB products were manufactured," as also mentioned in the IARC report quoted above.[17]

Although only a dozen countries have produced PCBs, these substances have been used virtually all over the world. Moreover, since the 1930s, PCBs have been present in innumerable places through the multiplicity of their usages—and correlatively the multiplicity of spaces, objects, and products into which they have been put. Indeed, their high thermal and chemical stability,

Table 6.1. Volume and duration of PCB production in countries with known production (by production volume). Adapted from IARC, *Polychlorinated Biphenyls* (2016), 72.

Producer	Country	Duration Start	Stop	Volume (metric tons)	Reference
Monsanto	USA	1929*	1977	641,246	de Voogt & Brinkman (1989); Spears (2014)
Bayer AG	Germany, western	1930	1983	159,062	de Voogt & Brinkman (1989)
Orgsteklo	Russian Federation	1939	1990	141,800	AMAP (2000)
Prodelec	France	?**	1984	134,654	de Voogt & Brinkman (1989)
Monsanto	United Kingdom	1954	1977	66,542	de Voogt & Brinkman (1989)
Kanegafuchi	Japan	1954	1972	56,326	Tatsukawa (1976)
Orgsintez	Russian Federation	1972	1993	32,000	AMAP (2000)
Caffaro	Italy	1958	1983	31,092	de Voogt & Brinkman (1989)
2.8 Vinalon and the Sunchon Vinalon Complex	Democratic Republic of Korea	1960[a]	2012[b]	30,000[c]	NIP Korea DPR (2008)
SA Cros	Spain	1955	1984	29,012	de Voogt & Brinkman (1989)
Chemko	Former Czechoslovakia	1959	1984	21,482	Schlosserová (1994)
Xi'an	China	1965	1980	10,000	Jiang et al. (1997); NIP China (2007)
Mitsubishi	Japan	1969	1972	2,461	Tatsukawa (1976)
Electrochemical Co.	Poland	1966	1970	1,000	Sułkowski et al. (2003)
Zaklady Azotowe Tarnow-Moscice	Poland	1974	1977	679	Sułkowski et al. (2003)
Geneva Industries	USA	1972	1974	454	EPA (2008b)
Total		1929	2012	1,357,810	

* The Swann Chemical Company plant in Anniston, Alabama, began manufacturing PCBs in 1929; it was acquired by Monsanto Chemical Company in 1935.
** During the 1930s or during the 1940s; sources disagree (see Bletchly, *Report*, Annex 1, p. 1; Fournié and Peyrichou, "L'emploi," 14; Meunier, *Rapport*, 15–16).
a. During the 1960s.
b. "The Ministry of Chemical Industry will, by 2012, take measures to dismantle the PCBs production process and establish a new process of producing an alternative."
c. Estimated from Republic of Korea 2008, National Implementation Plan for the Stockholm Convention on Persistent Organic Pollutants.

their nonflammability, and the fact that they constitute a very good electrical insulator led to their use in many technoindustrial applications. PCBs were first employed as dielectric fluids in capacitors and transformers, and this, according to many sources, was their predominant use. They have also been used in hydraulic systems, cooling systems, and heat transfer systems (especially in the food industry), and as a major component in various sealants and coatings (on account of its plasticizing and flame-retardant properties). Finally, they have also been included in inks, paints, adhesives, plasticizers, and rubber products; in cutting and lubricating oils and as immersion oil for microscopes; in fluorescent lamp ballasts; as pesticide extenders; for wire and cable insulation; and for the "microencapsulation" of dyes used in carbonless copy papers.[18]

As early as the 1930s, in the United States, medical surveys and industrial hygiene research were conducted to study the health problems faced by workers regularly exposed to PCBs. Their results revealed the toxicity of PCBs in animals and humans. However, these findings did not achieve public visibility. They were discussed in a few private meetings and remained confined to a few documents, often classified as confidential in companies that had financed the studies.[19]

Foodstuffs and the Making of a Public Environmental Health Problem

During the second half of the 1960s, the status of PCBs changed from miracle product exploitable in innumerable technical applications to environmental and health nightmare; food risks were central to this shift.

In 1966, Sören Jensen, a Danish chemist conducting ecotoxicology research in the Stockholm archipelago, first identified PCBs among organochlorine contaminants found in the bodies of wild animals.[20] Expanding his investigations, he found PCBs not only in pike, eagles, and seals, but also in his own hair and in that of his wife and his five-month-old daughter. In a text he published that year, he suspected food to be a relatively important route of contamination, and given the age of his child, he assumed that PCBs could also be transmitted through breast milk.[21] Subsequent scientific work confirmed these hypotheses.

Two years later, in 1968, in Japan, approximately 1,600 people were intoxicated when they consumed rice oil contaminated with PCBs used as a heat transfer agent during oil processing. Five deaths were immediately associated with this contamination, along with severe skin problems known as "chloracne," sometimes going so far as to cause disfigurement. "Follow-up studies in the 1970s revealed birth defects in babies born to mothers who had been exposed to contaminated rice oil," writes historian Ellen G. Spears.[22]

Sören Jensen's discoveries and this mass intoxication, commonly referred to as the "Yusho incident," came at a time when questions relating to the harmful effects of chemicals were capturing the attention of the scientific community, the public, politicians, and authorities. The atmospheric fallout of nuclear tests, particularly in the form of strontium-90, the contamination by pesticides of fruits such as cranberries, and the devastating birth defects caused by prenatal use of thalidomide all contributed to public debates on the toxicity of certain chemicals, including the invisible dispersion of toxicants in the environment and the seeming inability of authorities to protect public health.[23] Jensen's discoveries and the Yusho incident came just a few years after Rachel Carson's best-selling *Silent Spring* sounded an alarm concerning the health and environmental effects of certain chemicals dispersed in the environment, at the forefront of which was DDT (dichloro-diphenyl-trichloroethane), a substance which, like PCBs, belongs to the family of chlorinated hydrocarbons. Like radioactivity from fallout ending up in milk, Jensen's discovery showed that PCBs could end up in both human and animal bodies.[24] More specifically, by making highly visible the toxicity of PCBs for humans and the role food can play, the Yusho incident contributed to turning the problem of global environmental pollution by PCBs into an environmental health problem.

Knowledge Production and the Transnational Management of the PCB Problem

Scientific Knowledge about the Critical Role of Dietary Exposure and PCBs' Toxicity

After 1966, scientific studies on PCB pollution and its (eco)toxicological effects multiplied. They continued to show these substances as an omnipresent and chronic threat. By the late 1960s, PCBs had been described as "bioaccumulative" for their ability to accumulate in the tissues of living organisms and particularly in fat.[25] During the 1970s, investigations that found PCBs in the Arctic—that is to say, in a place particularly far-removed from the areas of production and use—and other similar studies strengthened the thesis of PCB's ubiquity in the environment. In the same decade, other scientific works began to document the toxic effects of their bioaccumulation.[26] This knowledge became even more important as scientists came to consider food to be the most important exposure pathway for the general population due to the conjunction of two properties of the PCB molecules: persistence (or virtual nondegradability) and liposolubility.[27] Two phenomena result from the conjunction of these properties. First, PCBs accumulate in fats *throughout the life* of an organism (this is a long-term corollary of the notion of "bioaccumulation"). Second, the higher an organism is located in a food chain, the more it accumulates PCBs because it collects the PCB burden of every other organism it eats. Scientists call this phenomenon "bioamplification" or "biomagnification."[28] As humans are one of the highest organisms in food chains, they are among those who are the most exposed to PCBs.[29]

Over time, the scientific community has attributed to PCBs many different toxic effects in mammals, including in humans. These effects range from carcinogenicity, immunotoxicity, and neurotoxicity to reproductive, developmental, and neurobehavioral effects.[30] Since the 1970s, PCBs have also climbed up the World Health Organization's classification of carcinogenic substances. In 1978, the International Agency for Research on Cancer stated that PCBs "should be regarded as if they were carcinogenic to humans."[31] When the classification was defined in 1979, PCBs were categorized as "possibly carcinogenic to humans" (Group 2B), after which they were included in Group 2A ("probably carcinogenic to humans") in 1987, and finally one congener (PCB-126, which belongs to the congeners known as "dioxin-like") was listed in Group 1 ("carcinogenic to humans") in 2012.[32] In addition, because of their ability to influence hormonal mechanisms, PCBs are also to be found among the substances that formed the basis for the endocrine disruptors hypothesis developed in the early 1990s.[33]

First Signs of Political Recalcitrance

In the 1970s, many countries started to regulate the use of PCBs, or even to ban their production. The objects and schedules of these prohibitions were different, however, depending on the country. While Japan banned production and new uses of PCBs in 1972, many countries, including France, chose to restrict applications by forbidding their use in so-called "open systems,"—that is to say, in applications that were considered to allow direct diffusion of PCBs into the environment, such as paints, caulk, glues, etc.[34] Chronologically, several waves of regulation can be distinguished, which more or less correspond to transnational steps in the mounting restriction of PCB use.[35]

Finland, Sweden, Norway, Switzerland, and Germany banned PCB use in "open systems" in 1971 and 1972, and a 1973 OECD decision required all member states to enact similar restriction.[36] This decision was made "considering the use of [PCBs] should be controlled by international action in order to minimize their escape into the environment pending the realization of the ultimate objective of eliminating entirely their escape into the environment."[37] Among other things, the OECD guidelines also required the control of production and import and export flows, and demanded that the articles, products, and devices containing PCBs be labeled so as to indicate the presence of these substances. In short, these measures were supposed to give member states the capacity to control their PCB releases into the environment. The first PCB regulatory measure promulgated in France, in 1975,[38] enforced OECD injunction, as did the first European regulation in 1976.[39]

A second wave of regulation was initiated in the mid-1980s. In 1985, a European directive prohibited the selling of "closed systems" containing PCBs (such as transformers or capacitors). It stated that "despite the restriction on the use of PCBs and PCTs introduced by Directive 76/769/EEC . . . , as last amended by Directive 83/478/EEC . . . , there is generally no indication that pollution of the environment by PCBs and PCTs has lessened significantly."[40] European norms were transposed into French law in 1986 and 1987. That year, the OECD published a "decision-recommendation" that formulated the same argument and the same requirements as the European directive.[41] The supranational and international organizations who pushed for a second regulatory step in the mid-1980s thus recognized the failure of the first wave of transnational measures taken in the 1970s.

Moreover, various affairs and controversies had occurred between these two waves, proving that the first regulatory step—focused on production, use, and disposal—was not a sufficient response to deal with the diversity of PCB problems and hazards. In France, in the mid-1980s, associations, scientists, and authorities addressed at least four types of PCB problems: (1) safety and public health hazards related to explosions and fires of PCB-containing transformers (incidents resulting in the release not only of PCBs but also of other toxic substances, including dioxins and furans); (2) PCBs as an environmental health hazard not only for the general population but especially for babies (knowledge about PCB body burden being partly transferred to the child via breast milk had become public, and raised questions about the balance between the benefits and risks of breastfeeding); (3) a problem of toxic waste management; (4) an abnormal local pollution of the Rhône River.[42]

These examples not only show that the first step of regulation was not enough to get the PCB problem under control, but also underline the extent to which food issues contributed to making the PCB pollution more visible and of greater concern. First, the issue of PCBs in breast milk transformed the average level of contamination in adults and the practice of breastfeeding into potential health hazards for babies.[43] Second, in certain places, the establishment of plants for the disposal of PCB-contaminated wastes caused many actors to fear the local contamination of land, water, and agricultural products. Third, the "abnormal" local pollution on the Rhône River mentioned here had been revealed through the discovery of high PCB contamination levels in fishes. Focusing on this last problem, which first arose in the mid-1980s and then resurfaced in the mid-2000s, the rest of the chapter discusses more precisely the role that food risk controversies played at two different moments in putting the problem of river pollution by PCBs on the local, regional, and national political agendas.

Food Risks, the French Rhône River, and the Multiple Emergences of PCB Problems

In 1985, France had not yet regulated the presence of PCBs in food.[44] While in other countries such as the United States, PCB limit values had been promulgated for various food products, in France, such regulatory thresholds did not yet exist.[45] The first decree to remedy this situation in France was adopted in 1988, providing an administrative response to one among the many questions the affair described below brought to light.

Episode 1 (1985–90)

In 1985, ecotoxicologists participating in a multidisciplinary research program on the ecology of the Rhône River found particularly high concentrations of PCBs in the fish of one of the areas under study, upstream of the city of Lyon and downstream of a PCB disposal facility.[46] Even if at that time there was no threshold in France beyond which fish were considered unfit for human consumption, the ecotoxicologists considered these concentrations to be abnormal for several reasons. First, they were ten times higher than those reported in the scientific literature for fish caught in Lake Geneva and in American rivers. Second, they were higher than regulatory thresholds already existing in other countries, namely in Switzerland and in the United States.

Along with other scientists with whom they were collaborating, the group of ecotoxicologists alerted national authorities in charge of environmental issues (Ministère de l'Environment) and the local/regional administration in charge of health (Direction Départementale des Affaires Sanitaires et Sociales de la Préfecture du Rhône). In their report, the researchers explained that these levels of PCBs made them fear "problems for public health and risks for fish populations."[47] They feared not only that the maintenance of fish populations and the balance of the entire local ecosystem would be at risk, but also that people consuming fish from this area would be particularly exposed to health hazards associated with PCBs.

The scientists were particularly concerned about one specific community they had heard about. A professional fisherman who worked in this area, whom they knew well because he sometimes collaborated in their research, had informed them that among his clients were many people of Asian origin, whose consumption of fish from the contaminated area was particularly high. The researchers called for further studies to be launched in order to better diagnose the pollution and better analyze the risks faced by fish populations and fish consumers. They also proposed a way to investigate the considerable difference between the levels of PCB measured in the zone deemed to be problematic and those of other nearby areas they had studied. They measured PCB concentrations in freshwater mollusks upstream and downstream of the effluent discharge point of an industrial zone where a PCB disposal facility was located. These analyses revealed much higher rates downstream than upstream, leading the ecotoxicologists to formulate (in this same report) the hypothesis that the PCB disposal plant might be responsible for the contamination found in fish. At the same time, a regional federation of environmental associations and the above-mentioned professional fisherman (who had stopped selling fish immediately after being informed by the scientists of the contamination) asked the prefecture to take charge of the problem, and brought the case to court.

The prefecture organized several meetings, bringing together some of the researchers who had sounded the alert and several of its services. They finally set up a technoscientific and administrative plan to monitor the river's PCB contamination levels in the area where the problem had arisen. Regular analyses began in 1988 and continued until 1999. In response to the researchers' concerns about a possible health risk for the fisherman's customers, the local authorities conducted a study that did not confirm the hypothesis that the amount of contaminated fish eaten by these persons could expose them to health risks. However, these conclusions were highly contested: the researchers, as well as the professional fisherman, strongly criticized the conceptual framing and the methods used in the study.

In February 1988, one and a half years after the researchers had sent their report to the authorities, the French Ministry for Agriculture and Food promulgated a decree fixing a threshold above which fish were considered unfit for human consumption (2 milligrams per kilogram of fresh weight). The contamination level that was initially considered abnormal by the ecotoxicologists thus became officially higher than the regulatory threshold. Fishing was prohibited in the area concerned and the professional fisherman who used to work in this area moved and found work elsewhere, without assistance from the authorities.

The case following the complaints made by the regional federation of environmentalist associations and the professional fisherman was dismissed an additional eighteen months later. "The order notifying dismissal of the case is based on the fact that although the pollution certainly exists, it was not possible to determine who was responsible, and the waste-disposal plant ... in question has since improved its facilities," reported a newspaper.[48] However, the problem of Rhône River pollution and fish contamination by PCBs got back on the agenda of the local authorities some fifteen years later, and then became a problem that concerned all the river basins of the national territory.

Episode 2 (2005 to the Present Day)

Between 1990 and 2005, the regulatory and administrative context changed significantly. Condemned in 2002 by the European Court for failing to fulfill its obligations in regard to controlling PCBs, France took what can be considered a third step toward the elimination of objects containing these substances.[49] In 2003, the government published a "national plan" defining a schedule for the decontamination and elimination of equipment containing PCBs.[50] At the same time, a new European directive concerning the presence of PCBs in food was in preparation. Its aim was to "[set] maximum levels for certain contaminants in foodstuffs," and the "contaminants" concerned were dioxins and certain types of PCBs ("dioxin-like PCBs").[51] The directive was promulgated in February 2006—that is, after the events described below started.

By the time of this second affair, France also had for some years a new administrative body specializing in food risk issues. At the end of the 1990s, following several "health crises," France, like other European countries, set up several health security agencies, including a French Food Safety Agency created in 1999. In 2005, this public body was working under the authority of three ministries (agriculture, health, and consumer affairs), with several missions in the field of animal health and food hygiene (risk assessment, scientific and technical support to administrations, and research).[52]

The PCB pollution problem reappeared in the Rhône River in 2005 when high levels of PCBs were measured in fish caught in an area very close to the one where the ecotoxicologists had found abnormal contamination in 1985.[53] The French Food Safety Agency, to whom the local authority transmitted the analysis results, recommended further investigation on the pollution extent and pointed out that the consumption of fish contaminated at such levels could present a health risk. The prefecture prohibited the consumption of fish from the concerned area and launched the investigation suggested by the agency. Research was carried out in that area and gradually moved away toward others. For two years, more and more prohibitions on fishing and eating fish were enacted in new areas along the Rhône River as the results of analyses became available.

Several elected representatives and various environmental associations mobilized to urge the state to accelerate investigations into the scope of the contamination and to evaluate and manage the health risks to which the population was exposed with regard to fish consumption. With these mobilizations reaching national proportions toward the end of 2007, a three-year "action program" was set up at a regional level by the prefecture in charge of the French Rhône River basin, from the border with Switzerland to the Mediterranean Sea. A few months later, the ministries in charge of fisheries and agriculture, environment, and health launched a similar national "action program."

From 2005 to 2013, approximately 140 areas throughout France were the object of local regulatory measures prohibiting fishing and fish consumption (sometimes specific species only, sometimes all species). The first of these measures, in September 2005, was introduced in accordance with the precautionary principle and with the opinion of the French Food Safety Agency. Then, from February 2006, the fishing and fish consumption prohibitions were promulgated in accordance with the European directive mentioned above, which set thresholds for dioxins and "dioxin-like PCBs" in foodstuffs, including fish.[54] Investigations based on the analysis of sediment cores also confirmed that the PCB incineration plant located on the Rhône riverside (or at least the industrial zone where it stands), already suspected circa 1985, had played an important role in the pollution of this area.

These events also compelled the French authorities to introduce a new way of managing the sanitary risks. In 2008, two national health agencies (respectively in charge of food safety and public health monitoring) began a PCB-contamination study among freshwater fish consumers. They took blood samples from more than six hundred fishermen or members of their household. The results showed that the PCB contamination of some fish consumers exceeded the critical contamination values defined by the WHO, even if their consumption of highly bioaccumulative fish was below the general recommendations for fish consumption. Finally, three years after the beginning of the study, the agency in charge of food safety issued recommendations (more specifically maximum frequencies) to "enable consumers to eat strong PCB bio-accumulator fish without risks in the long term," to quote the agency.

Thus, in these two episodes of a (re)emerging contamination problem in the Rhône River, the authorities concluded that sanitary requirements were not being met: the PCB levels found in fish in 1985 and in 2005 exceeded the regulatory threshold values set by the French government in 1988 and by Europe in 2006. From a more general perspective, the two cases revealed—at two different times—that the technical and political devices designed by Europe and the French government to bring PCB pollution levels under control had not worked.

Conclusion: Food Risks, the Government of PCBs, and Political Recalcitrance

Food risks have played an important role in the political trajectory of the PCB problem: they contributed to spotlighting a global environmental health problem at the end of the 1960s and then brought it back to the authorities' attention on many occasions over the years. The 1968 mass intoxication in Japan alerted the international community to certain health risks associated with exposure to these substances. In France, first in 1985 and then again in 2005, in an area of the Rhône River slightly upstream of Lyon, PCB levels measured in fish and considered abnormal were a warning signal that led scientists and the authorities to work on what they agreed to define as a problem of contamination of the river environment and its biota. Furthermore, after the 2005 alert in France, the gradual expansion of the analysis campaigns, initially in the Rhône Basin and then at national level, revealed the extent (nationally at least) of freshwater fish contamination in multiple areas that exceeded European regulatory thresholds. In all three incidents, food security concerns highlighted the situation as a problem in the eyes of many actors. They also played an important role in making the problem legitimate in the eyes of the authorities, to the extent that they have undertaken various actions to manage it.

Faced with more and more difficulties—or, in other words, faced with the recalcitrance of the PCB problem—the authorities over time reinforced existing actions and introduced new ways of dealing with the PCB pollution. As illustrated above with the mention of the first two steps of transnational regulation and a third one in France, many countries phased out PCBs uses in several stages. In fact, others have followed. In France, evolution of the regulation that has restricted uses and has implemented the phasing out of PCB-containing objects can be schematically broken down into four steps: (1) the use of PCBs in "open systems" was prohibited in 1975; (2) selling new equipment containing PCBs was banned in 1987, but the use of devices already in service remained authorized; (3) a program to decontaminate and dispose of PCB-containing devices was introduced in 2003, setting deadlines to end the use of most equipment by 2010; and (4) new deadlines were promulgated in 2013 to oversee the disposal of equipment containing quantities of PCB lower than those that had been the subject of the 2003 phase-out program. Moreover, regulations on the use and disposal of PCBs were not the only normative measures put in place by the public authorities to try to manage the PCB pollution problem. Over time, environmental and health requirements led to the definition of regulatory thresholds, such as levels of PCB not to be exceeded in industrial facility effluents, or maximal concentrations above which a foodstuff is considered unfit for consumption.[55] In this context, foodstuffs have played a central role not only in the (re)emergence and (re)definition of the issues, but also as a target of risk management. Both emergences of the pollution problem on the Rhône River via the abnormal level of fish contamination, and the subsequent ban on catching and eating fish, embody these two dimensions—food as an indicator and as a target for risk management.

The changing ways in which public authorities have addressed the PCB pollution problem illustrates perfectly a dynamic that Soraya Boudia and Nathalie Jas have discussed in their work about expertise on and regulation of toxicants.[56] As they suggest, different "modes of government" have been developed over time to deal with the problems posed by toxicants, but these various modes have been implemented successively (and now coexist) without fully resolving issues. Eventually, as problems have appeared increasingly difficult to manage—and sometimes impossible to solve—tools and ways of managing hazards have been developed not only to try to control the pollution problem, but to live *with* the contamination.

PCBs have been the subject of a mode of action that remains rare in the regulation of toxic substances: the implementation of phasing out, which has involved the progressive ban of certain uses and the planning on an international scale of the disposal of products and appliances containing the substance. But the management of the problems posed by this family of chemicals has also been carried out through the other "modes of government" discussed by Boudia and Jas: management by control, by risk, and by adaptation.[57] As regulations organizing the phase-out of PCB-containing objects were moving forward, norms were introduced, defining, for example, the levels of PCB content at which objects had to be processed or eliminated, or fixing authorized PCB concentration levels in emissions from waste-disposal facilities. All of these norms aimed to take control of the pollution problem (by controlling PCB releases into the environment). Other regulatory measures promulgated at a later date aimed to regulate the management of the "quality" of environmental media and foodstuffs, and these were the embodiment of a government by risk. Threshold values were defined on the basis of tools for assessing and managing environmental and health risks; when contamination levels exceeding these thresholds were detected, this required action by administrations in charge of managing the dangers that contaminated media or foodstuffs represent. Finally came government by adaptation, which does not consist of regulatory measures, but of sanitary authorities publishing recommendations concerning foodstuff consumption frequencies, as in other long-term polluted areas such as Chernobyl or the French Indies (particularly polluted by chlordecone, an insecticide, miticide, and fungicide). Authorities have issued information that is supposed to "enable" people to deal with risk-taking at an individual level.

Thus, the political trajectory of PCB problems is part of a broader history in which the different "modes of government" implemented to deal with environmental (health) problems have neither fully resolved them nor prevented new problems from arising. Nevertheless, there are also specificities that help account for the particular recalcitrance of the PCB problem. As mentioned above, certain physical and chemical properties of PCBs and the resulting environmental behavior provide some explanatory answers—and this is knowledge about PCBs that has been the subject of a very broad consensus within the scientific community for decades. These molecules are persistent and semi-volatile. Consequently, PCBs in the environment circulate for decades in air, soil, living tissues, etc. Moreover, many places known to be particularly polluted have not been cleaned up, and thus constitute particularly important PCB reservoirs.[58] More broadly, global contamination has been around for decades. These "background levels" also constitute a PCB reservoir, which is far more diffuse but still omnipresent and more or less circulating from one environmental medium to another.

Furthermore, by detailing *how* PCB problems have re-emerged and analyzing the political responses to recurring affairs, this chapter shows that *political recalcitrance* is not a mere translation of "biochemical persistence." As
explained, the notions of "biochemical persistence," "bioaccumulation," and
"biomagnification" are important to take into account in order to understand
why food issues played a critical role in making the PCB pollution (period-
ically) visible. Nevertheless, this chapter highlights—in three respects—that
it is also crucial to consider sociopolitical context, structures, and dynamics
to understand why the PCB problem has kept reemerging over the last five
decades through food issues. First, according to scientists, international organizations, and the public authorities involved, the first international step taken
to regulate PCB use proved to be ineffective with respect to its stated objective,
which was to reduce the level of environmental contamination. In addition,
the ban of many different uses of PCBs as well as the phase-out of products
and equipment that contain them have been very slowly implemented in many
countries. The legal disposal of objects representing potential sources of PCB
releases into the environment has lasted for more than forty years and still goes
on.[59] Second, the elements presented above concerning the two cases in France
also show that the recalcitrance of the PCB problem is linked to the evolution
of health risk assessment tools and food safety requirements. In these cases,
new tools and requirements have contributed to a broader relegitimization of
the contamination problem of the river environment and its biota in the eyes
of the authorities. Third, this chapter also shows the importance of the mobilization of actors external to the administrative and regulatory arena, who have
helped define problems and present them to other actors, whether in the late
1960s with Sören Jensen, or later on the researchers studying fish populations
in the Rhône River in the mid-1980s. In sum, both the importance and the recalcitrance of PCBs as an environmental (health) problem are entangled with
the detection and management of food risks.

Acknowledgments

Special thanks for helpful comments and suggestions go to Angela Creager,
Jean-Paul Gaudillière, Soraya de Chadarevian, and Xaq Frohlich. The French
Région Île-de-France supported part of the research for this chapter under the
PhD thesis project "Le problème PCB (polychlorobiphényles) des années 1960
à 2010: enquête socio-historique sur une pollution visible, massive et dura-
ble." This thesis was carried out at the CERMES3 (Centre de recherche, méde-
cine, sciences, santé, santé mentale, société) and the EHESS (École des Hautes
Études en Sciences Sociales).

Aurélien Féron is a postdoctoral researcher associated with the Laboratoire Interdisciplinaire Sciences Innovations Sociétés (LISIS) in France and a part-time invited lecturer at the Department of History and Philosophy of Sciences of the Université de Paris (France). He studied biochemistry and social sciences, with a specialization in history of sciences and science and technology studies. His scholarly work addresses production of knowledge, mobilizations, and transnational government in the field of environmental and health hazards in the twentieth and twenty-first centuries.

Notes

1. Koppe and Key, "PCBs"; Eckley and Selin, "All Talk"; Féron, "Persistance."
2. This notion of *political recalcitrance of a problem* could be discussed in relation to the notion of "wicked problems," which is widely used in political science. (See Peters, "What is So Wicked," for an example of a recent discussion of this notion.) This would, however, take us away from the purpose of this chapter. In short, I use the phrase *political recalcitrance* to insist on the distinction between this notion and that of "biochemical persistence" and because my primary purpose in this chapter is less to contribute to policy analysis than to present a sociohistorical investigation.
3. For texts summarizing the regulatory measures for PCBs, see, for example, IARC, *Polychlorinated Biphenyls* (2016); and Meunier, "Réglementation."
4. SSC, *Stockholm Convention.*
5. Wagner, "Inventories."
6. SSC, *Stockholm Convention.*
7. Wagner, "Inventories," 9.
8. INERIS, *Tierce expertise*, 9.
9. See for example Gusfield, *Contested Meanings*; Gusfield, *The Culture*; Gilbert and Henry, *Comment se contruisent les problèmes.*
10. By "*global* definitions of PCB problems" I refer here to the way in which international and supranational organizations define these problems.
11. For a recent essay that suggests considering "residues," including PCBs, as both material and political entities, see Boudia et al., "Residues."
12. Erickson, "Introduction," xxvii.
13. See Eckley and Selin, "All Talk," especially 88; Féron, Persistance, 135–65.
14. Spears, *Baptized in PCBs*; IARC, *Polychlorinated Biphenyls* (2016); Koppe and Keys, "PCBs"; Amiard et al., *PCB.*
15. Spears, *Baptized in PCBs.*
16. IARC, *Polychlorinated Biphenyls* (2016), 71.
17. Ibid., 70–74
18. Spears, *Baptized in PCBs*; IARC, *Polychlorinated Biphenyls* (2016); Koppe and Keys, "PCBs"; Amiard et al., *PCB.*
19. Spears, *Baptized in PCBs*, 66–74; Rosner and Markowitz, "Persistent Pollutants."
20. Spears, *Baptized in PCBs*, 133–35; Jensen, "The PCB Story"; "Report," 612.
21. "Report," 612.

22. Spears, *Baptized in PCBs*, 139–40; Kuratsune et al., "Epidemiologic Study on Yusho." For the broader history of public health problems in Japan due to industrial pollution, see Walker, *Toxic Archipelago*.

23. Spears, *Baptized in PCBs*, 131–33.

24. See de Chadarevian, "Radioactive Diet," this volume.

25. Koppe and Keys, "PCBs."

26. Ibid.

27. Amiard et al., *PCB*, 3, 463; IARC, *Polychlorinated Biphenyls* (2016), 90.

28. This phenomenon had been demonstrated first in radioactive waste; Creager, *Life Atomic*, 368–77.

29. Ibid., 424; Amiard et al., "Expositions."

30. Koppe and Keys, "PCBs"; Dargnat and Fisson, *Les PolyChloroBiphényles*; Amiard et al., *PCB*.

31. IARC, *Polychlorinated Biphenyls* (1978), 84.

32. IARC, *Polychlorinated Biphenyls* (2016), 34 (table 1); Lauby-Secretan et al., "Carcinogenicity."

33. Krimsky, *Hormonal Chaos*; Langston, *Toxic Bodies*; Gaudillière and Jas, "Introduction."

34. Kim and Masunaga, "Behavior"; Meunier, "Réglementation," 89; Eckley and Selin, "All Talk," 86.

35. In the United States, very little time elapsed between the two *steps of restriction* distinguished here with regard to transnational regulation: with the Toxic Substances Control Act promulgated in 1976, "PCBs could only be manufactured, processed, distributed and used in a 'totally enclosed manner,'" and only eighteen months later "all manufacture, processing and distribution of PCBs was prohibited" (with exemptions on a case-by-case basis). Koppe and Keys, "PCBs," 67. See also McGurty, "Transforming Environmentalism," 26; and Spears, *Baptized in PCBs*, 163, 167.

36. Meunier, "Réglementation," 89; Eckley and Selin, "All Talk," 86.

37. OECD, *Decision of the Council.*

38. MIR, "Arrêté du 8 juillet 1975. Inscriptions"; MIR, "Arrêté du 8 juillet 1975. Conditions."

39. CEU, "Council Directive 76/403/EEC"; CEU, "Council Directive 76/769/EEC."

40. CEU, "Council Directive 85/467/EEC."

41. OECD, *Decision–Recommendation.*

42. Féron, "Persistance."

43. See Langston, "Toxic Bodies" (preface), which relates the experience of an American woman confronted with these issues with even more concern because she grew up along the Fox River in Wisconsin, an environment particularly polluted by PCBs.

44. In 1985, in France as in many other countries, the use of PCBs in "open systems" had been banned for about ten years, and (as mentioned above) the European Community promulgated new regulatory measures regarding the use, sale, and disposal of PCBs, which were transposed into French law in 1986 and 1987. However, these regulatory measures did not address the issue of PCBs in food.

45. Meunier, "Réglementation," 108–9.

46. This section (*Episode 1*) is based on archives of two scientists (Gilles Monod, Henri Persat), a nonprofit organization (FRAPNA), and local authorities (Archives Départementales de l'Ain, 426W52), as well as interviews conducted with various actors.

47. *Contamination du Rhône par les PCB. Problèmes pour l'Hygiène Publique et Risques*

pour les Populations Piscicoles, November 1986, Gilles Monod's archives and Henri Persat's archives.

48. "La FRAPNA Arrête son Action en Justice sur la Pollution du Rhône par les PCB," *Le Monde*, 9 April 1990, Gilles Monod's archives; translation from French to English by the author.
49. European Court, "Judgment."
50. MEDAD and ADEME, *Plan National*.
51. CEC, "Commission Regulation."
52. Besançon, "L'institutionnalisation."
53. This section (*Episode 2*) is based on various documents (reports, press release, regulatory measures, etc.) collected on official websites of authorities (Agence de Bassin Rhône-Mediterranée, www.pollutions.eaufrance.fr; Office National de l'eau et des Milieux Aquatiques, www.onema.fr; Agence Française de Sécurité Sanitaire des Aliments, www.anses.fr) and nonprofit organizations (Robin des Bois, robindesbois.org; France Nature Environnement, www.fne.asso.fr; WWF France and Association Santé Environnement France, www.stopauxpcb.com), as well as interviews conducted with various actors.
54. CEC, "Commission Regulation."
55. Meunier, "Réglementation," 98–153.
56. Boudia and Jas, *Toxicants*; Boudia and Jas, *Powerless Science*; Boudia and Jas, *Gouverner*.
57. Ibid.
58. In France, for instance, knowing river sediments to be an important PCB reservoir contributing to the contamination of freshwater fish, the government funded research (carried out in partnership with private companies) on their depollution, but at the end no solution was considered to be appropriate.
59. In addition, many actors have subsequently emphasized the amount of equipment that has escaped the legal disposal pathway—that is to say, disposal in authorized facilities (through accidents with and vandalism of transformers, in service or stored; equipment abandoned "in the wild"; illegal discharges, etc.). Féron, "Persistance."

Bibliography

Archives

Archives Départementales de l'Ain, Bourg-en-Bresse, France
Fédération Rhône-Alpes de Protection de la Nature (FRAPNA), private archives, head office of the organization, Villeurbanne, France
Gilles Monod, private archives of Gilles Monod, ecotoxicologist, researcher at Institut National de la Recherche Agronomique, France
Henri Persat, private archives of Henri Persat, ecologist, researcher at University Claude Bernard—Lyon 1, France

Publications

Amiard, Jean-Claude, Thierry Meunier, and Marc Babut, eds. *PCB, Environnement et Santé*. Paris: Lavoisier, 2016.

Amiard, Jean-Claude, Jean-François Narbonne, and Paule Vasseur. "Expositions et Imprégnations Humaines aux PCB." In *PCB, Environnement et Santé*, edited by Jean-Claude Amiard, Thierry Meunier, and Marc Babut, 445–69. Paris: Lavoisier, 2016.

Besançon, Julien. "L'Institutionnalisation de l'Agence Française de Sécurité Sanitaire des Aliments comme Organisation-Frontière. Bureaucratisation de l'Expertise et Régulation des Risques Alimentaires." PhD diss., Institut d'Etudes Politiques de Paris, 2010.

Bletchly, J. D. *Report on a Study of Measures to Avoid Dispersion into the Environment of Polychlorinated Biphenyls (PCBs) and Polychlorinated Terphenyls (PCTs) from Existing Installations*. 1985.

Boudia, Soraya, Angela N. H. Creager, Scott Frickel, Emmanuel Henry, Nathalie Jas, Carsten Reinhardt, and Jody Roberts. "Residues: Rethinking Chemical Environments." *Engaging Science, Technology, and Society* 4 (2018): 165–78.

Boudia, Soraya, and Nathalie Jas. *Gouverner un Monde Toxique*. Versailles: Quae, 2019.

———, eds. *Powerless Science? Science and Politics in a Toxic World*. New York: Berghahn Books, 2014.

———, eds. *Toxicants, Health and Regulation since 1945*. London: Pickering & Chatto, 2013.

CEC (Commission of the European Communities). "Commission Regulation (EC) No 199/2006 of 3 February 2006 Amending Regulation (EC) No 466/2001 setting Maximum Levels for Certain Contaminants in Foodstuffs as Regards Dioxins and Dioxin-Like PCBs." *Official Journal of the European Union* L32 (4 February 2006): 34–38.

CEU (Council of the European Union). "Council Directive 76/403/EEC of 6 April 1976 on the Disposal of Polychorinated Biphenyls and Polychlorinated Terphenyls." *Official Journal of the European Communities* L108 (26 April 1976): 41–42.

———. "Council Directive 76/769/EEC of 27 July 1976 on the Approximation of the Laws, Regulations and Administrative Provisions of the Member States Relating to Restrictions on the Marketing and Use of Certain Dangerous Substances and Preparations." *Official Journal of the European Communities* L262 (27 September 1976): 201–3.

———. "Council Directive 85/467/EEC of 1 October 1985 Amending for the Sixth Time (PCBs/PCTs) Directive 76/769/EEC on the Approximation of the Laws, Regulations and Administrative Provisions of the Member States Relating to Restrictions on the Marketing and Use of Certain Dangerous Substances and Preparations." *Official Journal of the European Communities* L269 (11 October 1985): 56–58.

Creager, Angela N. H. *Life Atomic: A History of Radioisotopes in Science and Medicine*. Chicago: University of Chicago Press, 2013.

Dargnat, Cendrine, and Cédric Fisson. *Les PolyChloroBiphényles (PCB) dans le Bassin de la Seine et son Estuaire*. Rouen: GIP Seine Aval, 2010.

Eckley, Noelle, and Henrik Selin. "All Talk, Little Action: Precaution and European Chemicals Regulation." *Journal of European Public Policy* 11, no. 1 (2004): 78–105.

Erickson, Mitchell D. "Introduction: PCB Properties, Uses, Occurrence, and Regulatory History." In *PCBs: Recent Advances in Environmental Toxicology and Health Effects*, edited by Larry W. Robertson and Larry G. Hansen, xi–xxx. Lexington: University Press of Kentucky, 2001.

European Court. "Judgment of the Court (Fourth Chamber) of 6 June 2002. Commission of the European Communities v French Republic. Failure by a Member State to Fulfil its Obligations. Articles 4 and 11 of Directive 96/59/EC on the Disposal of Polychlorinated Biphenyls and Polychlorinated Terphenyls (PCB/PCT). Case C-177/01." *European Court Reports* ECLI:EU:C:2002:352 (2002): I-05137.

Féron, Aurélien. "Persistance Biochimique et Récalcitrance Politique. Enquête Socio-Historique sur les Résurgences Multiscalaires d'un Problème Environnemental et Sanitaire (1966–2013)." Ph.D. Diss., École des Hautes Études en Sciences Sociales, Paris, 2018.

Fournié, Robert and François Peyrichou, "L'emploi des PCB dans les Transformateurs Électriques présente-t-il des risques ?" *RGE. Revue Générale de l'Electricité*, no. 8 (September 1987): 13–22.

Gaudillière, Jean-Paul, and Nathalie Jas. "Introduction: la Santé Environnementale Au-Delà du Risque? Perturbateurs Endocriniens, Expertise et Régulation en France et en Amérique du Nord." *Sciences Sociales et Santé* 34, no. 3 (2016): 5–18.

Gilbert, Claude, and Emmanuel Henry, ed. *Comment se Contruisent les Problèmes de Santé Publique*. Paris: La Découverte, 2009.

Gusfield, Joseph R. *Contested Meanings: The Construction of Alcohol Problems*. Madison: University of Wisconsin Press, 1996.

———. *The Culture of Public Problems: Drinking-Driving and the Symbolic Order*. Chicago: University of Chicago Press, 1981.

IARC (International Agency for Research on Cancer), ed. *Polychlorinated Biphenyls and Polybrominated Biphenyls*. IARC Monographs on the Evaluation of the Carcinogenic Risk of Chemicals to Humans, 18. Lyon: IARC, 1978.

———, ed. *Polychlorinated Biphenyls and Polybrominated Biphenyls*. IARC Monographs on the Evaluation of Carcinogenic Risks to Humans, 107. Lyon: IARC, 2016.

INERIS (Institut National de l'Environnement Industriel et des Risques). *Tierce Expertise de l'Étude d'Interprétation de l'État des Milieux et Études Complémentaires. Site APROCHIM de Grez-en-Bouère*. Rapport d'étude INERIS-DRC-15-154613-09277A. Verneuil-en-Halatte: INERIS, 2015.

Jensen, Sören. "The PCB Story." *Ambio* 1, no. 4 (1972): 23–131.

Kim, Kyoung S., and Shigeki Masunaga, "Behavior and Source Characteristic of PCBS in Urban Ambient Air of Yokohama, Japan." *Environmental Pollution* 138, no. 2 (2005): 290–98.

Koppe, Janna G., and Jane Keys. "PCBs and the Precautionary Principle." In *Late Lessons from Early Warnings: The Precautionary Principle, 1896–2000*, edited by Poul Harremoës, David Gee, Malcolm MacGarvin, Andy Stirling, Jane Keys, Brian Wynne, and Sofia Guedes Vaz, 64–75. Copenhagen: European Environment Agency, 2001.

Krimsky, Sheldon. *Hormonal Chaos: The Scientific and Social Origins of the Environmental Endocrine Hypothesis*. Baltimore: Johns Hopkins University Press, 2000.

Kuratsune, Masanori, Takesumi Yoshimura, Junichi Matsuzaka, and Atsuko Yamaguchi. "Epidemiologic Study on Yusho, a Poisoning Caused by Ingestion of Rice Oil Contaminated with a Commercial Brand of Polychlorinated Biphenyls." *Environmental Health Perspectives* 1 (1972): 119–28.

Langston, Nancy. *Toxic Bodies: Hormone Disruptors and the Legacy of DES*. New Haven: Yale University Press, 2010.

Lauby-Secretan, Béatrice, Dana Loomis, Yann Grosse, Fatiha El Ghissassi, Véronique Bouvard, Lamia Benbrahim-Tallaa, Neela Guha, Robert Baan, Heidi Mattock, and Kurt

Straif. "Carcinogenicity of Polychlorinated Biphenyls and Polybrominated Biphenyls." *Lancet Oncology* 14, no. 4 (2013): 287–88.

McGurty, Eileen. *Transforming Environmentalism: Warren County, PCBs, and the Origins of Environmental Justice.* New Brunswick: Rutgers University Press, 2009.

MEDAD (Ministère de l'Ecologie du Développement et de l'Aménagement Durables) and ADEME (Agence de l'Environnement et de la Maitrise de l'Energie). *Plan National de Décontamination et d'Elimination des Appareils Contenant des PCB et des PCT.* 2003.

Meunier, Philippe, ed. *Rapport d'Information sur le Rhône et les PCB: Une Pollution au Long Cours.* Paris: Assemblée Nationale, 2008.

Meunier, Thierry. "Réglementation des PCB en France." In *PCB, Environnement et Santé,* edited by Jean–Claude Amiard, Thierry Meunier, and Marc Babut, 89–182. Paris: Lavoisier, 2016.

MIR (Ministère de l'Industrie et de la Recherche). "Arrêté du 8 juillet 1975. Inscriptions aux Tableaux des Substances Vénéneuses (section I)." *Journal Officiel de la République Française* (26 July 1975): 7600.

MIR (Ministère de l'Industrie et de la Recherche). "Arrêté du 8 juillet 1975. Conditions d'Emploi des Polychlorobiphényles." *Journal Officiel de la République Française* (26 July 1975): 7600.

OECD (Organisation for Economic Co-operation and Development). *Decision of the Council on Protection of the Environment by Control of Polychlorinated Biphenyls.* OECD/LEGAL/0108. Paris: OECD, 1973.

———. *Decision-Recommendation of the Council on Further Measures for the Protection of the Environment by Control of Polychlorinated Biphenyls.* OECD/LEGAL/0230. Paris: OECD, 1987.

Peters, Guy B. "What Is So Wicked about Wicked Problems? A Conceptual Analysis and a Research Program." *Policy and Society* 36, no. 3 (2017): 385–96.

"Report of a New Chemical Hazard." *New Scientist,* no. 32 (15 December 1966): 612.

Rosner, David, and Gerald Markowitz. "Persistent Pollutants: A Brief History of the Discovery of the Widespread Toxicity of Chlorinated Hydrocarbons." *Environmental Research* 120 (2013): 126–33.

Spears, Ellen G. *Baptized in PCBs: Race, Pollution, and Justice in an All-American Town.* Chapel Hill: University of North Carolina Press, 2014.

SSC (Secretariat of the Stockholm Convention), ed. *Stockholm Convention 10th Anniversary: Major Achievements in 10 Years.* Geneva: SSC, 2011.

Wagner, Urs K. "Inventories of PCBs: An Expert's Point of View." *PCBs Elimination Network. PEN magazine* 1 (2010): 9–11, 24–27, 38–41, 56–59.

Walker, Brett. *Toxic Archipelago: A History of Industrial Disease in Japan.* Seattle: University of Washington Press, 2010.

 PART II

Ordering Risks

 CHAPTER 7

Trace Amounts at Industrial Scale
Arsenicals and Medicated Feed in the Production of the "Western Diet"

Hannah Landecker

Arsenic-based drugs were widely used in meat production in twentieth-century America, and their story runs its course over approximately a hundred years, from the early twentieth century to our present moment. While the conjunction of "arsenic" and "food" in the same sentence will for many readers evoke accidental exposure or food fraud, this is an account about intention: systems of chemical dosing designed, due to their organization around food production, to get into the maximum number of bodies. It is the very intentional engineering and regulatory facilitation of the ubiquitous mass distribution of small amounts of potent chemicals into the bodies of food animals that is at the crux of this story.

Arsenical medications were used in feed animals at industrial scale between 1944 and 2015 in the United States with the full blessing of relevant federal agencies, and remain in use in many other countries around the world for the purposes of growth promotion, meat coloration, disease prevention, and parasite treatment. Here I am concerned with two main questions: how did arsenicals become a standard component in the feed of animals grown as human food? And, how could it have been regarded for so long as being without consequence for human health? In this chapter, I trace the history of the particular substances—arsenicals—but also address the significance of the rather obscure world of medicated feed revealed therein. Through arsenicals we may understand better the role of animal feed as a massive chemical redistribution mechanism, one specifically designed to travel the metabolic routes of ingestion, incorporation, and excretion that link microbes, plants, animals, and humans.

The remarkable efficacy of this chemical redistribution via animal feed means that this chapter is not about the risk of exposure to dietary or drinking water arsenic. That is already given to us by history and extensively studied in public health epidemiology.[1] Rather, it is to use the story of arsenicals to elucidate the twentieth-century origins of the historically particular biochemical ecologies, created through food systems, in which we are now immersed. The arsenic in the urine of the toddler fed rice snacks today is a historical indicator of the character of poultry feed in mid-century America, linked to the history of copper smelting and the concept of specific toxicity and the development of chemical pesticides.[2] Today, the main source of arsenic intake for most people is indeed dietary, while in those areas where arsenic in water is high, that source is more pertinent for them.[3] Not all of the arsenic in the diet (or the water) comes from anthropogenic agricultural sources, yet this does not diminish the importance of this contributor to our contemporary chemical landscape.[4]

Arsenical medications played an important role in establishing an infrastructure for moving growth-promoting chemicals through animal metabolism. Arsenicals were an important early element in the establishment during the 1940s of *medicated feed* as a ubiquitous and workaday element of animal husbandry.[5] Medicated feed (also sometimes called manufactured, formulated, commercial, scientific, or condimental feed) refers to small amounts of medicines or nutritional supplements mixed with a larger bulk of fodder, which itself is usually designed to provide rational management of the macronutrient and fiber intake of the animal according to its role as producer of eggs or milk or more animals, or as meat.[6] In the mid-twentieth century, farmers would buy these feeds ready-made to supplement grazing or farm-grown inputs. Today in large concentrated animal feeding operations (CAFOs), concentrates and fodder ingredients are purchased from suppliers and then mixed on site, often in proprietary combinations. The importance of animal feed in both the past and the present should not be underestimated; it remains the primary cost of raising animals for food. The economy of the ratio of food input to growth output, or *feed efficiency*, has been a central preoccupation for American animal husbandry for more than a century.

The story of arsenicals in agriculture has been largely overlooked, overshadowed in both the historical and the safety literature by antibiotics and hormones. Because the history of arsenicals in animal husbandry has been frequently mentioned but not examined in depth in the scholarly literature in the history of science, public health, agriculture, or food, the greater part of this article is dedicated to tracing out a thorough account of their origin in American chemistry and agriculture: how these medications came into use, why they seemed a good idea at the time, why they were approved for agricultural use in the 1940s by the FDA, and how they became firmly established as a ubiquitous part of animal diets for decades, and eventually came to be withdrawn from the American market between 2013 and 2015.

An important element of this history is analysis of the scientific knowledge originally generated around arsenicals concomitant with their introduction for growth promotion and meat pigmentation. The relatively few studies conducted by a handful of scientists working within research divisions of chemical and pharmaceutical companies in the animal nutrition market between 1945 and 1965 stood for more than five decades as the only evidence by which arsenicals were judged safe to use in animals destined for human consumption. The questions dominating this literature were predominantly those of efficacy in producing more rapid growth or better feed efficiency; concerns about toxicity were framed in terms of reduced growth or laying productivity, not questions of effects on consumers via meat or eggs or the broader environment.

I conclude by turning to the present, from the history of the biology of arsenic toward an understanding of the biology of this history: the specific biological and medical sequelae of this socially, economically and culturally shaped distribution of matter.[7] As the role of the so-called "Western diet" in causing metabolic disorders is taken up by contemporary science, the interacting legacies of the acceleration of meat production and the arsenical accelerants used to achieve it are coming into view. Shifting from research on toxicity at high levels to disease predisposition at low levels, dietary arsenic is now being considered as a contributing factor in diseases such as diabetes.[8] Thus the story of feed arsenic is also that of the risk it is thought to hold for human health. Its story takes us from threshold-controlled poison to belatedly banned substance, and out the other side to the emergence of a new kind of risk domain, that of chronic disease predisposition. In this domain, danger lies not in the toxicant in itself, but in the chronic temporality of exposure and the synergisms between arsenic and other dietary components and microorganisms that have also been configured by twentieth-century food systems.

How to "Feed Away Disease": Arsenicals in the Feed Bag

Arsenical medicines became feed additives in the 1940s. Exploration of arsenical compounds for the treatment of intestinal parasites of birds was driven by the enormous growth in the broiler industry and changes to the way chickens were hatched, raised, and grown in the 1920s and 1930s. Conditions were fundamentally altered by the discovery of vitamin D's role in the prevention of rickets in humans and "weak leg disease" in chickens. As William Boyd notes of the incorporation of chicken biology into American "circuits of capital," the introduction of cod liver oil in the early 1920s and then synthetic vitamin D after 1927 ushered in intensive confinement of birds from hatching to slaughter, replacing the seasonality of sunlight exposure.[9] Boyd writes that "continuous-flow production" unperturbed by the seasons was a "kind of biological time-space compression, creating a platform upon which intensification and breeding could proceed."[10] The number of commercial hatcheries in the United States expanded from 250 in 1918 to over ten thousand in 1927, and increasing rural electrification allowed extensive management of indoor environments for growing broilers. In short, in the 1920s and 1930s, chickens came inside, and in increasingly large numbers. By the 1940s, 1,400 million chicks a year were being hatched.[11]

With these changes came crippling outbreaks of infectious disease. Coccidiosis has been one of the most enduring problems dogging industrial poultry production over the twentieth century; it was (and is) a common affliction of turkeys as well.[12] An illness caused by several species of protozoa, it is marked by loss of appetite, intestinal bleeding, diarrhea, and death. Parasite oocytes (eggs) shed in feces quickly affected whole flocks. New flocks could be infected from the growing floor even after diseased ones were replaced.

Coccidiosis was not limited to chickens, but also caused extensive morbidity and mortality in young cattle, lambs, and animals raised for fur, such as mink. It therefore became a matter of urgent investigation in agricultural research settings in land-grant colleges, federal research institutions, and in pharmaceutical and chemical companies active in the animal nutrition market. "The frankly injurious species of coccidia are characterized by an uncanny ubiquity," wrote a representative of the Zoological Division of the Bureau of Animal Industry in 1949, noting that "producers of poultry, dairy and range calves, and feeder lambs appear everywhere to have suffered loss from coccidiosis."[13] He estimated the economic losses to lie, conservatively, in the range of $21 million dollars a year in the United States.

Between 1940 and 1950, the race to produce new remedies for this new scourge intensified. Pharmaceutical and fine chemical companies that had in the previous decades aggressively entered the animal nutrition market with vitamin and mineral concentrates looked to expand their portfolios. While we tend to think of vitamins, minerals, and amino acids as nutritional supplements, and drugs such as sulfonamides as medicines, such separation of nutrition and medicine would be misleading in terms of the continuity between these categories in the historical scene.[14] Vitamins and amino acids paved the way for chemotherapeutic drugs via the lucrative commercial and technical infrastructure established to get small concentrated quantities of metabolically active chemicals into the much larger volume of bulk fodder and then distribute it back out to individual animals en masse.

In addition, vitamins cured deficiency diseases and therefore were already understood as therapeutic. Trials to test feed components such as buttermilk concentrates or vinegar for the treatment of coccidiosis were conducted; a disconsolate list of failed therapies was published in 1934.[15] Then came drugs meant for pathogens that also acted on host metabolism, further blurring the line between medication and metabolic catalyst. Having profited enormously from the provisioning of the nutrients that drove the massive expansion in the number of animal bodies requiring supplementation, these companies then stood to benefit further from providing the medications to treat the diseases of intensification; "officers of pharmaceutical companies were enticed by prospects of selling drugs by the ton rather than by the ounce."[16]

According to industry observers, "the chemical that gave the tremendous impetus to feed medication was sulfaquinoxaline and the disease was and is coccidiosis. Other chemicals for the same and other disease conditions have followed in rapid succession."[17] The technical aspects of mixing concentrates with fodder, the economic and transport aspects of manufactured feed distribution, and the sociocultural work of establishing scientific feeding as the norm for farmers were already in place.[18] Therefore, the first companies to put modern chemotherapeutic drugs against coccidiosis on the market (displacing earlier more toxic remedies such as powdered inorganic sulfur) already sold vitamin concentrates and premixes in the animal nutrition market. Lederle Laboratories marketed sulfaquinidine and sulfathiazole for periodic administration in feed in 1946. Merck followed suit in 1948 with sulfaquinoxaline, marketed as SQ.[19] Yet SQ had its own drawbacks, and its success only spurred the concomitantly unfolding research with another class of synthetic antiparasitics, in this case the arsenicals.

Arsenic is a metalloid, having properties of both metals and solid nonmetals. As a transitional element, it can form alloys with other metals and form bonds with nonmetal elements. When bonded to carbon, it is an organo-arsenic, much less toxic to humans than inorganic arsenic (iAs), and thus the basis of the first chemotherapeutic agents used in medicine. Paul Ehrlich famously developed the original "magic bullet" drug—arsphenamine—as a treatment for syphilis. On the logic that some chemical dyes that had high affinity for fibers might have "selective toxicity" for pathogens and not for the patients' cells, Ehrlich set out to systematically test variations on an older arsenic-based medicine, atoxyl, in the search for one specific to the spirochete bacteria causing syphilis.[20] This systematic sequential tinkering with the various groupings and bonds in the molecule and therefore affinities and toxicities led to Salvarsan—"the arsenic that saves"—launched in 1910, followed by Neosalvarsan in 1914. While rather toxic and hard to administer, it was the first effective therapy for syphilis, and came into wide use in the United States with two million doses produced yearly by 1920.[21]

During World War I, an agent of chemical warfare called Lewisite was made by combining acetylene and arsenic trichloride, manufactured at a plant constructed for that purpose in Ohio. With the onset of World War II, efforts to find an antidote to this extremely lethal weapon expanded, sparking intensive research into the metabolism of arsenic.[22] By the late 1940s, research on arsenic was both chemical and biochemical: "For an understanding of the nature of the action of arsenic on pathological microorganisms such as *Treponema pallidum*, or its toxic effects in man [sic], it is clear that knowledge is needed not only of the nature of the biochemical defect which it produces in the life of the cell, but also of the chemical grouping or groupings with which it reacts."[23] At midcentury, more than 12,500 organoarsenicals had been tested for therapeutic uses, and "probably more work has been done to correlate the chemical structure and parasitical action of organic arsenicals than for any other group of chemicals."[24] World War II thus reanimated research into both the toxic and the therapeutic properties of organic and inorganic arsenical compounds in humans, agents that entered into cellular metabolism of both host and parasite.[25]

In animal medicine, the use of atoxyl from 1907 for treating spirochetosis in chickens did not result in widespread use in farming. Treatments of individual animals after they fell ill, with medications administered by injection, were expensive and often involved a veterinarian. Use of arsenicals diminished with the advent of sulfonamides in the 1930s, though "Fowler's Solution" (potassium arsenite) continued as a treatment for leukemia, and was also used to brighten show animals' coats and as a "pick-me-up" for horses and cows.[26] The return to arsenicals for animals was driven by the historical confluence of large flocks, expensive epidemics of parasitical diseases, low profit margins, fine chemical and pharmaceutical companies already deeply embedded in the business of making animal feed supplements, and intensive research into the metabolism and toxicity of arsenicals in the context of war. Last but not least, the growth of the manufactured feed industry greatly facilitated "mass medication," or treating whole herds or flocks all at once to "feed away disease," instead of individual injections or tonics.[27]

Treating an individual animal after the appearance of disease was quite different from treating the flock or herd before any disease was apparent. As historian Susan Jones has observed, "Larger-scale production units yielded more profit on a commodity with a tight profit margin: they also rendered animals who were worthless individually (such as poultry) valuable as a group."[28] The chemical properties of the new therapies also played a part, as toxicity was key to mode of application. If a therapeutic dose curative of infection also reduced weight gain or suppressed egg production, there was no sense in paying to use it. Thus a smaller, low-cost prophylactic dose that could prevent disease without disturbing growth became a driver of experimentation.

Researchers settled on food or drinking water as the most stable and labor-saving means for administering medication to large groups of animals all at once. Dr. Salsbury's Laboratories, a small company in Charles City, Iowa, dedicated to developing veterinary medications for use in chickens, began to screen large numbers of arsenicals in the late 1930s, with the first patent appearing in 1940. These substances were tested at various levels for their effect on growth of chicks and young birds and egg production of laying hens. Maintaining feed efficiency was central to any drug's success.

Salsbury's initiated marketing of Ren-O-Sal, drinking-water tablets in which the active ingredient was 3-nitro 4-hydroxyphenylarsonic acid, in 1944. But the research around Ren-O-Sal brought further surprises: the drug had positive effects on growth over and above disease prevention at subtherapeutic levels. A brief research report given in 1944 by Salsbury's scientists at the Symposium on Parasitology in Relation to the War reported that "when the concentration of 3-nitro 4-hydroxyphenylarsonic acid was 0.00256 percent in [drinking] water marked growth-stimulating properties were noted."[29] This abstract was followed by papers on arsenicals for control of coccidiosis in 1946 and confirmation of the growth promoter effect in 1949; the company also filed a patent on their formulation as both a coccidiostat and growth promoter in 1946.[30]

These "surprising" findings were taken up immediately by a broader community of animal nutrition researchers.[31] Writing in 1949, Frans Goble of the chemotherapy section of the Sterling-Winthrop Research Institute reflected that "notwithstanding the anticoccidial usefulness of certain sulfonamides," it was still desirable to search for "an effective chemoprophylactic agent which is sufficiently non-toxic to be fed throughout the 'starting' period."[32] Testing twelve phenylarsonic acids or their sodium salts, Goble suggested that the perfect balance between prophylaxis and toxicity could be found: "the innocuousness of the 0.0075 per cent level . . . suggests that safe and effective dosages might be established between 0.006 and 0.009 per cent."[33] Questions of toxicity were directed at the question of the outcome for the chicken, not its consumers.

Others thought arsenicals might be the unknown but intensively sought-after "animal protein factor" (APF) that seemed necessary for the growth of animals fed on corn or soy. This mysterious factor could come from fishmeal or even cow manure. The animal production industry had grown prodigiously, and the very materials that farmers had come to rely on as feed supplements in previous decades—fishmeal, buttermilk, tankage from slaughterhouses—were becoming more expensive and harder to source, particularly during the war. The APF was isolated and named vitamin B_{12} late in 1948, but only after decades of wide-ranging efforts to find it.[34] In 1948, investigators at USDA's Bureau of Animal Industry reported that organoarsenicals did not substitute for the unknown factor, but seemed to work in concert with it.[35] Thus news of low-dose arsenicals as growth promoters was in the scientific community and the literature several years before antibiotics were used in this way. They set the stage for the idea of drugs in feed as "chemoprophylaxis" rather than disease treatment; indeed, "it was clear before 1950 that many compounds, related only by their capacity to inhibit microbial growth, were all capable of stimulating growth in chickens."[36]

Dr. Salsbury's Laboratories patented, promoted and marketed arsenical medicines from 1944 onward. The company widely distributed *Dr. Salsbury's*

Poultry Health Messenger as well as poultry disease "manuals" for farmers, offering a mix of information and advertising illustrated with colorful "how to do it pictures" and graphic representations of the insides of birds and disease processes (figure 7.1). It is important to emphasize that consumers of these publications were unlikely to realize that they were feeding chicks arsenic. They were being taught to recognize certain brand name remedies, such as Ren-O-Sal, as efficacious and profitable measures to take against costly diseases in their flocks.

Here and in the allied official scientific papers, the benefits of continuously feeding the drug were touted. Studies and popular missives from Salsbury's claimed reduced mortality in chicks, faster growth of 10 to 15 percent, no adverse effect on egg hatchability, plus faster maturation, better laying, increased weight gain, and control of coccidiosis in chickens and turkeys alike—more profit with no down sides. Emphasis was placed on the improved "look" of the birds, promising higher prices:

> A three pound broiler chicken which has gained 10 per cent faster than its
> control as a result of arsonic compound in its feed, has gained 0.3 pounds

Figure 7.1a and b. Front cover of *Dr. Salsbury's Manual of Poultry Diseases*, 1951 (a); and back cover, proudly illustrating the company's chemical manufacturing plant at Charles City, Iowa (b).

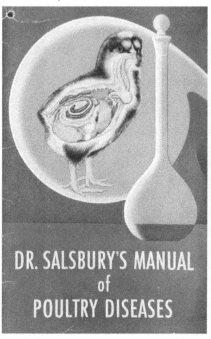

more weight. At 30 cents per pound, this bird is worth 10 cents more than its control weighing 2.7 pounds. Yet it is possible to buy enough arsonic to supply this bird from hatching to market for about one-half to two cents. . . . Better feathering, better pigmentation, and greater uniformity in the flock often induce poultry buyers to pay one or two cents premium for arsonic treated flocks.[37]

Feed mixers who bought arsenical concentrate premixes to sell onward as components of their proprietary feeds were also enthusiastic about the uptake of these new additives. The assistant manager of the poultry department of the Hales and Hunter Company (maker of Red Comb Broiler Mash) reported cheerfully in 1952 to the readers of *Feed Age*, a business magazine for feed manufacturers, "Over 80 per cent of the broilers we are feeding receive feeds containing an arsonic compound. Considering that these feeds have been offered on an optional basis for only one year, with no great sales effort, we feel the growers are sold on the product."[38]

Low cost relative to gain was key to the success of arsenical medications as feed additives. Plentiful raw materials for the derivation of arsenicals came from the copper smelting industry. The American Smelting and Refining Company (ASARCO) owned eleven smelters, transforming arsenic from a nuisance byproduct in smelting smoke to an important commercial product in its own right at the beginning of the twentieth century. Arsenious oxide was condensed from smelter smoke and refined to make arsenic trioxide (As_2O_3). Arsenic recovery operations were concentrated in Tacoma, Washington, seeding a robust industry in arsenic-based herbicides, pesticides, and wood treatment products just next door.[39] Lead or calcium arsenate was sprayed on apples for codling moth and on cotton to control boll weevil, as well as used in the citrus industry as a growth regulator in grapefruit to accelerate maturity and alter the sugar/acid ratio in the fruit. Supplanted by DDT during the war, the market for arsenic pesticides diminished just as arsenical feed additives came into production.

Antibiotics for both disease control and growth promotion broke on the animal production industry in 1951, but this did not so much displace as complement arsenicals. Arsenicals in particular were used "for improvement of feathering and increase of pigmentation of shanks, skin, comb and wattles. Their effect on appearance is quite consistent."[40] Investigation of possible synergistic effects of low-dose antibiotics and arsenicals led to combinations of these drugs with various vitamins and minerals. Abbott Laboratories in Chicago patented new methods for synthesizing arsanilic acid, and began to market it as a growth promoter called Pro-Gen in 1952 in competition with Salsbury's products. Chemist Douglas Van Anden Frost, who had attained a "missionary zeal concerning the trace elements" during his graduate training with a giant of the era of vitamin discovery, Conrad Elvehjem at the University of Wisconsin, was Abbott's champion for arsanilic acid.[41] Frost quickly became the most outspoken advocate for the use of arsenicals in animal feed, maintaining until his death that it was likely to be an essential nutrient.

A Useful Toxin Becomes a Potential Carcinogen

This brings us to questions of safety and regulation. Today one might be rightly inclined to ask what on earth made it seem like a good idea to put any arsenic—even relatively less toxic organic arsenic—in food animals at industrial scale for seventy years. Yet the discussion surrounding the introduction of arsenicals into animal husbandry was dominated by an overwhelmingly positive assessment of their value and an attitude of certainty as to the control of toxicity, perhaps because all of the relevant knowledge was being produced within the companies themselves or on the basis of industry funding. Assumptions as to the stability of the different forms of arsenic over time reinforced confidence about control; organic and inorganic arsenic were not seen as at all equivalent. Positive evidence of the safety of arsenical growth promoters and medicines in terms of both their regulatory approval and their apparent acceptance by consumers was derived from a small number of studies done between 1945 and 1965. This handful of papers continued as the reference point for arsenical use into the twenty-first century, despite profound shifts in the understanding of the biochemistry of arsenic metabolism, and mounting epidemiological, clinical, and experimental evidence as to the range and severity of the health impacts of arsenic exposure.

Studies on arsenical toxicity prior to the 1958 food additive amendment to the Food, Drug and Cosmetic Act focused on establishing dose response curves, toxicity thresholds in several different species, and lethal dose.[42] Questions of residues were addressed by measures of arsenic in muscle tissue, liver, eggs, and excreta to ascertain how much arsenic stayed in animals fed at different doses, and how much was excreted.[43] In the 1940s when arsenicals first emerged as coccidiostats, the pharmaceutical producers of arsenical compounds had to file a New Drug Application with the Food and Drug Administration (FDA), which included providing information about the safety of the drug in these terms: toxicity thresholds, lethal dose, and residues. In turn, any feed manufacturer who wished to incorporate arsenicals into a feed formulation and ship it across state lines had to fill out a form that referenced the original supplier's approval.

These early studies were taken by the FDA as satisfactory evidence of safety. Arsenicals were approved for use in disease prevention and control, growth promotion, and meat pigmentation without controversy. Like other substances seen as necessary compromises for the sake of economic prosperity and meeting market demand, their risks were judged as minimized by regulation and the setting of thresholds.[44] Feed containing these drugs had to be labeled properly with clear instructions about contents, use, precautions, and dosage. Allowable grams per ton of feed were stipulated, and maximum levels of arsenic content for tissues were set at 0.5 parts per million in meat and 1 part per million in edible by-products such as liver.[45] Arsenical feeds had to be withdrawn five days prior to slaughter.

It was not inevitable that arsenical medicines would be so easily accepted and regarded as safe. Arsenic pesticide residues in food figured large in the 1933 exposé *100,000,000 Guinea Pigs: Dangers in Everyday Foods, Drugs and Cosmetics* because these sprays were widely used on apples, cranberries, and tobacco, as well as other staple vegetable crops such as cucumbers.[46] The precursor to the FDA, the Department of Agriculture's Bureau of Chemistry, set tolerance levels for arsenic residues on fruit and vegetables in 1927 following an incident in which a British family was poisoned by apples exported from the United States.[47]

Paradoxically, the case of arsenic pesticides may have helped rather than hindered the introduction of arsenical feed additives. Industry researchers emphasized the difference between inorganic sprays and organic feed additives, and the trace amounts necessary to attain the desired effects. The argument that the amounts used were too low to be harmful went unchallenged. According to historian Frederick Davis, pressure from apple growers in the late 1930s pushed Congress to transfer authority for the examination of insecticides to the Public Health Service "by revising the appropriation act of the FDA so that none of the funds could be used for the study of toxicity of lead or arsenic," which meant a shift from laboratory animal studies that might pick up chronic effects, to field studies surveying farmers about their health more likely to detect acute toxic events.[48]

The conceptual distinction between organic and inorganic arsenic was very firm, and researchers subscribed to a half-century-old theory inherited from Paul Ehrlich that "arsenicals may be active only to the degree that they are bound, and that the enormous differences in the toxicity of arsenic compounds may be due merely to their varying affinity for the body tissues."[49] This enabled an imagination (not actually directly demonstrated) of these less toxic substances as binding to parasites but not animal cells. Useful nontoxic versions of arsenic compounds were thought to pass through the environment with little "affinity" for human or animal tissue. In this doctrine of selective toxicity, "the cells that are to be affected by the drug are called *uneconomic* cells in contrast with the others, the *economic* cells, which have to remain completely unaffected."[50]

Confidence in the clear divisions between organic and inorganic, economic and uneconomic, derived in part from confidence in the stability of different arsenic compounds over time. In experimental tests, amounts of arsenic detectable in animal tissues or eggs came in well below the thresholds set for pesticide residues, particularly if the withdrawal period was observed. A typical assessment read, "Analyses of representative eggs, edible flesh and liver showed that the arsenic content was within a safe level (less than 0.2 micrograms per gram, as compared with the tolerance for animal tissues of 2.65 micrograms per gram tentatively established by the Food and Drug Administration)."[51]

Purveyors of arsenicals and their scientific advocates were forced to shift briefly into defensive mode after the food additive amendment containing the Delaney clause of 1958 declared zero tolerance for substances that had been shown to cause cancer in humans or animals. The FDA raised concerns in 1959 with the Animal Health Institute, an association of animal drug manufacturers, and the American Association of Feed Manufacturers, taking the position that "our scientific advisers cannot say that arsenic compounds (referring to organic arsenicals in feed) . . . are safe until it is demonstrated that such compounds do not have this carcinogenic potential."[52] Arthur Flemming, secretary of health, education and welfare, spoke of evidence showing many hog and poultry farmers were not observing the prescribed withdrawal period. While organic arsenic compounds had not been shown to cause cancer when fed to test animals, he said, "we do need more advice from the scientists who are most knowledgeable in the arsenical field."[53]

Manufacturers were clearly alarmed. General counsel for Abbott Laboratories, Paul Gerden, spoke at length in a congressional hearing in 1960 on proposed legislation to regulate color additives about the effect of "the cancer clause" on arsenicals, even though they were not the topic under consideration. He warned of impediments impossible to surmount:

> Thus the statutory language contained in the cancer clause has been administratively distorted to require manufacturers to attempt to prove that a compound which has not been found to induce cancer in fact cannot cause cancer. This is not possible to prove scientifically. In other words, as the cancer clause in the food additives amendment has been interpreted, any substance can be rendered suspect on the basis of inconclusive reports, even reports concerning different substances, and its manufacturers then given the impossible task of "proving" that the substance is not a carcinogen.[54]

Arsenicals thus became a line of argument in condemnation of the Delaney Clause as inimical to scientific judgment.[55] Their example demonstrated why cancer "should not be singled out for special legislative attention distinct from the many other diseases that afflict mankind" in the proposed color additives legislation being considered. According to Gerden, arsenicals in feed were under threat chiefly because of what he claimed was a "tenuous" relationship to potassium arsenite, which itself only caused hyperkeratosis (skin thickening) in patients given high doses of this inorganic arsenic medication. It was the skin thickening that caused the cancer, he said, not the arsenical.

The *Chicago Daily Tribune* reported on the hearing under the headline "Warns House on Too Strong Cancer Laws: Abbott Asserts Dinner Table Would Suffer." Gerden was quoted as saying that "if the flat ban on a cancer clause is to be applied on the basis of remote implications from inconclusive scientific reports," then "coffee, tea, milk, cream, cocoa, claret, fat, vitamins, eggs and sugar might have to be excluded from the food supply as carcinogenic."[56] This publicity brought a response from FDA administrator George Larrick in the *Washington Post* that approval was being withheld on new uses of arsenicals until drug makers could present "factual evidence that they don't turn into cancer-causing inorganic arsenicals in animal products."[57]

The controversy produced a second wave of experiments from Dr. Salsbury's Laboratories and Abbott Laboratories, as well as a public relations campaign waged in the scientific literature complete with indignant defenses of "scientific judgment" under attack by overzealous regulation. The discussion of arsenic as naturally present in the world and thus inevitably present in food regardless of the use of additives began to open industry-based research articles on the topic. The natural presence of arsenic in shellfish was regularly mentioned. Douglas Frost wrote an impassioned defense of arsenic as "both a tonic and a toxin" that could be used at "a level easily consistent with safety," and that while "concern about arsenicals is natural and wise, when properly used few other groups have served mankind better in so many ways."[58] "The idea," opined Frost, "that nature alone provides the only proper food for man is both short-sighted and incorrect," and he complained that the Delaney Clause precludes "scientific judgment in that it provides for no tolerance, even for essential nutrients which may be purely carcinogenic in huge dosage in one species or another."[59] Frost promoted arsenic as a beneficent gift of nature and a natural counteragent to selenium toxicity.

Indeed, despite decades of suspicion about links between arsenic and cancer, there was no unambiguous proof of its carcinogenicity in the 1950s. Even the scientists advocating caution had to admit that "no reliable experimental reproduction of arsenic cancers in animals [had] been accomplished."[60] Frost claimed that arsenic "has not been shown to cause cancer in experimental animals," citing a National Research Council report on carcinogens from 1959. Therein arsenic trioxide (the inorganic trivalent form of the molecule) and sodium arsenite were listed in the appendix of "compounds sometimes referred to as being carcinogenic for experimental animals but for which available evidence seems inadequate to support such conclusion."[61]

Thus arsenical animal feeds were not regulated under the Delaney Clause. Being poisonous was not equivalent to carcinogenicity and was deemed controllable using thresholds. Evidence from industry scientists seemed to show that organoarsenicals did not cause tumors in animals, particularly when compared to known carcinogens. Authors of this industry-derived literature assumed that the body of the animal served as a shield between the toxin and the end consumer, and they questioned "whether the effectiveness of even potent carcinogens can be transmitted from animal to animal."[62] Experiments in which liver from arsenical-raised pigs was fed to rats were interpreted as demonstrating that tissue arsenic of animals raised this way was "not available" to subsequent eaters, because the rats excreted 100 percent of the tissue arsenic they ingested, while a comparison group fed arsenic trioxide excreted only about 50 percent. The authors likened tissue arsenic to that found naturally in shrimp, hypothesizing that both were "metabolically inert" because it was "firmly bound, probably to protein, and is not easily solubilized."[63]

Importantly, there were no contrasting voices in the regulatory sphere. In the end, the drug approvals from the 1940s stayed in place, and no further limitations were placed on their use. It would take another forty years for questions to be seriously raised anew. If environmental or health concerns were raised, they focused instead on waste residues at production sites of arsenical drugs. In 1971, the EPA administrator William Ruckelshaus ordered Whitmoyer Laboratories of Myerstown, Pennsylvania, to halt a scheduled shipment of seventy tons of waste containing 1 percent arsenic, which was being readied in barrels for being dumped 150 miles out to sea, the company's mode of disposing of such wastes for the previous seven years.[64] Dumping at sea was apparently a solution to arsenic pollution of the facility's soil from dumping waste into unlined lagoons, detected in 1964 when Whitmoyer became a subsidiary of the firm Rohm & Haas. In 1984, the Whitmoyer Laboratories site was declared a superfund site.[65] Similarly, the Salsbury's Laboratories site in Charles City, Iowa, adjacent to the Cedar River, came to sit atop an estimated six million pounds of arsenic waste and was ranked ninth on the EPA's list of 419 sites prioritized for cleanup in 1982.[66] These, however, remained local concerns of little public note and were not associated with the products, particularly after the original companies were bought out by larger transnational ones.[67]

In short, this animal feed additive was little known to the public and seemed both out of sight and out of mind. Indeed, nations including the United States were slow to regulate even drinking water. The maximum contaminant level (MCL) for arsenic in drinking water was set at 50 µg/liter by the US Public Health Service in 1942, and was not revised by the Environmental Protection Agency downward to 10 µg/liter until 2001, despite mounting evidence of the cancer risks to the population identified in the interim.[68] Feed arsenicals caught no one's attention in this context. Paradoxically, the growth promoters that came into place slightly after arsenicals—antibiotics, nitrofurans, and hormones—dominated discussions around the safety of medicated feed. This attention tended to erupt substance by substance: concerns about antibiotic resistance drove focus on medically relevant antibiotics; mutagenicity worries meant a focus on nitrofurans; concerns about cancer clusters in humans exposed to diethylstilbestrol in utero meant a focus on DES.[69]

A brief review of the development of the public health view on arsenic and experimental findings on the biochemistry of arsenic metabolism shows a growing lag between developments in the scientific literature and regulatory standards for arsenicals in feed. This lag was already evident in the 1960s, widened in the 1990s, and finally became unsustainable in the 2010s after pressure from consumer advocacy groups. Occupational health studies of miners, ASARCO smelter workers, and the pesticide manufacturing industry linked prolonged inorganic arsenic exposure in the workplace to lung cancer in the 1960s.[70] Despite concerted opposition from industry, new arsenic air level standards were proposed by the Office of Occupational Health and Safety (OSHA) in the mid-1970s, coming into effect in the mid-1980s after the resolution of court challenges.[71]

In 1968, the first of a series of reports on the relationship between arsenic levels in drinking water in southwestern Taiwan and skin cancer was published; subsequent studies established a dose-response relationship between arsenic concentration in well water and mortality due to cancers of the internal organs in the 1980s.[72] The digging of tube wells in Bangladesh and the ensuing arsenic poisoning of populations made arsenic a central problem in global public health. In 1992, on the basis of this burgeoning epidemiological literature, scientists warned that the 50 μg/liter drinking water standard established in 1942 was insufficient, estimating the combined cancer risk at 1 in 100 in individuals being exposed to inorganic arsenic at that level.[73] Indeed, the tracing of health effects of arsenic in drinking water became a kind of subindustry within the world of public health. Arsenic has since the 1990s become a "top ten" chemical of major health concern globally for the World Health Organization. Arsenic is number one on the substance priority list for the US Agency for Toxic Substances and Disease Registry, a ranking achieved by concern about a substance's frequency, toxicity, and potential for human exposure at National Priority List Superfund sites.

Despite growing certainty that inorganic arsenic in drinking water was an important causative element in human cancer, the element in its multitudinous forms remained elusive in terms of mechanistic understandings of carcinogens and mutagens for decades. There was no animal model in which exposure to this "paradoxical" carcinogen reliably produced tumors; arsenic came up as nonmutagenic in bacterial and mammalian cell mutation assays that measured mutation at a single gene locus, did not show reactivity with DNA, and did not appear to cause point mutations, although it was clearly "clastogenic" in terms of chromosome damage and thus included in a list of compounds by which "the human gene pool can become insidiously polluted."[74] As measures of mutagenicity of environmental pollutants were increasingly important in state regulation of industrial hazards in the 1960s and 70s in Europe and North America, arsenic's scientific uncertainty also meant regulatory uncertainty.[75]

A contrasting example helps illustrate the impact of this uncertainty: nitrofurans, a class of synthetic antimicrobials introduced to human medicine and animal feed from 1953, were like arsenicals in that they were anti-protozoal, and were likewise quickly incorporated into the battle against intestinal parasites in poultry and hog farming. However, nitrofurans were clearly mutagenic and genotoxic in bacterial and cell culture assays. The food preservative AF-2 was banned in 1979, and then nitrofurans were as a class banned by the FDA and the European Union from systemic use in animal feed in the early 1990s. Arsenicals took far longer to come under renewed regulatory scrutiny despite much more obvious epidemiological data linking arsenic to cancer outcomes in humans.[76]

In the 1990s, the teasing apart of what happened to different species of arsenic compounds in their course through the body showed a complex path from adsorption to excretion marked by a series of chemical transformations. Of great significance to the question of carcinogenicity were the metabolic intermediates formed in the sequence of reduction and oxidative methylation processes to which cells subject arsenic compounds in order to detoxify and excrete them. The generation of monomethylated and dimethylated arsenicals (MMAs and DMAs) in metabolism was intensively studied, and these metabolites proved to be variously teratogenic, nephrotoxic, tumor promoting, carcinogenic, cytotoxic, genotoxic, and enzyme inhibiting, suggesting that "the methylation of iAs should be regarded as an activation process that forms more reactive species which exert unique toxic effects . . . key to understanding its actions as a toxicant and a carcinogen."[77] Arsenic is predominantly metabolized in the liver and excreted in urine, and the ability to test for these arsenic metabolites in urine changed the accuracy with which individual exposures could be measured and therefore the accuracy of epidemiological studies on the subject.

It may seem that these developments far afield from agriculture were hardly relevant to the question of organoarsenicals in animal feed because early safety studies demonstrated that the majority of the drug fed to animals is excreted unchanged. Moreover, much of the public health literature was focused on arsenic in drinking water or the workplace. Nonetheless, within this context of greatly heightened concern about environmental arsenic pollution, new attention to what happened to excreted arsenical drugs began to finally shift assumptions that went untested for decades, still being repeated in the 2000s.[78] In particular, a 1969 study by Joseph Morrison of the biochemistry department of Dr. Salsbury's Laboratories stood as the sole investigation of the fate of arsenic in chicken manure.[79] He had studied the effect on soil arsenic fertilized with poultry litter, using chemical techniques developed in the 1950s for the detection of arsenic. On the basis of three samples each from long-term treated, short-term treated, and untreated fields and crops around a single commercial broiler operation in Arkansas, Morrison found no evidence that use of poultry litter affected inorganic arsenic content in soil, or crops grown on litter-treated soil, or in drainage water from such fields.

However, the problem of animal waste management had become a pressing environmental problem in its own right by 2000. Expansion in the national and international appetite for chicken and pork also meant more manure and therefore greater volume of excreted drugs: trace amounts add up. Poultry litter (droppings, feathers) from growing operations is often disposed of by spreading it on fields, incineration, or refeeding. When the question of the fate of arsenicals after excretion was taken up again, the answers were considerably different than those of 1969. Researchers from the National Water Quality Laboratory of the US Geological Survey reaffirmed that the arsenical drugs given to chickens were excreted unchanged and remained stable in dried litter, but with the addition of water and the passage of thirty days, the "speciation" of the organoarsenic shifted to arsenate (inorganic arsenic in its pentavalent state).

Researchers noted that if the amount of roxarsone excreted by a single broiler fed at industry standard over its forty-two-day growth period is 150 mg, and if one assumed that 70 percent of the 8.3 million broilers grown in the United States in 2000 were fed with this arsenical feed, "the manure would contain about 9×10^5 kg of roxarsone or 2.5×10^5 kg of arsenic."[80] Clostridium bacteria in chicken litter, a dominant species in the microbial flora of the chicken cecum, transforms the drug chemically, releasing inorganic arsenic.[81] In 2003, a lawsuit was brought by a family with a child with leukemia against Alphapharma, a subsidiary of Pfizer that was by then the primary supplier of poultry arsenicals, and poultry producers in Arkansas. The attempt to hold the companies accountable for the leukemia case was defeated when the court found for the defendant, but *Green v. Alphapharma* was nonetheless an important development in awareness about the environmental impact of arsenical drugs.

Concern about environmental arsenic was also prompted by studies of arsenic in rice. Sampling of polished white rice from ten countries over four continents showed that rice from the United States contained more inorganic arsenic, on average, than that from almost any other country (exceeded only by some from France).[82] Within the United States, rice grown in Texas and Arkansas had the highest mean arsenic concentrations.[83] In 2012, *Consumer Reports* set out to sample apple and grape juice for arsenic, and then took on the question of arsenic in rice and products made with rice, using their find-ings to call for a federal standard limit on arsenic in food, just as there is for drinking water.[84] Subsequently, a group of Arkansas rice farmers (unsuccess-fully) filed suit against Arkansas poultry producers alleging the chicken pro-cessors knew of high arsenic levels in chicken manure that could potentially contaminate rice crops.

It took somewhat longer to dispel the notion that no meaningful amounts of arsenic reached consumers through meat. Serious retesting of assumptions about tissue affinity and bioavailability did not occur until very late in arsenic's emergence as a public health hazard. Indeed, it was only because of "growing interest in various sources of arsenic exposure" that a biostatistics master's stu-dent was set on examining data collected on meat and poultry liver samples by the Food Safety and Inspection Service National Residue Program between 1989 and 2000. Reportedly it came as a profound surprise that arsenic levels in young chickens were at least three times that found in other meat.[85] The sub-sequent analysis took into account American consumption patterns, records of apparent violations of withdrawal periods for arsenicals, and ratios of liver arsenic to muscle tissue arsenic to model the amount of arsenic in meat that would reach consumers.[86]

These estimates raised considerable alarm when published in 2004. Subse-quent market basket surveys showed conventionally raised chicken purchased from American supermarkets to contain both arsenical drug residues and in-organic arsenic, while organic samples did not; moreover, arsenic speciation was altered in the samples according to whether they were cooked or not.[87] In 2009, the Center for Food Safety and the Institute for Agriculture and Trade Policy filed a citizen petition with the FDA requesting the withdrawal of ap-proval for arsenical drugs as animal feed additives: a subsequent lawsuit in 2013 pressed the agency to act on the petition. This pushed the FDA to conduct its own analysis of arsenical feeding and arsenic residues in chickens. Zoetis, at that time the animal health subsidiary of Pfizer, voluntarily suspended sales of arsenical medications for chickens in 2013, and the FDA withdrew approval for medications containing arsanilic acid, roxarsone, and carbasone.[88] Ap-proval for turkey arsenicals was withdrawn in 2015.

Conclusion: From Toxicity to Disease Promotion

In the history recounted above, the mass dissemination of nutrients such as vitamin D through animal feed resulted in novel disease conditions including widespread parasitic intestinal diseases in chickens. The manufactured feed in-dustry provided the infrastructure for ameliorating the very conditions it had enabled, delivering billions of trace doses of various drugs in the feedbag or the water dish. Arsenicals were the first modern growth-promoter promising both disease prevention and acceleration of growth and maturation; antibiotics and hormones followed close on. They were established within a culture of industry-led toxicity assessment in the 1940s and 1950s, and the inertia of established practice carried them forward through the twentieth century. The idiosyncratic complexities of arsenic metabolism and carcinogenicity meant that this element escaped regulatory scrutiny in the food supply for decades. It was only rising concern about arsenic exposure in public health and concerted pressure from consumer advocates and federal scientists that eventually led to the reexamination of questions of direct transfer of arsenic to consumers through meat, and indirect contamination of soil and water.

Having reached this point in a narrative arc that has moved from enthusiastic launch to quiet shelving, a point that feels very much like a denouement, one might expect this is the end of the story. However, the biomedical science of arsenic is still unfolding along with the biological legacy of the decades described here. While this is not the place to delve into the depths of a complex contemporary biomedical literature, a few key trends speak to the themes of this book, in particular the transition in what kind of risk is involved when arsenic is on the table.

Notably, what is emerging is a calculation of risk to human health at the intersection of several distinct legacies of twentieth-century food systems. Dietary transformation is one legacy; widespread low-level arsenic exposure is another, in the wake of decades of agricultural use of arsenicals, not just in animal feed, but also through the application of tons of arsenical pesticides, herbicides, and wood preservative. Even foods consumed in the name of metabolic health in a time of obesity, such as organic brown rice sweetener, carry the legacy of arsenicals in agriculture into the body.[89] The acceleration of animal product generation and the growth accelerant itself might have been physically separated in the 2010s with the withdrawal of arsenical agents from meat and egg production in the United States, but their material consequences continue to interact.

Risk at intersections of diet, metabolic disorder, and arsenic are being made legible by epidemiology, experiments with arsenic and the microbiome, and increasing understanding of cellular arsenic metabolism. As epidemiologists begin to study low-dose chronic exposure to arsenic, they note synergistic effects between body mass index (BMI) and arsenic exposure, in which the risk of type 2 diabetes upon arsenic exposure is exacerbated at higher BMI, indicating the two factors interact "in a greater than additive manner."[90] Nutritional factors such as "consumption of a Western-style diet can make the liver susceptible to low, normally non-hepatotoxic concentrations of arsenic."[91] Low-fiber diets and antibiotics compromise the buffering effect of gut microbes, which themselves ingest and metabolize arsenic as it is drunk or eaten.[92] Arsenic compromises DNA repair and epigenetic processes, making it more likely that DNA damage will go unfixed.[93] As a "metabolic disruptor," arsenic undermines the ability of cells and organs to respond to oxidative stress, an effect magnified in bodies perhaps already taxed by insulin insensitivity.[94]

This kind of risk to organs, cells, genomes, and enzymes is not of direct damage with linear causality—a toxin experienced in high quantities that comes in and causes a genetic lesion that then leads to cancer. Rather, it is the risk of the generation of conditions under which diseases are more likely to occur, such as a compromised gut wall, a remodeled liver, or DNA-binding proteins rendered dysfunctional by having their normal zinc ion displaced by an arsenic ion. Arsenic is being rethought as a metabolic disruptor that *potentiates* the impact of more traditional causative factors for diabetes, such as high-fat diets.[95] Thus one cannot think about arsenic by itself; it comes to us as a very specific historical bundle of circumstance. It is this historically specific biochemical ecology that matters more than the chemical substance on its own. Evolving conceptions of risk over the twentieth century shaped how and where arsenic was distributed through animal metabolism; now arsenic's ubiquitous low-dose presence and influence on human health is reshaping how we think about the risk not just of specific toxins in foods, but of food systems, whose interacting legacies form a complex ecology for human health and wellbeing.

Hannah Landecker holds a joint appointment in the Department of Sociology and the Institute for Society and Genetics at the University of California, Los Angeles. She is a historian and sociologist of the modern life sciences and biotechnology, with a particular focus on cell biology, metabolism, and epigenetics. Her work includes *Culturing Life: How Cells Became Technologies* (Harvard University Press, 2007), and many articles on the use of film technology in the biosciences, the rise of antibiotic resistance, and the history and sociology of metabolism and metabolic disorders in the biosciences.

Notes

1. Nigra et al., "Poultry Consumption."
2. Karagas et al., "Association of Rice."
3. Cubadda et al., "Human Exposure."
4. Nachman et al., "Mitigating Dietary Arsenic Exposure."
5. Welch and Martí-Ibáñez, *Symposium on Medicated Feeds.*
6. Sapkota et al., "What Do We Feed."
7. Landecker, "Antibiotic Resistance."
8. Kuo et al., "The Association."
9. Boyd, "Making Meat."

10. Ibid., 638.
11. Foster, "The Economic Losses."
12. Reid, "History of Avian Medicine."
13. Foster, "The Economic Losses," 435.
14. Schwerin, Stoff, and Wahrig, *Biologics.*
15. Becker, *Coccidia and Coccidiosis.*
16. Reid, "History of Avian Medicine," 514.
17. Welch and Martí-Ibáñez, *Symposium on Medicated Feeds.*
18. Landecker, "The Food of our Food."
19. Lesch, *The First Miracle Drugs.*
20. Albert, *Selective Toxicity.*
21. Wright, Seiple, and Myers, "The Evolving Role"; Brandt, *No Magic Bullet.*
22. Vilensky and Redman, "British Anti-Lewisite (Dimercaprol)."
23. Stocken and Thompson, "Reactions of British Anti-Lewisite," 168.
24. Frost, Overby, and Spruth, "Arsenicals in Feeds," 235.
25. The chemical warfare research and the anti-parasitic research were not separate endeavors. The search for anti-parasitic medications was a matter of wartime priority because of tropical theaters of war, and the combination of the Lewisite antidote BAL with arsenicals generated a new agent with tamped-down toxicity that was still effective against several strains of trypanosomes that had become almost completely resistant to earlier generations of arsenical therapies.
26. Cullen and Reimer, *Arsenic Is Everywhere.*
27. Welch and Martí-Ibáñez, *Symposium on Medicated Feeds*, 150.
28. Jones, *Valuing Animals*, 100.
29. Morehouse and Mayfield, "The Effect of Some Aryl Arsonic Acids" (1944), 6.
30. Morehouse and Mayfield, "Poultry Treatment Composition"; Morehouse and Mayfield, "The Effect of Some Aryl Arsonic Acids" (1946).
31. Frost, Overby, and Spruth, "Arsenicals in Feeds."
32. Goble, "Para-Substituted Phenylarsonic Acids," 533.
33. Ibid., 537.
34. Summons, "Animal Feed Additives."
35. Bird, Groschke, and Rubin, "Effect of Arsonic Acid Derivatives."
36. Frost, Overby, and Spruth, "Arsenicals in Feeds," 235.
37. Morehouse, "The Economic Value," 34.
38. Karrasch, "Arsonic Compounds in Feeds," 31.
39. Tepper and Tepper, "The Rise and Fall."
40. Bird, Groschke, and Rubin, "Effect of Arsonic Acid Derivatives," 30.
41. Oldfield, "Douglas Van Anden Frost."
42. Frost, "Safety First."
43. Evans et al., "The Arsenic Content of Eggs."
44. Jas and Boudia, *Toxicants.*
45. 21 CFR §121.1138. Arsenic. (1964, p. 477).
46. Kallet and Schlink, *100,000,000 Guinea Pigs.*
47. Parascandola, *King of Poisons.*
48. Davis, *Banned*, 29.
49. Hogan and Eagle, "The Pharmacologic Basis," 93.

50. Albert, *Selective Toxicity*, 1. Emphasis in original.
51. "Arsenicals as Growth Promoters," 208.
52. Quoted in statement of Paul Gerden, US Congress, *Hearings on Color Additives*, 206.
53. Letter from Arthur S. Flemming to Oren Harris, 18 April 1960, reproduced in US Congress, *Hearings on Color Additives*, 218.
54. Statement of Paul Gerden, US Congress, *Hearings on Color Additives*, 206.
55. Markowitz and Rosner, *Deceit and Denial*.
56. Hearst, "Warns House."
57. "Animal-Feed Arsenic Stirs FDA Dispute."
58. Frost, "Arsenic and Selenium," 129–30.
59. Ibid., 129.
60. Hueper, "Carcinogens and Carcinogenesis," 361.
61. National Research Council (US), Food Protection Committee, *Problems in the Evaluation*, 39. The appendix lists results from J. L. Hartwell's 1951 *Survey of Compounds Which Have Been Tested for Carcinogenic Activity*, in which arsenic trioxide applied to ten rabbits produced one carcinoma and three papilloma, applied to twelve rabbits produced three papilloma; sodium arsenite applied to forty rats produced one sarcoma—no information about dosage or time of exposure was provided.
62. Frost, "Arsenic and Selenium," 129.
63. Overby and Frost, "Nonavailability to the Rat," 42.
64. "Company Halts Plan."
65. United States Environmental Protection Agency, "Whitmoyer."
66. Jasen, "Toxic City."
67. Dr. Salsbury's Laboratories was acquired by Solvay Chemicals in 1979, and its trademarks on 3-nitro were eventually acquired by Zoetis, originally the animal science subsidiary of Pfizer that is now an independent company.
68. Smith et al., "Arsenic Epidemiology." The new 10 μg/L standards took effect in 2006.
69. Office of Technology Assessment, United States Congress, *Drugs in Livestock Feed*.
70. Ott, Holder, and Gordon, "Respiratory Cancer"; Lee and Fraumeni, "Arsenic and Respiratory Cancer."
71. Sullivan, "Contested Science."
72. Tseng et al., "Prevalence of Skin Cancer"; Hughes et al., "Arsenic Exposure and Toxicology."
73. Smith et al., "Arsenic Epidemiology."
74. Basu et al., "Genetic Toxicology"; Shaw, "Human Chromosome Damage," 409.
75. Creager, "The Political Life"; Schwerin, "Low Dose Intoxication."
76. McCalla, "Mutagenicity of Nitrofuran Derivatives."
77. Drobna, Styblo, and Thomas, "An Overview," 433.
78. Jones, "A Broad View of Arsenic."
79. Morrison, "Distribution of Arsenic."
80. Garbarino et al., "Environmental Fate," 1509.
81. Stolz et al., "Biotransformation."
82. Meharg et al., "Geographical Variation."
83. Zavala and Duxbury, "Arsenic in Rice."
84. Consumer Reports, "Arsenic In Your Food."
85. Lasky, "Arsenic in Chicken."

86. Lasky et al., "Mean Total Arsenic Concentrations."
87. Nachman et al., "Roxarsone, Inorganic Arsenic."
88. *Federal Register* 79, no. 39 (27 February 2014): 10963–65.
89. Jackson et al., "Arsenic, Organic Foods."
90. Castriota et al., "Obesity."
91. Watson, "Molecular Mechanisms," 37.
92. Dietert and Silbergeld, "Biomarkers"; Massey et al., "Oligofructose Protects."
93. Bailey and Fry, "Arsenic-Associated Changes."
94. Heindel et al., "Metabolism Disrupting Chemicals."
95. Kirkley and Sargis, "Environmental Endocrine Disruption."

Bibliography

Albert, Adrien. *Selective Toxicity with Special Reference to Chemotherapy.* London: Methuen, 1951.

"Animal-Feed Arsenic Stirs FDA Dispute." *Washington Post,* 12 February 1960, p. A3.

"Arsenicals as Growth Promoters." *Nutrition Reviews* 14, no. 7 (1956): 206–9.

Bailey, Kathryn A., and Rebecca C. Fry. "Arsenic-Associated Changes to the Epigenome: What Are the Functional Consequences?" *Current Environmental Health Reports* 1, no. 1 (2014): 22–34.

Basu, A., J. Mahata, S. Gupta, and A. K. Giri. "Genetic Toxicology of a Paradoxical Human Carcinogen, Arsenic: A Review." *Mutation Research* 488, no. 2 (2001): 171–94.

Becker, Elery R. *Coccidia and Coccidiosis of Domesticated, Game and Laboratory Animals and of Man.* Ames: Collegiate Press, 1934.

Bird, H. R., A. C. Groschke, and Max Rubin. "Effect of Arsonic Acid Derivatives in Stimulating Growth of Chickens." *Journal of Nutrition* 37, no. 2 (1949): 215–26.

Boyd, William. "Making Meat: Science, Technology, and American Poultry Production." *Technology and Culture* 42, no. 4 (2001): 631–64.

Brandt, Allan M. *No Magic Bullet: A Social History of Venereal Disease in the United States since 1880.* New York: Oxford University Press, 1987.

Castriota, Felicia, Johanna Acevedo, Catterina Ferreccio, Allan H. Smith, Jane Liaw, Martyn T. Smith, and Craig Steinmaus. "Obesity and Increased Susceptibility to Arsenic-Related Type 2 Diabetes in Northern Chile." *Environmental Research* 167 (November 2018): 248–54.

"Company Halts Plan to Dump Arsenic." *Chicago Tribune* (1963–1996). 13 March 1971.

Consumer Reports. "Arsenic In Your Food." Accessed 8 April 2020, https://www.consumer reports.org/cro/magazine/2012/11/arsenic-in-your-food/index.htm.

Creager, Angela N. H. "The Political Life of Mutagens: A History of the Ames Test." In *Powerless Science? Science and Politics in a Toxic World,* edited by Soraya Boudia and Nathalie Jas, 46–64. Environment in History, vol. 2. New York: Berghahn Books, 2014.

Cubadda, Francesco, Brian P. Jackson, Kathryn L. Cottingham, Yoshira Ornelas Van Horne, and Margaret Kurzius-Spencer. "Human Exposure to Dietary Inorganic Arsenic and Other Arsenic Species: State of Knowledge, Gaps and Uncertainties." *The Science of the Total Environment* 579 (1 February 2017): 1228–39.

Cullen, William R., and Kenneth J. Reimer. *Arsenic Is Everywhere: Cause for Concern?* Cambridge: Royal Society of Chemistry, 2016.

Davis, Frederick Rowe. *Banned: A History of Pesticides and the Science of Toxicology*. New Haven: Yale University Press, 2014.

Dietert, Rodney Reynolds, and Ellen Kovner Silbergeld. "Biomarkers for the 21st Century: Listening to the Microbiome." *Toxicological Sciences* 144, no. 2 (2015): 208–16.

Drobna, Zuzana, Miroslav Styblo, and David J. Thomas. "An Overview of Arsenic Metabolism and Toxicity." *Current Protocols in Toxicology* 42, no. 431 (2009): 4.31.1–4.31.6.

Evans, Robert John, Selma L. Bandemer, David A. Libby, and A. C. Groschke. "The Arsenic Content of Eggs from Hens Fed Arsanilic Acid." *Poultry Science* 32, no. 4 (1 July 1953): 743–44.

Foster, A. O. "The Economic Losses Due to Coccidiosis." *Annals of the New York Academy of Sciences* 52, no. 4 (1949): 434–42.

Frost, Douglas V. "Arsenic and Selenium in Relation to the Food Additive Law of 1958." *Nutrition Reviews* 18, no. 5 (1960): 129–32.

———. "Safety First in Use of Arsonic Acid Compounds in Feed." *Feed Age* 2, no. 12, (1952): 30–31.

Frost, D. V., L. R. Overby, and H. C. Spruth. "Arsenicals in Feeds, Studies with Arsanilic Acid and Related Compounds." *Journal of Agricultural and Food Chemistry* 3, no. 3 (1955): 235–43.

Garbarino, J. R., A. J. Bednar, D. W. Rutherford, R. S. Beyer, and R. L. Wershaw. "Environmental Fate of Roxarsone in Poultry Litter. I. Degradation of Roxarsone during Composting." *Environmental Science & Technology* 37, no. 8 (2003): 1509–14.

Goble, Frans C. "Para-Substituted Phenylarsonic Acids as Prophylactic Agents against *Eimeria Tenella* Infections." *Annals of the New York Academy of Sciences* 52, no. 4 (1949): 533–37.

Hartwell, Jonathan L. *Survey of Compounds Which Have Been Tested for Carcinogenic Activity*. 2nd ed. Bethesda: National Cancer Institute, National Institutes of Health, 1951.

Hearst, Joseph. "Warns House on Too Strong Cancer Laws: Abbott Asserts Dinner Table Would Suffer." *Chicago Daily Tribune*, 11 February 1960, part 5.

Heindel, Jerrold J., Bruce Blumberg, Mathew Cave, Ronit Machtinger, Alberto Mantovani, Michelle A. Mendez, Angel Nadal, et al. "Metabolism Disrupting Chemicals and Metabolic Disorders." *Reproductive Toxicology* 68 (March 2017): 3–33.

Hogan, Ralph B., and Harry Eagle. "The Pharmacologic Basis for the Widely Varying Toxicity of Arsenicals." *Journal of Pharmacology and Experimental Therapeutics* 80, no. 1 (1944): 93–113.

Hueper, W. C. "Carcinogens and Carcinogenesis." *American Journal of Medicine* 8, no. 3 (1950): 355–71.

Hughes, Michael F., Barbara D. Beck, Yu Chen, Ari S. Lewis, and David J. Thomas. "Arsenic Exposure and Toxicology: A Historical Perspective." *Toxicological Sciences* 123, no. 2 (2011): 305–32.

Jackson, Brian P., Vivien F. Taylor, Margaret R. Karagas, Tracy Punshon, and Kathryn L. Cottingham. "Arsenic, Organic Foods, and Brown Rice Syrup." *Environmental Health Perspectives* 120, no. 5 (2012): 623–26.

Jas, Nathalie, and Soraya Boudia, eds. *Toxicants, Health and Regulation since 1945*. London: Pickering & Chatto, 2013.

Jasen, Georgette. "Toxic City: Iowa Town Learns to Live with Chemical Waste: Economic Woes Overshadow Pollution Worries." *Wall Street Journal,* 27 April 1983, sec. 1.

Jones, F. T. "A Broad View of Arsenic." *Poultry Science* 86, no. 1 (2007): 2–14.

Jones, Susan D. *Valuing Animals: Veterinarians and Their Patients in Modern America.* Baltimore: Johns Hopkins University Press, 2003.

Kallet, Arthur, and Frederick John Schlink. *100,000,000 Guinea Pigs: Dangers in Everyday Foods, Drugs, and Cosmetics.* New York: Grosset & Dunlap, 1935.

Karagas, Margaret R., Tracy Punshon, Vicki Sayarath, Brian P. Jackson, Carol L. Folt, and Kathryn L. Cottingham. "Association of Rice and Rice-Product Consumption with Arsenic Exposure Early in Life." *JAMA Pediatrics* 170, no. 6 (2016): 609–16.

Karrasch, R. H. "Arsonic Compounds in Feeds: A Manufacturer's Viewpoint." *Feed Age* 2, no. 12 (1952): 31.

Kirkley, Andrew G., and Robert M. Sargis. "Environmental Endocrine Disruption of Energy Metabolism and Cardiovascular Risk." *Current Diabetes Reports* 14, no. 6 (June 2014): 494.

Kuo Chin-Chi, Moon Katherine A., Wang Shu-Li, Silbergeld Ellen, and Navas-Acien Ana. "The Association of Arsenic Metabolism with Cancer, Cardiovascular Disease, and Diabetes: A Systematic Review of the Epidemiological Evidence." *Environmental Health Perspectives* 125, no. 8 (2017): 087001-1–087001-15.

Landecker, Hannah. "Antibiotic Resistance and the Biology of History." *Body & Society* 22, no. 4 (2016): 19–52.

———. "The Food of Our Food: Medicated Feed and the Industrialization of Metabolism." In *Eating Beside Ourselves,* edited by Heather Paxson. Durham: Duke University Press, in press.

Lasky, Tamar. "Arsenic in Chicken: A Tale of Data and Policy." *Journal of Epidemiology & Community Health* 71, no. 1 (2017): 1–3.

Lasky, Tamar, Wenyu Sun, Abdel Kadry, and Michael K Hoffman. "Mean Total Arsenic Concentrations in Chicken 1989–2000 and Estimated Exposures for Consumers of Chicken." *Environmental Health Perspectives* 112, no. 1 (2004): 18–21.

Lee, Anna M., and Joseph F. Fraumeni. "Arsenic and Respiratory Cancer in Man: An Occupational Study." *JNCI: Journal of the National Cancer Institute* 42, no. 6 (1969): 1045–52.

Lesch, John E. *The First Miracle Drugs: How the Sulfa Drugs Transformed Medicine.* New York: Oxford University Press, 2007.

Markowitz, Gerald E., and David Rosner. *Deceit and Denial: The Deadly Politics of Industrial Pollution.* Berkeley: University of California Press, 2002.

Massey, Veronica L., Kendall S. Stocke, Robin H. Schmidt, Min Tan, Nadim Ajami, Rachel E. Neal, Joseph F. Petrosino, Shirish Barve, and Gavin E. Arteel. "Oligofructose Protects against Arsenic-Induced Liver Injury in a Model of Environment/Obesity Interaction." *Toxicology and Applied Pharmacology* 284, no. 3 (2015): 304–14.

McCalla, D. R. "Mutagenicity of Nitrofuran Derivatives: Review." *Environmental Mutagenesis* 5, no. 5 (1983): 745–65.

Meharg, Andrew A., Paul N. Williams, Eureka Adomako, Youssef Y. Lawgali, Claire Deacon, Antia Villada, Robert C. J. Cambell, et al. "Geographical Variation in Total and Inorganic Arsenic Content of Polished (White) Rice." *Environmental Science & Technology* 43, no. 5 (2009): 1612–17.

Morehouse, Neal F. "The Economic Value of Arsonic Compounds in Feeds." *Feed Age* 2, no. 12 (1952): 34.

Morehouse, Neal F., and Orley J. Mayfield. "The Effect of Some Aryl Arsonic Acids on Experimental Coccidiosis Infection in Chickens." *Journal of Parasitology* 30, no. 4, Supplement (1944): 6.

———. "The Effect of Some Aryl Arsonic Acids on Experimental Coccidiosis Infection in Chickens." *Journal of Parasitology* 32, no. 1 (1946): 20–24.

———. "Poultry Treatment Composition." United States US2450866A, filed 4 October 1946, and issued 5 October 1948. https://patents.google.com/patent/US2450866A/en.

Morrison, Joseph Louis. "Distribution of Arsenic from Poultry Litter in Broiler Chickens, Soil, and Crops." *Journal of Agricultural and Food Chemistry* 17, no. 6 (1969): 1288–90.

Nachman, Keeve E., Patrick A. Baron, Georg Raber, Kevin A. Francesconi, Ana Navas-Acien, and David C. Love. "Roxarsone, Inorganic Arsenic, and Other Arsenic Species in Chicken: A U.S.-Based Market Basket Sample." *Environmental Health Perspectives* 121, no. 7 (2013): 818–24.

Nachman, Keeve E., Gary L. Ginsberg, Mark D. Miller, Carolyn J. Murray, Anne E. Nigra, and Claire B. Pendergrast. "Mitigating Dietary Arsenic Exposure: Current Status in the United States and Recommendations for an Improved Path Forward." *Science of the Total Environment* 581–582 (2017): 221–36.

National Research Council (US), Food Protection Committee. *Problems in the Evaluation of Carcinogenic Hazard from Use of Food Additives*. National Research Council. Publication 749. Washington, DC: National Academies Press, 1959.

Nigra, Anne E., Keeve E. Nachman, David C. Love, Maria Grau-Perez, and Ana Navas-Acien. "Poultry Consumption and Arsenic Exposure in the U.S. Population." *Environmental Health Perspectives* 125, no. 3 (2017): 370–77.

Office of Technology Assessment, United States Congress. *Drugs in Livestock Feed*. Washington, DC: US Government Printing Office, 1979.

Oldfield, J. E. "Douglas Van Anden Frost (1910–1989)." *Journal of Nutrition* 122, no. 3 (1992): 405–11.

Ott, Marvin Gerald, Benjamin B. Holder, and Harold L. Gordon. "Respiratory Cancer and Occupational Exposure to Arsenicals." *Archives of Environmental Health: An International Journal* 29, no. 5 (1974): 250–55.

Overby, L. R., and D. V. Frost. "Nonavailability to the Rat of the Arsenic in Tissues of Swine Fed Arsanilic Acid." *Toxicology and Applied Pharmacology* 4, no. 1 (1962): 38–43.

Parascandola, John. *King of Poisons: A History of Arsenic*. Washington, DC: Potomac Books, 2012.

Reid, W. Malcolm. "History of Avian Medicine in the United States. X. Control of Coccidiosis." *Avian Diseases* 34, no. 3 (1990): 509–25.

Sapkota, Amy R., Lisa Y. Lefferts, Shawn McKenzie, and Polly Walker. "What Do We Feed to Food-Production Animals? A Review of Animal Feed Ingredients and Their Potential Impacts on Human Health." *Environmental Health Perspectives* 115, no. 5 (2007): 663–70.

Schwerin, Alexander von. "Low Dose Intoxication and a Crisis of Regulatory Models: Chemical Mutagens in the Deutsche Forschungsgemeinschaft (DFG), 1963–1973." *Berichte Zur Wissenschaftsgeschichte* 33, no. 4 (2010): 401–18.

Schwerin, Alexander von, Heiko Stoff, and Bettina Wahrig, eds. *Biologics: A History of Agents Made from Living Organisms in the Twentieth Century*. London: Pickering & Chatto, 2013.

Shaw, M. W. "Human Chromosome Damage by Chemical Agents." *Annual Review of Medicine* 21, no. 1 (1970): 409–32.

Smith, Allan H., Peggy A. Lopipero, Michael N. Bates, and Craig M. Steinmaus. "Arsenic Epidemiology and Drinking Water Standards." *Science* 296, no. 5576 (2002): 2145–46.

Stocken, Lloyd A., and R. H. S. Thompson. "Reactions of British Anti-Lewisite with Arsenic and Other Metals in Living Systems." *Physiological Reviews* 29, no. 2 (1949): 168–94.

Stolz, John F., Eranda Perera, Brian Kilonzo, Brian Kail, Bryan Crable, Edward Fisher, Mrunalini Ranganathan, Lars Wormer, and Partha Basu. "Biotransformation of 3-Nitro-4-Hydroxybenzene Arsonic Acid (Roxarsone) and Release of Inorganic Arsenic by Clostridium Species." *Environmental Science & Technology* 41, no. 3 (2007): 818–23.

Sullivan, Marianne. "Contested Science and Exposed Workers: ASARCO and the Occupational Standard for Inorganic Arsenic." *Public Health Reports* 122, no. 4 (2007): 541–47.

Summons, Terry G. "Animal Feed Additives, 1940–1966." *Agricultural History* 42, no. 4 (1968): 305–13.

Tepper, Lloyd B., and Jeffrey H. Tepper. "The Rise and Fall of the Tacoma Arsenic Industry." *IA. The Journal of the Society for Industrial Archeology* 39, no. 1/2 (2013): 65–78.

Tseng, W. P., H. M. Chu, S. W. How, J. M. Fong, C. S. Lin, and S. Yeh. "Prevalence of Skin Cancer in an Endemic Area of Chronic Arsenicism in Taiwan." *JNCI: Journal of the National Cancer Institute* 40, no. 3 (1968): 453–63.

US Congress. *Color Additives: Hearings before the Committee on Interstate and Foreign Commerce, House of Representatives, 86th Congress, 2nd Session*. Washington, DC: US Government Printing Office, 1960.

US Environmental Protection Agency. "Whitmoyer Laboratories Site Profile." Accessed 11 August 2018, https://cumulis.epa.gov/supercpad/SiteProfiles/index.cfm?fuseaction=second.Cleanup&id=0300643#bkground.

Vilensky, Joel A., and Kent Redman. "British Anti-Lewisite (Dimercaprol): An Amazing History." *Annals of Emergency Medicine* 41, no. 3 (2003): 378–83.

Watson, Walter H. "Molecular Mechanisms in Arsenic Toxicity." In *Advances in Molecular Toxicology*, edited by James C. Fishbein and Jacqueline M. Heilman, 9: 35–75. Elsevier, 2015.

Welch, Henry, and Félix Martí-Ibáñez. *Symposium on Medicated Feeds: Proceedings of the Symposium on Medicated Feeds Sponsored by U.S. Department of Health, Education, and Welfare, Food and Drug Administration, Veterinary Medical Branch*. New York: Medical Encyclopedia, 1956.

Wright, Peter M., Ian B. Seiple, and Andrew G. Myers. "The Evolving Role of Chemical Synthesis in Antibacterial Drug Discovery." *Angewandte Chemie (International Edition in English)* 53, no. 34 (2014): 8840–69.

Zavala, Yamily J., and John M. Duxbury. "Arsenic in Rice: I. Estimating Normal Levels of Total Arsenic in Rice Grain." *Environmental Science & Technology* 42, no. 10 (2008): 3856–60.

 CHAPTER 8

Between Bacteriology and Toxicology
Agricultural Antibiotics and US Risk Regulation (1948–77)

Claas Kirchhelle

Since their first use about eighty years ago, antibiotics have acquired great significance in US food production. In 2018, almost 11,567 metric tons of antibiotics were sold for food-producing animals.[1] On farms, antibiotics are used to treat and prevent disease, substitute labor previously devoted to the care of individual animals, and increase feed efficiency.[2] Using antibiotics in food production is not without its risks. Antibiotics can mask disease in animals and enable substandard welfare conditions. If there is insufficient time for drugs to clear animals' systems, residues can taint agricultural produce and sensitize or trigger allergies in humans. Antibiotic-laden excrement can also contaminate soil and water. In addition to welfare and residue hazards, agricultural antibiotic use selects for antimicrobial resistance (AMR). Now considered a major global health challenge, reducing bacterial resistance to antibiotics is an important public health goal. For agricultural regulators, achieving antibiotic reductions without jeopardizing farm productivity and animal welfare is a major challenge.[3]

This chapter traces US officials' response to the antibiotic dilemma between 1948 and 1977. It argues that, from the beginning, antibiotics proved an awkward fit for regulators' traditional division of labor. Antibiotics' status as human and animal drugs that were also being used as agricultural chemicals and food preservatives not only blurred bureaucratic responsibilities but also undermined the traditional risk scenarios guiding regulatory agencies. In the United States, responsibility for antibiotic regulation was divided between three agencies: (1) the Food and Drug Administration (FDA) was responsible for antibiotic licensing, enforcing guidelines, and protecting consumers from residues and AMR; (2) the United States Department of Agriculture (USDA) was responsible for promoting the safe but efficient use of new technologies like antibiotics; and (3) the Communicable Disease Center (CDC) was responsible for monitoring microbial hazards.[4] Often divided in their assessment of antibiotic risks, the three agencies also faced contradictory external pressures to either expand or reduce agricultural antibiotic use. The chapter explores the resulting "bureaucratic pluralism" in which competing agendas and risk interpretations repeatedly fragmented US antibiotic regulation.[5] Rather than develop a comprehensive strategy of drug licensing, stewardship, and access, individual agencies reacted to external pressure and nontraditional risk scenarios like AMR by delaying action, using bureaucratically "safe" but ineffective measures like monitoring residues, and fighting to expand or retain their bureaucratic authority over antibiotic regulation.

The final part of the chapter argues that the piecemeal and increasingly combative nature of official responses to the antibiotic problem should not surprise us and was not solely due to "capture" or regulatory "failure." AMR posed a grave nontraditional challenge to a regulatory system that had been created around the notion that food purity equated safety. Mirroring what contemporary sociologists described as professional bureaucracies' slow, fragmentary, and defensive responses to novel problems, the rupturing of this purity-safety equation in the late 1960s led to regulatory power struggles, delayed decision-making, and unsystematic responses to AMR.[6] In many ways, the resulting fragmentation of visions of safety and misalignment of regulatory priorities over AMR remains a major challenge for US stewardship efforts.

An Acceptable Risk: The Rise of Agricultural Antibiotics (1948–60)

The introduction of antibiotics into US food production started as a trickle and then became a flood. Building on their already substantial experience with arsenic additives and treatments, American farmers and veterinarians first used synthetic sulfonamides to treat individual animals and then to treat entire herds and flocks from the mid-1930s onward.[7] Described by Susan Jones and others, the US poultry industry was an early pioneer of mass medication. By the 1940s, producers across the US were incorporating standardized ratios of potentially toxic sulfa drugs into feed and water to control coccidiosis, fowl typhoid, and pullorum disease. The pharmaceutical industry was quick to capitalize on agricultural demand. In 1948, Merck's sulfaquinoxaline became the first officially licensed antibiotic for routine inclusion in animal feeds and proved an instant commercial success.[8] Poultry producers were not the only converts to mass medication. Starting in 1942, large migratory beekeeping operations began to experiment with sulfathiazole to control European and American foulbrood. Less toxic than previous treatments, sulfonamides—and later penicillin, chloramphenicol, tylosin, and oxytetracycline—were mixed in syrup and pollen substitutes and provided colonies with enough time to clear out infected brood chambers.[9]

Therapeutic applications were joined by nontherapeutic ones. In 1949, industry researchers announced that feeding low-dosed antibiotic supplements could promote animal growth by enhancing feed efficiency. Within a year, antibiotic growth promoters (AGPs) were being used to cut feed costs and prevent disease across the United States.[10] Over the next decade, further antibiotic applications emerged to protect plants against infection, delay bacterial spoilage in fish, shellfish, and poultry, and preserve whale meat. By 1960, antibiotics were ubiquitous throughout US food production. Of 2,130,000 kg of antibiotics produced in the United States, over one-third (770,000 kg) were added to feeds and other nontherapeutic applications.[11]

Initially, most US officials were unconcerned by agricultural antibiotics' meteoric rise. This was in part due to their strong reliance on industry science and in part due to traditional toxicological and microbial risk scenarios. In the case of the FDA, regulators primarily focused on containing toxic—and increasingly carcinogenic—hazards. Guiding regulators were risk models according to which chemical exposure was inevitable and became problematic only once it exceeded a certain threshold. This threshold focus was deeply rooted in FDA history. Since the interwar period, one of the FDA's main tasks had been to protect consumers from toxic pharmaceuticals and tainted food. As a result of the 1927 Hunt Committee, officials focused on defining when chemical exposure turned from "natural" to "unnatural" and evaluated already marketed products.[12] Ten years later, mass fatalities resulting from tainted sulphanilamide cough syrup reconfirmed this focus and led to the passage of the 1938 Federal Food, Drug, and Cosmetic Act (FDC). The FDC significantly boosted FDA powers by allowing officials to judge industry new drug applications (NDAs), which contained information on drugs' composition, manufacturing process, intended use, and evidence of safety. Although NDAs did not cover pesticides and food additives, FDA officials now had unique powers to act as gatekeepers for substances used in medicine and food production.[13]

Their traditional focus on toxic thresholds meant that postwar FDA regulators tended to view low-dosed antibiotic applications favorably. In 1948, Merck's sulfaquinoxaline was allegedly approved in less than twenty-four hours.[14] Only when dosages were potentially toxic did regulators impose restrictions. In 1949, the FDA reacted to reports of self-medication on farms and the poisoning of humans and animals by introducing mandatory safety labels and maximum concentrations for premixed sulfa feeds and concentrates. Trusting in consumer compliance, safety labels became a favorite FDA means to influence behavior without disrupting the pharmaceutical market and expanding policing. The rules established for sulfonamide feeds and concentrates were seamlessly applied to AGPs.[15] Retrospectively legalizing an already booming market, the FDA licensed a broad range of AGPs in 1951 and 1953. Ready-mix feeds were to include denatured drugs, labeled "for feeding use only," and include no more than 50 grams per ton—an already existing industry standard. Only drugs for which no reasonable usage guidelines could be developed were assigned prescription-only status from 1951 onward.[16]

In addition to licensing many other new products, the FDA also collaborated with the USDA to promote antibiotics' adoption on farms. Repeatedly opposing veterinary calls for prescription requirements, regulators removed bureaucratic hurdles for drug manufacturers and feed mills.[17] Meanwhile, USDA officials advertised antibiotic feeds and therapeutics in manuals, yearbooks, and agricultural journals. Building on a general sense of optimism about agriculture's "chemical revolution," antibiotics were heralded as a safe and progressive way to produce more with less labor and feed.[18]

Regulators' antibiotic optimism was soon challenged. In 1950, dairies were already complaining about the destruction of bacterial starter cultures by antibiotic residues in milk. The residues were caused by farmers using antibiotics to treat mastitis in cows and not discarding contaminated milk.[19] As described by Kendra Smith-Howard, FDA officials first reacted by informing farmers about the time that was required to allow antibiotics to clear cows' bodies, then by lowering the dosage of mastitis treatments, and finally by pioneering a residue monitoring program in 1960. However, surveillance and enforcement remained limited to culturally sensitive milk.[20] In 1954 and 1958, the Miller and Food Additives Amendments to the FDC had overturned earlier FDA attempts to define antibiotics in meat as adulteration (see below) and instead legalized thresholds below which residues could be deemed "safe."[21] Seven parts per million (ppm) residue tolerances subsequently enabled the licensing of antibiotic preservatives for raw poultry, fish, scallops, and shrimp in 1955, 1956, and 1959. Similar to their assessment of AGPs, FDA regulators viewed these residue levels as too low to pose toxic or carcinogenic hazards. Concerns about AMR selection were of secondary importance.[22]

While FDA regulators focused on toxicity, the newly founded CDC assessed antibiotics' effects on individual pathogens. In 1947, the CDC had established a specialized Veterinary Public Health Section (VPHS) and Laboratory (VPHL). The VPHS initially concentrated on evaluating well-known microbial threats like typhoid, brucellosis, bovine TB, and bacterial food poisoning. The fragmentation of US laboratory networks meant that this was no easy task. For much of the 1950s, CDC experts struggled to convince local and state health authorities to share and standardize data collection.[23] The lack of national surveillance systems had serious consequences for the evaluation of antibiotic risk. Instead of worrying about abstract scenarios of AMR selection in nonhospital settings, the CDC focused limited resources on surveying outbreaks of individual pathogens. This "organismal" focus entailed an awareness of nonhuman AMR insofar as it undermined treatments of specific microbes in individual animals and herds but not as a wider environmental hazard to human and animal health.

The CDC's "organismal" view of AMR became clear in August 1952 when the American Veterinary Medical Association (AVMA) asked the US Committee on Public Health to assess antibiotic additives. Headed by VPHS chief James H. Steele, a specialized veterinary public health committee studied the situation. From the beginning, AMR concerns were mostly limited to the presence or absence of drug residues, which might select for resistant pathogens in food. Experts cited FDA warnings that antibiotic residues and preservatives might "result in the emergence of strains of pathogenic microorganisms resistant to these drugs" but shied away from a wider environmental risk evaluation of agricultural AMR selection.[24] This was despite US reports on AMR selection in AGP-fed poultry and lab animals, as well as on AMR transfer from resistant to sensitive pneumococci.[25] In their final report, experts called for more research, the evaluation of human feed experiments being conducted in Guatemala, and awareness campaigns against "the promiscuous use of [higher-dosed mastitis] antibiotics which may cause serious emergencies."[26] Lower-dosed antibiotic feed additives were dangerous if they left residues in food and triggered consumer allergies but not necessarily if they selected for AMR in nonhuman settings away from consumers' tables. In the absence of contradicting official AMR data, this relatively benign assessment was endorsed by an international conference on the use of antibiotics in agriculture three years later.[27] Throughout the 1950s, most epidemiologists believed that natural or mutational AMR in individual organisms could be passed to others only hereditarily—that is, from one generation to the next (this is known as vertical transmission.) Although selection pressure aided this proliferation by suppressing sensitive competitors, most popular feed additives were deemed too low-dosed to cause serious problems. Even if individual pathogens had acquired resistance, it was thought that their proliferation could be contained by reducing selection pressure, using antibiotic combinations, and applying strict hygiene protocols.[28]

Should they clash, residue concerns trumped AMR warnings. In early 1957, public health officials warned that FDA plans to limit commercial mastitis treatments to 100,000 units per dose would not only fail to curb residues but also increase treatment times: "When the treatment time is extended, some of the organisms develop resistance or tolerance toward the antibiotic, and the bacterial flora of the udder changes from the usual high strep count to a predominantly high staph and pseudomonas count."[29] CDC concerns were seconded by researchers at the USDA's Beltsville experimental station. However, Antibiotics Division director Henry Welch "did not request comments on the plans, which the FDA expect[ed] to implement immediately."[30] The rules of 1950s antibiotic regulation were clear: spreading antibiotics' benefits was a priority as long as they did not contaminate milk or leave toxic residues in meat. AMR was an afterthought.

Standard Responses to Nonstandard Threats: Antibiotic Reform (1960–66)

The initial mode of US antibiotic regulation came under pressure from around 1960 onward. Faced with growing concerns about many of the new chemicals on farms and in households, officials were forced to revisit, and in some cases overturn, earlier regulations. Budgetary constraints and corruption allegations worsened regulators' position.[31] Attempting to regain public trust, the FDA, USDA, and CDC announced ambitious monitoring programs to protect US food purity by combatting salmonellosis and drug residues in 1964 and 1966. Investment in these traditional modes of food protection coincided with new risk scenarios centering on horizontal AMR transfer—that is, the sharing of the trait between organisms in contact with each other, not just progeny (vertical transmission) but neighbors (horizontal transmission). Between 1959 and 1965, research on bacteria's ability to "communicate" AMR not only within but also across species amplified the potential costs of agricultural antibiotic use, and of low-dosed AGPs in particular. Even AMR selection in harmless commensal bacteria could lead to the dissemination of mobile resistance genes (R-factors) to pathogens. This nontraditional ecological risk scenario undermined existing antibiotic policies and led to growing tensions between regulatory agencies.[32]

The situation was particularly uncomfortable for the FDA. Between 1959 and 1961, Senate hearings shed an unfavorable light on ties between regulators and industry and led to the sacking of Antibiotics Division chief Henry Welch. Still working with 1950s staffing levels, an increasingly overburdened FDA struggled to streamline contradictory policies and respond to allegations that it was "flying by the seat of [its] pants."[33] Morale plummeted. While senior officials surrounding Commissioner Larrick adopted a siege mentality, dissenters expressed growing alarm at industry influence on decision-making. In 1962, members of several FDA divisions issued memos criticizing close ties of some investigators to industry, vague regulatory guidelines, and attempts to legalize AGP rations for mature animals.[34] A parallel FDA program of farm and feed mill inspections added to concerns about producers' compliance with drug guidelines. Inspectors submitted sobering reports: substandard or banned antibiotics were being marketed; antibiotics were being fed to animal species they had never been licensed for; some mills lacked basic weighing equipment; mixed leftover medicated feeds were being mixed with new drugs; and owners were trying to intimidate inspectors. According to one report, "Some of the

weirdest things were going on in the way of irrational [drug] combinations being used."[35] On larger poultry farms, "convenience and economic factors" often
led to a total disregard of FDA guidelines.[36] Drug residues in meat were likely.[37]
Although Commissioner Larrick continued to promote voluntary measures
and educational measures in reaction to increasing data on noncompliance,
pressure for widespread reform was mounting.[38]

A significant reform impulse occurred ahead of Larrick's 1965 resignation.
Reacting to USDA reports of penicillin in red meat, FDA officials hastily commissioned an ad hoc expert group on antibiotic residues in food and the use
of antibiotics as food preservatives.[39] Parallel warnings about "infectious resistance" meant that the ad hoc committee also assessed AMR hazards.[40] However, in their 1966 report, assembled experts once again limited AMR concerns
to the presence or absence of residues. Prioritizing traditional risk scenarios,
experts called for updating several practices, including a critical review of antibiotic tolerances, national residue monitoring for antibiotics in meat, and new
labels or withdrawals of dangerous and residue-prone drugs. Although experts
expressed concern about AMR proliferation, they limited policy proposals to
financing more research and reducing selection pressure by curbing residues.[41]

Happy to sidestep AMR-inspired antibiotic restrictions, the FDA endorsed
nearly all aspects of the ad hoc report. In addition to commissioning a 1967
conference on the use of drugs in animal feeds, which was inconclusive, officials revoked antibiotic tolerances, reviewed inefficacious and persistent
antibiotic products, and collaborated with the USDA to establish national residue monitoring for meat.[42] USDA inspectors would collect random samples
to establish a statistical overview of contamination levels. Carcasses showing
suspicious injection marks would also continue to be sampled. Residue detections would lead to targeted FDA follow-up monitoring.[43] Although it was
impossible to inspect every carcass, officials hoped that the new mix between
random and targeted monitoring would put the "fear of God or FDA" in antibiotic abusers.[44] According to the FDA and USDA, consumers had a right to
chemically pure food; whether its production would select for AMR continued
to be of secondary importance.

The 1966 push to rid food of invisible agrochemical residues coincided
with new campaigns for eliminating microbes. Although CDC data remained
patchy, a steady decline of well-known diseases such as typhoid had been par-
alleled by a more than fourfold increase of reported salmonellosis-associated
food poisoning cases from 1,733 in 1951 to 8,500 in 1961.[45] Underreporting
meant that annual cases likely numbered in the hundreds of thousands. Similar to what Anne Hardy has described for Britain, the microbial ecology of
US outbreaks was also changing.[46] Since the 1940s, new production and food
processing technologies had created ideal environments for familiar and novel
microbial hazards. In the case of salmonellosis, previously unknown salmo-

nella serotypes were supplanting "native" serotypes. Although *Salmonella (S.) typhimurium* (as the strain was then called) remained the most common cause of foodborne salmonellosis, *S. reading* emerged as a major new source of salmonellosis in 1956. *S. blockley* isolates were rare until the bacterium rapidly spread throughout the US poultry industry from 1955 onward. *S. heidelberg* was not detected in the United States before 1953 but became the second most common type of salmonella isolated from humans and animals in 1963. In 1958, *S. dublin* emerged as a source of severe salmonellosis in cattle and humans west of the Rocky Mountains.[47]

For the CDC, curbing salmonellosis became both a priority and a welcome way to systematize national disease surveillance and gain influence. In 1961, it acquired responsibility for the collection of human salmonellosis reports from the National Office of Vital Statistics and established a Salmonella Surveillance Unit. By 1963, weekly surveillance reports were being compiled with information from forty-seven state health departments and five typing centers.[48] After hosting a national salmonella conference in 1964,[49] the CDC also trialed

Figure 8.1. Typhoid fever and other salmonellosis reported in the United States (1946–62). Adapted from *Proceedings: National Conference on Salmonellosis, March 11–13, 1964* (Washington, DC: US Government Printing Office, 1965), 7.

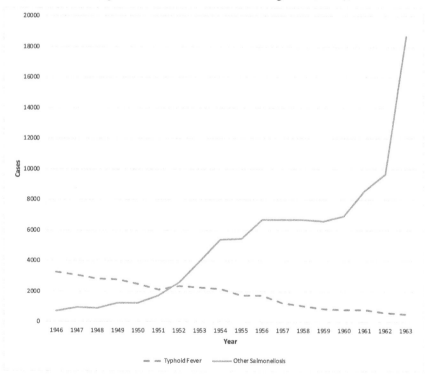

more active surveillance of salmonellae in human food and frequently contaminated animal feeds.[50] By 1966, officials were investigating 20,865 bacteriologically confirmed infections in humans (up from 1,700 reported infections in 1956).[51] Based on the parallel success of campaigns against S. *pullorum* and S. *gallinarum*, it was hoped that better data, industry compliance, 1965 pasteurization rules for frozen, liquid, and spray-dried egg and milk, and a 1968 federal (USDA/FDA) rendering plant certification program for states would curb other causes of salmonellosis.[52]

Surprisingly, systematized CDC salmonella surveillance did not lead to routine AMR monitoring. This was not because there was no interest. In 1958 and 1960, CDC researchers compared tetracycline and chloramphenicol resistance in contemporary S. *typhimurium* outbreaks with previous isolates. Whereas none of the two hundred cultures isolated prior to 1948 were resistant, 5 percent of one hundred human cultures and 9 percent of one hundred fowl cultures isolated between 1956 and 1957 were tetracycline resistant. Between 1959 and 1960, tetracycline resistance had risen to 13.9 percent among human cultures and 29 percent among animal cultures. Chloramphenicol resistance had also emerged in 1.9 percent of human isolates. The authors warned that "within a few years the percentage of salmonellae resistant to tetracyclines will reach sizable proportion" and called for more systematic surveillance.[53] In 1962, a further CDC study of 652 salmonellae cultures revealed another increase of chlortetracycline resistance amongst S. *typhimurium* isolates to 38 percent.[54] However, no AMR monitoring program emerged.

The seeming disinterest in nonhuman AMR can be explained by senior CDC officials' prioritization of familiar risks, which could be tackled using tried and tested means. Similar to the FDA's emphasis on residues, officials chose to focus on the traditional risk of microbial food contamination. Although improving food safety was important, their narrow focus on traditional risk scenarios made officials react slowly to mounting evidence of rising AMR on farms, which would have required novel regulatory interventions.

In addition to salmonellae, a good example of officials' narrow "organismal" prioritization of measures designed to curb the microbial hazards of well-known bacterial pathogens—and not by organisms' resistance profiles or by floating resistance genes ("infectious resistance")—were their slow reactions to rising AMR in staphylococci. In medicine, a contemporary pandemic of resistant *Staphylococcus aureus* phage type 80/81 had triggered campaigns for "rational" antibiotic use and boosted the status of infection control specialists.[55] For a while, the pandemic also heightened interest in potential animal reservoirs of Staph 80/81. In May 1960, Hawaii-based public health veterinarians surveyed dairy cows on Oahu. Using a technique called bacteriophage typing to differentiate between bacteria strains, researchers identified four cows as Staph 80/81 carriers. A dairy employee was subsequently found to be a carrier with "a large draining furuncle [boil] on his back at the waistline" and a history of furuncles on his finger.[56] Further examination revealed that his wife and eighteen-month-old child also suffered from boils and chronic sore throat. Human and animal isolates had a mostly identical antibiogram and were resistant to chlortetracycline, oxytetracycline, tetracycline, and dihydrostreptomycin. While it was possible that the dairy worker had infected the cows, frequent antibiotic mastitis treatments meant that it was also "possible that the resistant strains developed in bovine mammary glands."[57]

The Hawaii study triggered further research by the CDC's VPHL. In mid-1959, VPHL chief Mildred Galton had already agreed to support a study on bovine staph carriage proposed by the Los Angeles State Department of Public Health. Milk samples from a 180-cow operation employing convicts would be sent to the CDC for bacteriophage typing. An additional study of staph carriage among dairy workers was turned down: turnover was too rapid and the sheriff's department might baulk at external surveillance.[58] Between October 1959 and March 1960, the CDC typed 1,023 milk samples of which 74 percent contained coagulase-producing staphylococci. Although no Staph 80/81 was detected, 54 percent of 855 typeable cultures could be typed with human bacteriophages and showed varying levels of AMR.[59] Reflecting on the study and other contemporary investigations, CDC researchers acknowledged links between food poisoning and staphylococci in milk, cows, and their human handlers. Although it was impossible to distinguish between human and animal strains, resistant staph isolates indicated a potential public health threat. This threat was exacerbated by the "widespread use of antibiotic therapy in mastitis and the demonstration of antibiotic residues in milk . . . [and] by the addition of antibiotics to animal feeds as a prophylactic measure."[60] However, rather than propose novel policy measures, such as treatment restrictions, officials advocated traditional interventions, such as pasteurization and better veterinary health care. Described by Maryn McKenna, contemporary CDC investigations of Staph 80/81 outbreaks among abattoir workers similarly failed to lead to a re-evaluation of popular antibiotic preservatives.[61] Despite mounting evidence of resistant salmonella and staphylococci, the CDC did not yet view nonhuman AMR selection as a serious risk *sui generis*.

In contrast to the United Kingdom, where centralized epidemiological surveillance led to the first AMR-inspired AGP review in 1960, traditional food purity concerns remained foremost in the minds of US officials.[62] With the FDA, USDA, and CDC united in their quest to reduce salmonellosis and antibiotic residues, no agency was willing to "take ownership" of the new ecological problem of agricultural AMR selection.[63]

In Search of Safety: AMR and AGPs during the 1970s

Fragmented attempts to address nonhuman AMR selection only emerged during the 1970s. Under pressure to react to European regulators' decision to restrict the use of medically relevant "therapeutic" AGPs due to AMR hazards in the European Economic Community, the previously united front of US food regulation cracked. While the USDA defended agricultural antibiotics on economic grounds, CDC officials played an important role in raising AMR awareness. Caught between agro-industrial and public health interests, the FDA first delayed regulatory action by seeking external advice and then failed to implement AGP restrictions in the face of industrial opposition, internal divisions, and growing political attacks on alleged overregulation.

In the case of the CDC, internal AMR assessments shifted decisively following 1966 warnings about a return to the "pre-antibiotic middle ages."[64] Citing British research, an editorial in the *New England Journal of Medicine* explicitly attacked uncontrolled AGP use on farms. The impact on CDC rhetoric was significant. Whereas officials had previously described agricultural AMR selection as a secondary threat, they began to acknowledge it as a problem *sui generis*. During a 1966 lecture, Eugene Gangarosa, head of the CDC's Enteric Diseases Branch, explicitly addressed the "broad public health implications" of "extensive use of antibiotics in poultry and animal feed."[65] Although the practice had resulted in a "sharp decrease in overt salmonellosis," an unintended consequence had been "an increase in antibiotic resistant salmonellae to which man [*sic*] is now exposed."[66] According to the CDC, US agencies' earlier prioritization of traditional "organismal" salmonellosis control over antibiotic stewardship was no longer necessarily valid. However, more data was required to test this new risk prioritization.

Speaking at the FDA symposium on drugs in animal feeds one year later, VPHS chief James H. Steele set out an ambitious plan to study agricultural AMR selection. Although Steele reassured participants that nobody was contemplating blanket bans, he stressed the importance of evaluating antibiotics' impact on the microbial environment. Doing so would be a significant undertaking "and may even extend to the next century," he said.[67] Surveillance would have to be regional, national, and international and also include areas not using antibiotics, though, in Steel's words, "such areas are few today and will probably disappear in the next decade or two."[68] Steele proposed integrating clinical observations, lab-based studies, mathematical modeling of genetic change over time, and routine epidemiologic AMR surveillance. Extensive laboratory support and an additional fifty to a hundred CDC epidemiologists would achieve the "necessary surveillance, in the shortest time and at the least cost."[69] Additional laboratories could be contracted to establish "so-called normal microbial pathogens, and to note changes in resistance, susceptibility, etc."[70] Detailed surveys of national drug use should also be "initiated as soon as possible."[71]

Although such a program only emerged in the United States in 1996 in the wake of heightened concerns about re-emerging infections, Steele's prescient speech set out essential components underlying modern AMR surveillance. It also marked the beginning of systematic CDC research on nonhuman AMR selection. From 1967 onward, the CDC began to conduct in-house research on horizontal gene transfer.[72] In addition to requesting multiresistant *S. typhimurium* strains from Britain, CDC experts studied AMR in salmonellae outbreaks:[73] of four hundred representative strains from across the United States, 22 percent (89) were resistant to one or more of eleven antibiotics. *S. typhimurium* isolates were again significantly more resistant (41.1 percent) than other serotypes and frequently carried mobile R-factors conferring resistance against multiple antibiotics. While AMR in salmonellae seemed to have stabilized, it seemed to be increasing in other enteric organisms, such as *Escherichia coli*.[74] Between 1967 and 1971, CDC experts also reported a sudden increase of transferable ampicillin resistance in shigella.[75] In 1972, concerns increased following the nearly simultaneous detection of mobile multiple AMR—including chloramphenicol resistance—in virulent Central American strains of *S.* Typhi, *Shigella flexneri* 2a, and *Shigella dysenteriae* 1.[76] With over ten thousand confirmed cases occurring in Mexico, the multiple resistant typhoid outbreak in particular seemingly confirmed earlier warnings about the ecological fallout of mass agricultural antibiotic use by British experts and posed a concrete risk to US public health.[77] Although CDC experts remained cautious about publicly endorsing European AGP restrictions, there was no longer any doubt that they considered AMR a serious national threat in both human and nonhuman settings.

CDC action contrasted with FDA hesitancy. Although they had begun to withdraw inefficacious veterinary antibiotics after 1966, FDA officials had failed to develop an internal strategy regarding nonhuman AMR selection that went beyond curbing residues. As a consequence, they were surprised when Britain's prestigious Swann Committee on agricultural antibiotics recommended banning medically relevant AGPs in late 1969. On 21 November 1969, FDA's associate commissioner for science, Dale Lindsay, sent an urgent message to Bureau of Veterinary Medicine (BVM) director, Cornelius Donald Van Houweling. Van Houweling and Lindsay were to "develop an agency position on the use of antibiotics in animal feed" in response to press queries.[78] Van Houweling summarized the US situation: "The use of drugs, including antibiotics, in animal feeds has increased tremendously in recent years. It is reliably estimated that approximately 40 million tons of animal feed containing drugs was consumed in 1968. Also, that almost 80 percent of the meat, milk, and eggs consumed in the United States comes from animals fed medicated feeds."[79] Personally, Van Houweling saw no reason for quick action but noted that the FDA Bureaus of Science and Medicine were "acutely aware of the concern expressed in regard to the possible ecological effects of using these large amounts of antibiotics in animal feeds."[80]

Faced with conflicting risk assessments, FDA commissioner Charles Edwards decided to displace pressure by seeking external advice. Announced in April 1970 and sitting until February 1972, an FDA task force was charged with assessing antibiotic feeds. Meanwhile, interim regulations enabled ongoing AGP use. This decision proved controversial. Criticizing the BVM for insisting that AGPs remain available, Division of Microbiology deputy director, Robert Angelotti, warned, "Our course of action should be one of preventing an increase in the pool of resistant microorganisms."[81] Angelotti's warnings were eventually endorsed by the FDA task force. After heated clashes between its medical, agricultural, and industrial members, the task force recommended withdrawing tetracyclines, sulfonamides, dihydrostreptomycin, penicillins, and all other approved antibiotics from feeds if producers could not prove that they were safe. The safety proof had been inserted as a last-minute concession to industry to avoid a breakdown of deliberations. However, because of this concession, no restrictions occurred. Between February and August 1972, minority reports, hostile industry reviews, and critical NAS and USDA reports forced Commissioner Edwards to significantly water down testing requirements. In the end, no AGP was deemed unsafe.[82]

This development was greeted by USDA secretary Earl Butz, who had personally defended AGPs in 1970, but was criticized by CDC experts.[83] In late 1972, CDC Bacterial Diseases Branch chief John V. Bennett endorsed AGP restrictions in a tense letter exchange with antibiotics expert Maxwell Finland, who had led the hostile NAS-review of the task force recommendations.[84] According to Finland, the FDA task force had made a political rather than a scientific decision: "I think we can go on endlessly . . . . You say I (we) should not polarize my (our) opinion [*sic*], but the Task Force did in fact do so on the same basis by asking for a positive recall before definitive data were available."[85] Bennett responded: "How large does a risk:benefit ratio need to be to justify public health action? Personally, I feel the ratio is such that, given available data, the actions of the Task Force and Swann Committee were merited."[86]

In the absence of concrete proof of harm, official disagreement over AGP withdrawals continued. During the mid-1970s, the withdrawal faction was strengthened by studies on the spread of resistance from AGP-fed animals to farm families, as well as by official data on rising AMR in salmonellae.[87] However, FDA attempts to gain robust external endorsement of statutory bans remained unsuccessful. In early 1977, industry celebrated a last-minute victory over bans being proposed by an FDA-staffed subcommittee of the National Advisory Food and Drug Council. Appointed shortly afterward, the new FDA commissioner and former microbiologist Donald Kennedy unilaterally announced bans of penicillin and tetracycline AGPs in April 1977. Reactions highlight the extent to which US officials were divided over antibiotic stewardship: AGP bans were supported by the FDA's Division of Microbiology and the CDC but faced lackluster endorsement by the BVM and staunch opposition from industry-sponsored think tanks and USDA agronomists. Bogged down by lengthy withdrawal procedures, Kennedy's bans stalled. In 1978, Congress imposed a moratorium on FDA action and commissioned more NAS research. A frustrated Kennedy resigned in 1979, while vague terms of reference made the NAS call for further research in 1980. Democrat-led attempts to revive AGP-restrictions in 1981 and a 1985 withdrawal petition by the National Resources Defense Council suffered a similar fate.[88]

With the USDA, CDC, and FDA divided over the risks of agricultural AMR selection, it remained easy for lobbyists to sow further doubt and for industry-friendly politicians to delay regulatory action by calling for more research. Created in 1970, AGP interim regulations remained in place until new FDA guidelines on voluntary withdrawals were announced in 2013.

Treating Symptoms: Residues and Salmonellosis (1966–78)

Public divisions over AMR stood in contrast to US agencies' ongoing commitment to chemical and microbial food purity. In 1966, the USDA already analyzed 885 meat samples of which 141 (15.9 percent) tested positive for antimicrobial residues.[89] One year later, random monitoring was expanded to 3,900 red meat and 1,300 poultry samples.[90] Federal inspectors were also attempting to control salmonellae in food and feeds. However, it soon became clear that enhanced monitoring would accomplish little without wider reforms of enforcement, drug licensing, and withdrawal procedures. It was here that regulatory coordination faltered.

In April 1969, the BVM already warned that compliance on US feed mills remained problematic. A total of 8,567 registered medicated feed manufacturers had to be inspected every two years.[91] However, handing over inspections to states had backfired. The FDA had all but withdrawn from feed mill inspections in commissioned states, but states had inspected only 799 mills in the fiscal year (FY) 1968.[92] Without controls, noncompliance was widespread: over 45 percent of inspected manufacturers were not assaying feeds, and, in 25 percent of those that were conducted, the feed did not meet federal standards.[93] When such violations were detected, enforcement was toothless. In 1968, the FDA Division of Case Guidance complained that one manufacturer had been selling feeds containing an unapproved drug since 1961. The firm had stated that it would rather ignore FDA warnings "than lose the customer."[94]

According to the BVM, the situation was dangerous and residues likely: "We are becoming increasingly concerned that . . . our approving medicated feed applications represents little more than a "rubber stamp" operation. . . . In the apparent absence of a reasonably vigorous regulatory program we are witnessing an overall relaxation of controls in the manufacture of medicated feeds."[95] Amounting to just under 4 percent of FDA staffing and just over 3 percent of the FDA's budget, the 166 labor-years and $2.4 million allotted to the BVM were clearly insufficient to adequately control the vast quantities of veterinary drugs and feeds used in the United States.[96]

Scarce resources continued to undermine federal controls throughout the 1970s. Despite an ongoing rise of US salmonellosis reports, the USDA-FDA-industry program of enhanced controls of hygiene and heat treatments in rendering plants producing animal protein supplements like meat and bone meal was dropped in 1972. More successful programs against fowl typhoid and pullorum disease were stopped in 1974 and 1976 and replaced with a "sanitation monitored" certification program.[97] Antibiotic abuse also remained widespread. Although feed mill inspections increased slightly, they remained inadequate to curb noncompliance. Between 1970 and 1971, 12.7 percent of 347 FDA-examined (financial year 1971), and about 25 percent of approximately 10,000 state-examined, feed samples (financial year 1970) were violative.[98] It also emerged that nine of thirty-two states participating in feed mill controls had an inspection coverage of less than 5 percent.[99] The situation was not better on US farms, where inspectors reported nonadherence to withdrawal times and the destruction of drug records after animals had been sold.[100] The small size of residue monitoring in relation to the billions of animals slaughtered every year made it unlikely that offenders would be caught.[101]

Problematic FDA enforcement exacerbated the situation. Already identified as a problem in the 1950s, the FDA remained unable to assay all of the drugs it had licensed. In 1971, Denver district inspectors complained that major feed producer Elanco was illegally selling tylosin as a "'not new drug' for growth promotion."[102] Inspectors accused headquarters of willful ignorance:

> For the last few years we have been given the impression that the administration is "disinterested" in tylosin, because of assay problems. To conveniently ignore the tylosin problem does not reduce its potential for mis-use and only tends to further our lack of effective control of this drug. If we cannot rely on assays for regulatory consideration, what is the procedure that an inspector in the field should follow? Denver works with commissioned state officials . . . . Are they to consider it part of their "training" to ignore questions when the answer is difficult or inconvenient?[103]

Even if assays had been available, punishing noncompliance remained difficult. Withdrawing manufacturing licenses was an ineffective and "time-consuming process" requiring "due notice and an opportunity for a hearing . . . following failure to comply within a reasonable time after written notice."[104] Regulators could mostly only threaten withholding future approvals. The situation was not better regarding residue offenders. Between January 1972 and 1978, the FDA conducted only 282 follow-up investigations as a result of oxytetracycline, chlortetracycline, tetracycline, and/or penicillin residue detections. Successful prosecutions were rare.[105] In early 1977, the USDA was forced to inform a woman who was allergic to antibiotics that it was best to avoid conventional produce and buy from local alternative farmers.[106]

Enforcement problems were partially due to FDA licensing decisions. In 1975, the FDA's BVM had become embroiled in a major scandal after an internal report revealed that officials were unable to fulfill legal assay requirements for twenty-seven licensed drugs. During subsequent congressional hearings by Senator Edward Kennedy, it was investigated whether BVM director Van Houweling had pressured staff to license drugs for which there were no assays.[107] Although there was insufficient evidence to prosecute, parallel carcinogenicity allegations against nitrofuran antibiotics revealed how harmful the lack of official assays could be.[108] According to a 1976 General Accounting Office (GAO) report, the FDA had first been warned about nitrofurans' carcinogenicity in 1965 and twice (!) by manufacturers in 1967 and 1968.[109] However, despite zero-tolerance provisions against carcinogens in 1958, Van Houweling had not directly moved against the popular animal drugs. Following contradictory FDA reviews, he considered nitrofurans safe as long as residues did not reach humans.[110] This position was criticized by the FDA's Bureau of Food (BOF). Calculating safe exposure at less than two parts per billion, a BOF veterinarian argued that nitrofurans "have no place in the treatment of food-producing animals."[111] BOF pressure made the FDA issue notices of hearings on nitrofuran withdrawal procedures in 1971. Informed by Van Houweling, industry, however, claimed that extended withdrawal times would eliminate hazardous residues. By 1976, the FDA had neither accepted nor denied industry hearing requests.[112] In the absence of official assays with which to detect whether and how many carcinogens were reaching consumers, the FDA could not legally prove imminent harm.[113]

BVM leniency extended beyond the nitrofurans. Between 1973 and the second quarter of 1976, USDA residue monitoring detected an increase of violative sulfa samples from 1.4 to 11.1 percent. Follow-up investigations revealed that most violations were caused by feeds containing sulfamethazine (SMZ). The swine industry was particularly affected because of SMZ's inclusion in a popular feed mix called ASP-250 and the drug's tendency to contaminate later batches of feed produced in the same mill or used on the same farm. The FDA briefly considered withdrawing all sulfa-chlortetracycline-penicillin combinations but ultimately only extended mandatory withdrawal times.[114] This relatively mild response was again partly due to BVM director Van Houweling, who initially asked the BOF to raise SMZ tolerances.[115] Van Houweling remembered the situation in a 1990 interview. Although he had been tough on offending farmers, he considered a certain degree of cross-contamination in larger feed operations almost unavoidable.[116] Reflecting on the fact that his decisions had not always erred on the side of caution, he remembered "a kind of backhanded compliment" by FDA general counsel Peter Hutt: "[Hutt] said I'd have been a good executive secretary for a livestock association (laughter)[;] I always was concerned about the economic end. Coming from a farm background, farming is not, you know, a great big wonderful deal. Usually operating on a pretty close margin, and I think they ought to have every advantage they can get."[117] BVM deputy Fred Kingma was less complimentary about this period: "Congressional interest was at a peak insofar as veterinary medicine was concerned, some of it brought on by people working in the bureau taking actions which were questionable to some of us."[118]

Undermining the FDA's ability to combat noncompliance and residues, BVM licensing and enforcement came under pressure around 1978. This was in part due to the reforming zeal of new FDA commissioner Donald Kennedy and in part due to expanded and improved USDA residue monitoring. Whereas the USDA had analyzed around a thousand samples in 1966, it tested approximately twenty-five thousand samples in 1979.[119] In 1978, the USDA also introduced a new Swab Test On Premise (STOP), which was quicker and more sensitive to a broader range of antibiotics than previous tests. In 1977, STOP trials had already revealed that around 10 percent of slaughtered dairy cows had antibiotic residues in their tissues. For the FDA, this sudden rise in residue detections was bad news. According to a 1978 memorandum, the FDA would "be faced with a number of violative residue reports far above its current capability to conduct follow-up investigations."[120] In addition to overburdening inspectors, FDA prosecutors still lacked assays that were reliable enough to withstand precision cross-examination in court. Following an internal review, Commissioner Kennedy tried to improve controls over the veterinary drug market by focusing on securing successful prosecutions, stripping the BVM of its power to overrule the BOF, and prompting Van Houweling to retire.[121]

However, Kennedy's measures were short-lived. Following his 1979 resignation, ongoing funding problems and the Reaganite rollback of federal enforcement undermined reforms. In 1985, Congress's Intergovernmental Relations and Human Resources Subcommittee published a scathing review of medicated feed oversight. According to internal FDA estimates, "as many as 90 percent or more of the 20,000 to 30,000 new animal drugs estimated to be on the market" had not been approved as safe and effective.[122] Although systematized hygiene inspections and public shaming have since reduced problems, the disconnect between 1960s reform goals and outcomes remains stark.[123] Similar to AMR, one of its causes was the tendency to monitor symptoms like drug residues but not redress underlying problems like uncontrolled antibiotic access and licensing decisions. Although surveillance played an important role in keeping residues in the spotlight, US officials proved unable to self-correct problematic decision-making by individual agencies and to formulate coherent agendas for agricultural drug policy. This undermined their public health goals.

Conclusion: A Normal Breakdown?

The history of antibiotics confirms and complicates central observations from the rapidly growing research body on risk and substance regulation. Actions by the USDA and Van Houweling's BVM seem to lend weight to accounts prioritizing narratives of twentieth-century regulatory "capture" by industry for substances ranging from lead to cigarettes.[124] Meanwhile, US authorities' early focus on establishing "safe" levels of antibiotic use and residue tolerances speaks to the large body of academic publications focusing on how regulatory and scientific "boundary work" played an important role in legitimizing risk from around 1800 onward.[125] However, the described dilemmas of regulating a group of substances whose risks only gradually became evident and continually transgressed regulatory boundaries also point beyond familiar narratives and challenge us to historicize not just notions of risk but underlying ideals of safety.

Antibiotics entered the US agricultural market in the midst of a significant recalibration and extension of state authority that was designed to provide consumers with abundant, safe, and "pure" food. This equation of safety with chemical and microbial purity structured the policies of the USDA, the still young FDA, and the newly founded CDC. It also led to a form of risk management on the part of all three agencies that prioritized established policies against residues and bacterial contamination, but mostly ignored AMR hazards and made the rapid expansion of antibiotic use across US food production seem "safe." The purity-safety equation also structured initial responses to new ecological scenarios of horizontal AMR proliferation. Continually struggling to attract resources and under pressure to maintain public trust in their authority, 1960s regulators' first reaction was to amalgamate AMR politics with well-established and culturally acceptable policy modes. Following what Xaq Frohlich has described as the marketplace-focused mode of postwar food regulation, all three agencies intensified monitoring for chemically or microbially "impure" food that was about to reach consumers, but failed to reform the production practices and supply chains driving antibiotic use and AMR.[126]

This "inappropriate" initial response to AMR bears remarkable resemblance to mid-century sociologists' observations on bureaucracies' "learned inability" to adapt to nontraditional challenges.[127] It also foreshadowed significant infighting once the interagency purity-safety equation broke down. The fact that AMR could not be managed according to familiar risk protocols but required a wider recalibration of agricultural production systems exposed competing agendas both between and within government agencies. These agendas were shaped by agencies' core missions and mirrored increasingly polarized public disagreements between medical and agricultural experts over whether or not nonhuman AMR selection posed a significant hazard.[128]

Tasked with protecting the nation's health from microbial threats, the CDC had initially failed to view agricultural AMR selection as a hazard *sui generis*. This assessment changed around 1965 when CDC experts embraced horizontal AMR scenarios. Officials subsequently attempted to use AMR's ecological dimensions to capture new fiscal and bureaucratic resources by lobbying for national AMR surveillance and targeting agricultural production practices, which had traditionally been under the purview of the FDA and USDA. What was an opportunity for the CDC infringed upon the interests of the USDA. Although its joint mission was to promote agricultural productivity and protect consumers' health, the postwar period saw the USDA repeatedly prioritize productivist agendas aimed at maximizing the efficiency of US food production.[129] This tendency became particularly pronounced under USDA secretary Earl Butz between 1971 and 1976. Having defended AGPs as an expert witness in 1970, Butz and other USDA officials launched defensive research initiatives and warned about the economic costs of potential bans.[130] In contrast to stalled CDC demands for expanded AMR monitoring, the USDA successfully emphasized the need for more limited produce-focused residue regulation, drug monitoring in meat and milk, and industry self-control.

CDC and USDA disagreements complicated FDA regulation. Despite being uniquely placed to take action due to its combined power to regulate agricultural drug use, production methods, human medicines, and food safety, the FDA struggled to develop a coherent antibiotic policy. FDA commissioners not only had to compromise between USDA and CDC positions but also had to navigate increasingly fierce internal factionalism with the agriculture-focused BVM repeatedly prioritizing productivity and farmer interests and the BOF and other divisions prioritizing data on AMR-related human health hazards. FDA commissioners initially tried to displace inner- and inter-agency tensions by installing external committees. However, resulting reports and policy initiatives were quickly undermined by external attacks, as well as by ongoing internal antagonism over policies' interpretation and enactment. Divided and under nearly constant criticism from other agencies and consumer and producer organizations, the FDA responded to AMR, feed sector noncompliance, and residue problems in a fragmented way. Even when the balance of power briefly shifted in favor of more decisive AMR-focused action following the 1977 nomination of Commissioner Kennedy, opposition from industry, USDA agronomists, and an increasingly regulation-wary Washington toppled AGP restrictions and watered down FDA initiatives. The subsequent stagnation of US antibiotic reform was only gradually overcome. European antibiotic bans and enhanced AMR monitoring reenergized antibiotic stewardship efforts from around 2008 onward. However, familiar regulatory fault lines continue to characterize policy making. In 2013, CDC, USDA, and FDA officials clashed over plans to enact statutory drug withdrawals. Under pressure from the USDA, FDA regulators prioritized a program of initially voluntary industry guidelines over the CDC-supported congressional enactment of statutory bans.[131]

This ongoing misalignment of regulatory priorities and bureaucratic agendas should not surprise us. It is a "normal" consequence of life in a world in which the—both literally and figuratively—evolving ecology of AMR has exploded a regulatory system shaped by still powerful cultural equations of food safety with food purity. In contrast to mid-century hopes, nonlinear AMR hazards could not be made "safe" by regulatory frameworks focusing on the physical and microbial qualities of the food reaching US consumers' tables. By refocusing attention from the table to the practices used to produce this food, 1960s reports on horizontal AMR posed a significant and not completely resolvable dilemma to regulators: using antibiotics to realize productivist promises of plenty entailed a constant trade-off between gains in the present and potentially permanent and uncontainable shifts in the microbial environment. Which forms of use remained appropriate and whose interests should be prioritized by the resulting weighing of antibiotic benefits against AMR risks divided 1970s US regulators and remains a point of contention. Recent years have seen many high-income countries successfully reduce non-human antibiotic use to curb AMR. However, restricting farmers' direct access to low-dosed growth promoters has often proved far easier than restricting economically more important forms of therapeutic and prophylactic drug use. Meanwhile, the increasingly AMR-focused antibiotic stewardship equation structuring high-income policies does not necessarily hold true in parts of the world that lack access to effective drugs, face higher infectious disease burdens, or face growing consumer demand for cheap animal protein.[132] Over eighty years after antibiotics' advent in food production, it remains unclear how we will manage resulting risks on and off our tables.

Claas Kirchhelle is assistant professor of the history of medicine at University College Dublin. Supported by a Wellcome Trust University Award, his research fuses approaches from history and microbiology to explore the global history of infectious disease surveillance and control. *Pyrrhic Progress*, his monograph on the history of antibiotic use, resistance, and regulation in food production, was published by Rutgers University Press in 2020 and was awarded ICOHTEC's 2020 Turriano Prize.

Notes

1. FDA, *2018 Summary Report*, 12.
2. Lawrence, Fowler, and Novakofski, *Growth of Farm Animals*, 325–27.
3. Review on Antimicrobial Resistance, *Tackling Drug-Resistant Infections*.
4. The CDC was named National Communicable Disease Center in 1967, Center for Disease Control in 1970, Centers for Disease Control in 1980, and Centers for Disease Control and Prevention in 1992.
5. Häußermann, *Politik der Bürokratie*, 125.
6. Bonazzi, *Geschichte des organisatorischen Denken*s, 191; 203–5; 247–51.
7. Landecker, this volume.
8. Jones, *Valuing Animals*, 96–104; Lesch, *The First Miracle Drugs*, 197, 203, 287; Campbell, "History of the Discovery of Sulfaquinoxaline."
9. Oertel, "History of Beekeeping in the United States"; Haseman, "Sulfathiazole Control of American Foulbrood."
10. Finlay, "Hogs"; Finlay and Marcus, "'Consumerist Terrorists.'"
11. NRC, "Effects on Human Health of Subtherapeutic Use of Antimicrobials," table 1.
12. Sellers, *Hazards of the Job*, 198–201.
13. Carpenter, *Reputation and Power*, 73–75; Langston, *Toxic Bodies*, 26–27.
14. Campbell, "Sulfaquinoxaline," 938.
15. Kirchhelle, *Pyrrhic*, 33–53.
16. Ibid.
17. Smith-Howard, "Healing Animals in an Antibiotic Age," 724, 727.
18. Kirchhelle, *Pyrrhic*, 33–74.
19. Ibid.
20. Smith-Howard, "Antibiotics and Agricultural Change," 327–51.
21. On the impact of the FDC Amendments, see Vogel, *Is It Safe?*, 15–42, 45–46; Langston, *Toxic Bodies*, 77–82.
22. Donald Grove to John Foster, 2 October 1961, Folder 432.1 June–Dec, Box 3040, FDA General Subject/ Correspondence Files [GS] 432.1, DF A1/Entry 5, Record Group [RG] 88, National Archives and Records Administration [NARA]; §121.1014 Tolerances for residues of chlortetracycline, Subpart D—Food Additives Permitted in Animal Feed or Animal–Feed Supplements, Reissued Mar 20, 1962, Folder PA 190#95, Box 3245, FDA GS 432.1, DF A1/Entry 5, RG 88, NARA, 7.
23. Mildred M. Galton, Amherst Paper, 18 June 1963, Folder Recent Emphasis on Reporting and Surveillance of Salmonellosis, Box 6, Account Number [ANr] 69A750, RG 442, NARA Atlanta.

24. *Federal Register* 18, no. 37 (25 February 1953): 1077.

25. "Report on the current practice of adding drugs, antibiotics, and similar materials to feed for poultry and livestock and the possible effect on consumers of foods of animal origin in cases where such animals have been fed medicated feeds or in instances where they have been treated with therapeutic levels of the aforementioned drugs," Manuscript and Publications Steele 3, ANr 74A591, Box 2, RG 442, NARA Atlanta, 12–13.

26. Ibid., 16.

27. NAS, *Proceedings: First International Conference on the Use of Antibiotics in Agriculture.*

28. Kirchhelle, *Pyrrhic*, 63, 71.

29. Associated Chief for Program Bureau of State Services: Comments on FDA Plan to Reduce Amounts of Penicillin in Milk (11 January 1957), Folder EB VPH Section Dr. Steele, ANr 69A0020, Box 4, RG 442, NARA Atlanta.

30. Ibid.

31. On the 1960s "toxicity crisis," see Vogel, *Is It Safe?*, 44–46.

32. Bud, *Penicillin*, 175–76; Podolsky, *The Antibiotic Era*, 154.

33. Paul Sanders, "Summary of Some Differences and Sources of Confusion within the [FDA] and Their Jurisdiction over Medicated Feeds," 8 April 1959, enclosed in S. F. Kern to Commissioner FDA, 14 April 1959, Folder 432.1–432.1–11, Box 2668, FDA GS, DF A1/Entry 5, RG 88, NARA.

34. Charles Durbin to Bureau of Enforcement (Atten: C. Armstrong), 27 April 1962, Folder PA 190#95, Box 3245, FDA GS 432.1, DF A1/Entry 5, RG 88, NARA; J. F. Robens, "Memorandum of Conference," 6 December1962; K. J. Davis, A. A. Nelson and B. J. Vos to Bureau of Enforcement, "Re-Guideline 44—Animal Feeds Containing Certifiable Antibiotics at Growth Promoting Levels," [undated], Folder 70A190#95, Box 3245, FDA GS 432.1, DF A1/Entry 5, RG 88, NARA.

35. S. E. Jamison to Charles Durbin, 17 October 1961, Enclosed in J. N. Geleta to Jamison, 25 January 1962, Folder PA 190#95, Box 3245, FDA GS 432.1, DF A1/Entry 5, RG 88, NARA.

36. Daniel DeCamp to Durbin, "Memorandum—Current Poultry Feeding Practices," 18 September 1962, Folder 70A190#96, Box 3246, FDA GS, DF Entry A1, RG 88, NARA, 2.

37. Robert V. Marrs to A. Harris Kenyon, 26 September 1962, Folder 70A190#95, Box 3245, FDA GS 432.1, DF A1/Entry 5, RG 88, NARA, 2–4.

38. George Larrick, "Memorandum of Conference," 26 March 1965, see also enclosed memorandum on medicated feeds compliance by beef feeders, Folder 88-73-5#42, Box 3701, FDA GS, DF A1/Entry 5, RG 88, NARA.

39. *Federal Register* 31, no. 163 (23 August 1966): 11141.

40. "Infectious Drug Resistance," 277.

41. Kirchhelle, *Pyrrhic*, 54–74.

42. NAS, *Proceedings of a Symposium.*

43. Kingma, "Establishing," 33; "Use Medicated Feeds"; NAS, *Proceedings of a Symposium*, 45.

44. Subcommittee, *Regulation of Food Additives*, 585.

45. Mildred Galton, Recent Emphasis on Reporting and Surveillance of Salmonellosis, 18 June 1963, ANr 69A750, Box 6, RG 442, NARA Atlanta, 4–5.

46. Hardy, *Salmonella Infections*.
47. Galton, Recent Emphasis on Reporting and Surveillance of Salmonellosis, 7–13, NARA Atlanta. In the late twentieth century, *Salmonella typhimurium* and several other *Salmonella* strains were reclassified from being species to serotypes; it is now called *Salmonella enterica* serotype Typhimurium, or just *Salmonella* Typhimurium. This chapter uses the older nomenclature of its historical sources for these strains, in contrast to Hardy, this volume.
48. Ibid., 1–6; Mildred Galton and Philip S. Brachman, Salmonella Surveillance in the US 1965, ANr 69A0020, Box 2, RG 442, NARA Atlanta.
49. CDC, *Proceedings, National Conference on Salmonellosis*.
50. Surveillance of Salmonella in Foods and Feeds, 16 July 1964, Salmonella Surveillance Report Material for thru 1964, Anr 69A0020, Box 3, RG 442, NARA Atlanta.
51. Mildred Galton, "Joint WHO/Expert Committee on Zoonoses," 14 November 1966, ANr 69A750, Box 6, RG 442, NARA Atlanta, 1–2.
52. Ibid., 18–19; MS Epidemiology of Salmonellosis in the US, Manuscripts & Publications 3 Steele, James H.—1962, ANr 74A591, Box 2, RG 442; Alexander Langmuir to Galton, 19 June 1964, Salmonella Surveillance Report Material for thru 1964, ANr 69A0020, Box 3, RG 442, NARA Atlanta; Purchase, "Are We Ready for a National Salmonella Control Program?," 603.
53. Ramsey and Edwards, "Resistance of Salmonellae Isolated in 1959 and 1960 to Tetracyclines and Chloramphenicol," 390–91.
54. McWhorter, Murrell, and Edwards, "Resistance of Salmonellae."
55. Hillier, "Babies and Bacteria"; Condrau and Kirk, "Negotiating Hospital Infections."
56. Wallace, Quisenberry, and De Harne, "Preliminary Report," 457–58.
57. Ibid., 459.
58. Robert Huffaker to Mildred Galton, 21 July 1959; N. B. Gale/Robert J. Schroeder to Galton, 23 December 1959, California Cows Folder, ANr 69A0020, Box 3, RG 442, NARA Atlanta.
59. Draft (14 September 1960), Staphylococcal Distribution (by Bact. Unit), ibid.
60. First Draft, 10 April 1960, Epidemiological Aspects of Mastitis as Related to Public Health, Steele, Galton, Zinn, ANr 69A750, Box 6, RG 442, NARA Atlanta, 6.
61. McKenna, *Big Chicken*, 83–89.
62. Kirchhelle, "Swann Song."
63. On problems of determining "regulatory ownership" over complex problems like salmonellosis see Gusfield, "Constructing the Ownership of Social Problems," 433; Haalboom, "Who Owns Salmonella?"
64. "Infectious Drug Resistance."
65. Presentation, Lecture Public Health LA, Implications of Bacterial Persistence, 1966, ANr 69A1480, Box 2, RG 442, NARA Atlanta, 3.
66. Ibid.
67. James Steele, "Surveillance of Effects of Drugs in Animal Feeds," Manuscripts and Publications, Steele, 1967, ANr74A108, Box 3, RG 442, NARA Atlanta, 4.
68. Ibid.
69. Ibid., 9.
70. Ibid., 10.
71. Ibid., 9.

72. Assignment Descriptions Epidemiology Branch NCDC April 1968, EIS Conference Information 1968, ANr 72A1426, Box 1, RG 442, NARA Atlanta, 11.

73. Carolyn Dunn to M. J. Lewis, 22 October 1968, Correspondence Salmonella-Shigella Unit Dr. Morris, 1967–1968, ANr 71A1857, Box 1, RG 442, NARA Atlanta; George K. Morris to Philip Brachman—Projects in Progress, 29 November 1967; Morris to Brachman—Projects in Progress, 22 January 1968, Correspondence Salmonella-Shigella Unit Dr. Morris, 1967–1968, ibid.

74. Schroeder, Terry, and Bennett, "Antibiotic Resistance."

75. Jack Weissman, Eugene Gangarosa, and Herbert DuPont, "Changing Needs in the Antimicrobial Therapy of Shigellosis," ANr 75A106, Box 1, RG 442, NARA Atlanta.

76. Manuscript—An Epidemic-Associated Episome, 5 June 1972, ibid.; a simultaneous outbreak of MDR-Typhi occurred in India.

77. Kirchhelle, "Swann Song"; Kirchhelle, Dyson, and Dougan, "Biohistorical Perspective."

78. Dale Lindsay to Van Houweling, 21 November 1969, Folder 88-76-80 (Box 31), Box 4214, GS 432.1, DF A1/Entry 5, RG 88, NARA.

79. Van Houweling to Lindsay, 26 November 1969, Folder 88-76-80 (Box 31), Box 4214, FDA GS 432.1, DF A1/Entry 5, RG 88, NARA, 1.

80. Ibid.

81. Robert Angelotti to Herman F. Kraybill and through Keith Lewis, 27 March 1970, Folder #128 1975 May–July 432.1, [Accession 88-81-27], Box (FRC) 25, FDA GS 432.1, DF UD–WW/Entry 4, RG 88, NARA [all individual documents in this folder are attached to each other/ correspondence is continuation of FDA Task Force Folder], 2.

82. Kirchhelle, *Pyrrhic*, 185–213.

83. Butz, "Statement before FDA."

84. Kirchhelle, *Pyrrhic*, 185–213.

85. Countway Library of Medicine [CLM], FP Series VI, B. Veteran's Administration Committees and Projects Records, 1950–1953, Box 12, Folder 8, Maxwell Finland to John V. Bennett (20 December 1972).

86. Ibid., Bennett to Finland (15 January 1973).

87. Levy, Fitzgerald, Macone, "Changes in Intestinal Flora Of Farm Personnel after Introduction of a Tetracycline-Supplemented Feed on a Farm," 588; "Tab A—Scientific Summary with References"—likely prepared by Susan Feinman, enclosed in C. D. Van Houweling to Acting Commissioner, 07.03.1977, Folder 111 1977 432.1 Jan–May, Box (FRC) 22 (1977), FDA GS, DF UD–WW/ Entry 8, RG 88, NARA, 2–6; Ryder et al., "Increase in Antibiotic Resistance."

88. Finlay and Marcus, "Consumer Terrorists"; Kirchhelle, *Pyrrhic*, 185–213.

89. "Biological residue (supplement)," 6 March 1969, Folder 88-76-80, Box 4215, GS, DF A1/Entry 5, RG 88, NARA.

90. Kingma, "Establishing," 33.

91. Van Houweling to Herbert Ley, 17 April 1969, Folder 88-76-80 (Box 31), Box 4214, FDA GS 432.1, DF A1/Entry 5, RG 88, NARA, 1.

92. Ibid., 2.

93. Ibid., 3.

94. Daniel Kleber to General Counsel, 28 February 1968, Folder 88-76 5V (Box 34), Box 4103, FDA GS 432.1, DF A1/Entry 5, RG 88, NARA.

95. Van Houweling to Herbert Ley, 17 April 1969, Folder 88-76-80 (Box 31), Box 4214, FDA GS 432.1, DF A1/Entry 5, RG, NARA, 1, 3–4.

96. Subcommittee on Public Health and Welfare, *FDA Reorganization*, 478.

97. John McDowell to E. T. Mallinson, 22 June 1978, Folder #92 432.1 1978 June–July, Box (FRC) 18 [Accession 88-84-34], FDA GS, DF UD–WW/Entry II, RG 88, NARA; Purchase, "Are We Ready," 603–4.

98. F. D. Cazier to Van Houweling, 6 August 1971, Folder 432 By Products (Jan–Aug), Box 4480, FDA GS 432.1, DF A1/Entry 5, RG 88, NARA.

99. Ibid., 8.

100. "Inspection Report Kingfisher County Feed Yard, Inc.," 10 February 1971, Folder 88-77-28#30, Box 4481, FDA GS 432.1, DF A1/Entry 5, RG 88, NARA.

101. For sampling data, see Subcommittee, *Regulation of Food Additives*; Mussman, "Drug and Chemical Residues."

102. "Message C83206PAAUIJAZ RUWLROS002 0812134-UU," 22 March 1971, enclosed in Dennis Linsley to attention of Herbert Friedlander, 12 April 1971, Folder 88-77-28#30, Box 4481, FDA GS 432.1, DF A1/Entry 5, RG 88, NARA.

103. Ibid., 2–3.

104. E. J. Ballitch to Den F25 Attn: Dr. Steiner, "Memorandum—Medicated Feed Approvals," 6 September 1973, Folder #143 432.1 July–Sept, Box 4819, FDA GS 432.1, DF A1/Entry 5, RG 88, NARA.

105. Catherine W. Carnevale to Philip J. Frappaolo, 3 November 1978, Folder #90 1978 432.1 1978 Oct–Dec, Box (FRC) 17 (1978) [Accession 88-84-34], FDA GS 432.1, DF UD-WW/Entry II, RG 88, NARA.

106. L. V. Sanders to Evelyn Levy, 7 February 1977, Folder 111 1977 432.1 Jan–May, Box (FRC) 22 [Accession 88-83-62], FDA GS 432.1, DF UD-WW/Entry 8, RG 88, NARA, 2.

107. Subcommittee on Health, *FDA Practice and Procedure, 1975, Hearings 28.–29.01.1975—BVM & BOF*.

108. Oral History Interview of CD Van Houweling, 18 June 1990, 16.

109. GAO, *Use of Cancer-Causing Drugs*, i–iii, 7–8.

110. Ibid., 5–11.

111. Ibid., 12.

112. Ibid., 13–17, 27–28, 46.

113. Ibid., i–iii; 47; nitrofurans were eventually banned for systemic use in 1991 and for ophthalmic use in 2002.

114. "Synopsis of Meeting," 18 November 1976, enclosed in Maryln Perez to Honorable Berkley Bedell, 7 February 1977, 1; Homer R. Smith, "Memorandum of Phone Call," 27 May 1977, Folder 111 1977 432.1 Jan–May, Box (FRC) 22 [Accession 88-83-62], FDA GS 432.1, DF UD-WW/Entry 8, RG 88, NARA.

115. Maryln Perez to Honorable Berkley Bedell, 7 February 1977, Folder 111 1977 432.1 Jan–May, Box (FRC) 22 [Accession 88-83-62], FDA GS 432.1, DF UD-WW/Entry 8, RG 88, NARA, 1.

116. Interview Van Houweling, 31.

117. Ibid., 40.

118. Oral History Interview of Fred Kingma, 28 February 1990, 9.

119. Subcommittee on Health and the Environment, *Antibiotics in Animal Feed*, 460–68.

120. Lester M. Crawford to Commissioner, 18 August 1978 Folder #91 432.1 1978 Aug–Sept, Box (FRC) 18 (1978) [Accession 88-84-34], FDA GS 432.1, DF UD–WW/Entry II, RG 88, NARA, 2.

121. "FDA Talk Paper," 15 December 1978, Folder #90 1978 432.1 1978 Oct–Dec, Box (FRC) 17 [Accession 88-84-34]; "Action Plan for the [BF] Animal Drugs Program," 13 February 1978, enclosed in Joseph P. Hile and Van Houweling to the Commissioner, 7 June 1978, Folder #92 432.1 1978 June–July, Box (FRC) 18 [Accession 88-84-34], 5; Crawford to Commissioner, 18 August 1978, Folder #91 432.1 1978 Aug–Sept, Box (FRC) 18 [Accession 88-84-34], FDA GS 432.1, DF UD-WW/Entry II, RG 88, NARA, 4; Van Houweling subsequently had a successful career as an industry consultant; see Interview Van Houweling.

122. Committee on Government Operations, "Human Food Safety and the Regulation of Animal Drugs," 2.

123. *2000 FSIS National Residue Programme Data*, 4–6.

124. See, for example, Markowitz and Rosner, *Lead Wars*; Proctor, *Golden Holocaust*; Oreskes and Conway, *Merchants of Doubt*.

125. For overviews, see Sánchez and Llobat, "Following Poisons"; Kirchhelle, "Toxic Tales."

126. Frohlich, this volume.

127. See, for example, Merton, *Social Theory and Social Structure*, 25; Bonazzi, 203–5, 247–54; Selznick, "Foundations of the Theory of Organization," 29–30, 35; Gusfield, 433; Crozier, *Le Phénomène Bureaucratique*.

128. Kirchhelle, *Pyrrhic*, 163–213.

129. Schutter, "The Specter of Productivism."

130. Butz, "Statement before FDA."

131. Kirchhelle, *Pyrrhic*, 185–213.

132. Kirchhelle, Chandler, et al., "SETting the Standard."

Bibliography

Archives

National Archives and Records Administration, CDC, Atlanta, Georgia (NARA Atlanta)
National Archives and Records Administration, FDA and USDA, College Park, MD (NARA)

Publications

Bonazzi, Giuseppe. *Geschichte des organisatorischen Denkens*, 2nd ed. Edited by Veronika Tacke. Wiesbaden: Springer, 2014.

Bud, Robert, *Penicillin: Triumph and Tragedy*. Oxford: Oxford University Press, 2009.

Butz, Earl, "Statement before FDA Task Force on Antibiotics in Feeds Moline, Illinois, October 6, 1970." In *Proceedings of the Antibiotic Presentations to the US Food and Drug Administration Task Force on the Use of Antibiotics in Animal Feeds*, 309–13. Animal Health Institute, 1970.

Campbell, William C. "History of the Discovery of Sulfaquinoxaline as a Coccidiostat." *Journal of Parasitology* 94, no. 4 (2008): 934–45.

Carpenter, Daniel. *Reputation and Power: Organizational Image and Pharmaceutical Regulation at the FDA*. Princeton: Princeton University Press, 2010.

CDC (US Centers for Disease Prevention and Control). *Proceedings, National Conference on Salmonellosis, March 11–13, 1964*. Atlanta: US Health, Education, and Welfare, 1965.

Committee on Government Operations. "Human Food Safety and the Regulation of Animal Drugs." In *Union Calendar*, 2. Washington, DC: GPO, 1985.

Condrau, Flurin, and Robert Kirk. "Negotiating Hospital Infections: The Debate between Ecological Balance and Eradication Strategies in British Hospitals, 1947–1969." *Dynamis* 31, no. 2 (2011): 385–405.

Crozier, Michel. *Le Phénomène Bureaucratique*. Paris: Seuil, 1963.

FDA (US Food and Drug Administration). *2018 Summary Report on Antimicrobials Sold or Distributed for Use in Food-Producing Animals*. Washington, DC: 2019.

Finlay, Mark, "Hogs, Antibiotics, and the Industrial Environments of Postwar Agriculture." In *Industrializing Organisms: Introducing Evolutionary History*, edited by Philip Scranton and Susan R. Schrepfer, 237–60. London: Routledge, 2004.

Finlay, Mark, and Alan Marcus. "'Consumerist Terrorists': Battles over Agricultural Antibiotics in the United States and Western Europe." *Agricultural History* 90, no. 2 (2016): 146–72.

FSIS (USDA Food Safety and Inspection Service). *2000 FSIS National Residue Programme Data*. Washington, DC: FSIS, 2001.

GAO (US Government Accountability Office). *Use of Cancer-Causing Drugs in Food-Producing Animals May Pose Public Health Hazard: The Case of Nitrofurans; Report of the Comptroller General of the U.S.* Washington, DC: GAO, 1976.

Gusfield, Joseph R. "Constructing the Ownership of Social Problems: Fun and Profit in the Welfare State." *Social Problems* 36 (1989): 431–41.

Haalboom, Floor. "Who Owns Salmonella?" *BMGN – Low Countries Historical Review* 132, no. 1 (2017): 83–103.

Hardy, Anne. *Salmonella Infections, Networks of Knowledge, and Public Health in Britain 1880–1975*. Oxford: Oxford University Press, 2015.

Haseman, Leonard. "Sulfathiazole Control of American Foulbrood." *University of Missouri Agricultural Experiment Station Circular* 341 (1949): 1–10.

Häußermann, Hartmut. *Die Politik der Bürokratie. Einführung in die Soziologie der staatlichen Verwaltung*. Frankfurt: Suhrkamp, 1977.

Hillier, Kathryn. "Babies and Bacteria: Phage Typing Bacteriologists, and the Birth of Infection Control." *Bulletin of the History of Medicine* 80, no. 4 (2006): 733–61.

"Infectious Drug Resistance." *NEJM* 275, no. 5 (1966): 277.

Jones, Susan D. *Valuing Animals: Veterinarians and Their Patients in Modern America*. Baltimore: Johns Hopkins University Press, 2003.

Kingma, Fred J. "Establishing and Monitoring Drug Residue Levels." *FDA Papers* 1, no. 6 (1967): 8–9, 31–33.

Kirchhelle, Claas. *Pyrrhic Progress: Antibiotics in Anglo-American Food Production 1935–2013*. New Brunswick: Rutgers University Press, 2020.

———. "Swann Song: British Antibiotic Regulation in Livestock Production (1953–2006)." *Bulletin of the History of Medicine* 92, no. 2 (2018): 317–50.

———. "Toxic Tales: Recent Histories of Pollution, Poisoning, and Pesticides (ca. 1800–2010)." *NTM Zeitschrift für Geschichte der Wissenschaften, Technik und Medizin* 26, no. 2 (2018): 213–29.

Kirchhelle, Claas, Zoe Anne Dyson, and Gordon Dougan. "A Biohistorical Perspective of Typhoid and Antimicrobial Resistance." *Clinical Infectious Diseases* 69, Suppl. 5 (2019): S388–94.

Kirchhelle, Claas, Clare Chandler, et al. "SETting the Standard—Multidisciplinary Hallmarks for Structural, Equitable, and Trackable Antibiotic Policy." *BMJ Global Health* (2020); 5:e003091. doi: 10.1136/bmjgh-2020-003091.

Langston, Nancy. *Toxic Bodies: Hormone Disruptors and the Legacy of DES*. New Haven: Yale University Press, 2010.

Lawrence, Tony, Vernon Fowler, and Jan Novakofski. *Growth of Farm Animals*, 3rd ed. Wallingford: CABI, 2012.

Lesch, John E. *The First Miracle Drugs: How the Sulfa Drugs Transformed Medicine*. Oxford: Oxford University Press, 2007.

Levy, Stuart B., George B. Fitzgerald, and Ann B. Macone. "Changes in Intestinal Flora of Farm Personnel after Introduction of z Tetracycline-Supplemented Feed on a Farm." *NEJM* 295, no. 11 (1976): 583–88.

Markowitz, Gerald, and David Rosner. *Lead Wars: The Politics of Science and the Fate of America's Children*. Berkeley: University of California Press, 2014.

McKenna, Maryn. *Big Chicken: The Incredible Story of How Antibiotics Created Modern Agriculture and Changed the Way the World Eats*. Washington, DC: National Geographic, 2017.

McWhorter, Alma C., Mary C. Murrell, and P. R. Edwards. "Resistance of Salmonellae Isolated in 1962 to Chlortetracycline." *Applied Environmental Microbiology* 11, no. 4 (1963): 368–70.

Merton, Robert. *Social Theory and Social Structure*, 3rd ed. New York: Free Press, 1968.

Mussman, H. C. "Drug and Chemical Residues in Domestic Animals." *Federation Proceedings* 34, no. 2 (1975): 197–201.

NAS (US National Academy of Sciences). *Proceedings: First International Conference on the Use of Antibiotics in Agriculture*. Washington, DC: NAS-NRC, 1955.

NAS (US National Academy of Sciences). *Proceedings of a Symposium: The Use of Drugs in Animal Feeds*. Washington, DC: NAS, 1967.

NRC (US National Research Council). *The Effects on Human Health of Subtherapeutic Use of Antimicrobials in Animal Feeds*. Washington, DC: NRC, 1980.

Oertel, Everett. "History of Beekeeping in the United States." *Agriculture Handbook* 355 (1980): 2–9.

Oral History Interview of C. D. Van Houweling by Ronald T. Ottes, 18 June 1990. In *History of the US FDA*, 16.

Oral History Interview of Fred Kingma by Robert G. Porter, 28 February 1990. In *History of the US FDA*, 9.

Oreskes, Naomi, and Erik M. Conway. *Merchants of Doubt: How a Handful of Scientists Obscured the Truth on Issues from Tobacco Smoke to Global Warming*. New York: Bloomsbury Press, 2011.

Podolsky, Scott. *The Antibiotic Era: Reform, Resistance and the Pursuit of a Rational Therapeutics*. Baltimore: Johns Hopkins Press, 2015.

Proctor, Robert N. *Golden Holocaust: Origins of the Cigarette Catastrophe and the Case for Abolition*. Berkeley: University of California Press, 2011.

Purchase, H. Graham. "Are We Ready for a National Salmonella Control Program?" *Reviews of Infectious Disease* 1, no. 4 (1979): 600–6.

Ramsey, Carolyn H., and P. R. Edwards. "Resistance of Salmonellae Isolated in 1959 and 1960 to Tetracyclines and Chloramphenicol." *Applied and Environmental Microbiology* 9, no. 5 (1961): 390–1.

Review on Antimicrobial Resistance (O'Neill Review). *Tackling Drug-Resistant Infections Globally: Final Report and Recommendations.* London, 2016.

Ryder, R. W., P. A. Blake, A. C. Murlin, G. P. Carter, R. A. Pollard, M. H. Merson, S. D. Allen, and D. J. Brenner. "Increase in Antibiotic Resistance among Isolates of Salmonella in the US, 1967–1975." *Journal of Infectious Diseases* 142, no. 4 (1980): 485–91.

Sánchez, José Ramón Bertomeu, and Ximo Guillem Llobat. "Following Poisons in Society and Culture (1800–2000): A Review of Current Literature." *Actes d'història de la ciència i de la tècnica* 9, no. 1 (2016): 9–36.

Schutter, Olivier De. "The Specter of Productivism and Food Democracy." *Wisconsin Law Review* 2014, no. 2 (2014): 199–234.

Selznick, Philip. "Foundations of the Theory of Organization." *American Sociological Review* 13, no.1 (1948): 29–30, 35.

Schroeder, Steven A., Pamela M. Terry, and John V. Bennett. "Antibiotic Resistance and Transfer Factor in Salmonella, United States, 1967." *JAMA* 205, no. 13 (1968): 903–6.

Sellers, Christopher C. *Hazards of the Job: From Industrial Disease to Environmental Health Science.* Chapel Hill: North Carolina Press, 1997.

Smith-Howard, Kendra. "Antibiotics and Agricultural Change: Purifying Milk and Protecting Health in the Postwar Era." *Agricultural History Society* 84, no. 3 (2010): 327–51.

———. "Healing Animals in an Antibiotic Age: Veterinary Drugs and the Professionalism Crisis, 1945–1970." *Technology and Culture* 58, no. 3 (2017): 722–48.

Subcommittee on Public Health and Welfare of the Committee on Interstate and Foreign Commerce. *FDA Consumer Protection Activities—FDA Reorganization.* Washington, DC: GPO, 1970.

Subcommittee of the Committee on Government Operations. *Regulation of Food Additives and Medicated Animal Feeds.* Washington, DC: GPO, 1971.

Subcommittee on Health, Committee on Labor and Public Welfare. *Food and Drug Administration Practice and Procedure, 1975, Hearings 28.–29.01.1975—BVM & BOF.* Washington, DC: GPO, 1975.

Subcommittee on Health and the Environment of the Committee on Interstate and Foreign Commerce. *Antibiotics in Animal Feed. Hearings before the Subcommittee on Health and the Environment on H.R. 7285.* Washington, DC: GPO, 1980.

"Use Medicated Feeds Carefully and Wisely." *FDA Papers* 1, no. 6 (1967): 45.

Vogel, Sarah A. *Is It Safe? BPA and the Struggle to Define the Safety of Chemicals.* Berkeley: University of California Press, 2013.

Wallace, Gordon, Walter Quisenberry, and Maurice A. De Harne. "Preliminary Report of Human Staphylococcal Infection Associated with Mastitis in Dairy Cattle." *Public Health Reports* 75, no. 4 (1960): 457–60.

 CHAPTER 9

Conflicts of Interest, Ignorance, Capture, and Hegemony in the Diethylstilbestrol US Food Crisis

Jean-Paul Gaudillière

In April 1973, the US Food and Drug Administration (FDA) took the radical step of banning diethylstilbestrol (DES) implants in cattle and other livestock after having prohibited the use of DES in animal feeds one year earlier.[1] Significant risk of cancer in humans was the official motive for this unusually rash action, which caused an additive that had been employed as a growth enhancer in animal feeds for more than twenty years to disappear from the agricultural scene. The decision was immediately contested, and even though a federal judge confirmed the ban in 1978, illegal uses of DES in agriculture remained an issue until the mid-1980s.

The story of DES in the United States has been addressed in many ways. Most accounts have analyzed the uses of the drug in medicine, focusing on the controversy that surfaced in the 1970s when it was discovered that the massive prescription of DES to pregnant women in the 1950s and 1960s to prevent spontaneous abortion had resulted in cancer and anomalous reproductive tract development in the daughters of treated women. One group of studies that has looked at the unfolding of this medical drama focuses on the lessons to be drawn from the affair, either from the perspective of medical practice or of its regulation.[2] Another group, mostly by historians and sociologists of medicine, has investigated the medical uses of DES by situating them in the long history of sex hormones and gynecology, while more recent studies on gender have analyzed the role played by the then-emerging Women's Health Movement in the political crisis.[3] A much smaller group of scholars have dealt with the story central to this chapter—namely, the "meat and DES" controversy. Such work has emphasized its two distinct contexts: the rise of industrial agriculture on the one hand, and, on the other, the development of the environmental and consumer movements in the 1960s and 1970s.[4] This literature on agricultural DES shares with that on medicine a strong interest in how political and economic interests have made possible both the hormone's extended use and—when social movements grew in importance—its final ban. Departing from the last of these scholarly trends, this chapter's aim is to take the DES and meat crisis as a point of entry into the question of industrial influence—of how it was perceived during the 1970s affair and the debates that led to the 1978 ban; how we, as analysts, should analyze it; and how the analytical categories we choose matter to the possibility of regulatory reform.

The case is all the more interesting because the massive prescription of DES to pregnant women was an off-label use of this molecule. Indeed, the risk of miscarriage was not initially among the pathological indications accepted by the FDA when it authorized the marketing of DES; this was an off-label extension of prescription indications. Although in the 1950s and 1960s the research backing such use had never been sufficient to build consensus, the practice was nonetheless deemed safe because DES was deemed an analogue of *natural* estrogens. In practice, its use became so widespread that many women without any specific history of miscarriage were being prescribed DES as preventive medication.[5] In parallel, the uses of DES in agriculture originated, together with antibiotics to accelerate the growth of farm animals from chicken to cattle, in a technological race to increase productivity and, indirectly, to favor the consumption of meat by humans through its cheap mass production. In both instances, industrial practices and their relations to patterns of uses (actual and/or legitimate) were central. The DES affair was therefore not only a collective drama about the medicalization of women's health and the regulation of pharmaceuticals, but also a drama about agricultural industrialization and its relations to the dangers of food and to food safety experts; a drama opposing on the one hand those who believed, with the feeding industry, that the medical crisis was irrelevant to the agricultural situation since the doses of DES administered on the farms were incomparably lower than those administered in the clinic, and on the other hand those who believed, with Ralph Nader and a growing number of consumer activists and concerned citizens, that the very same industry that had pushed for overconsumption in medicine, confused issues, and lobbied the FDA, was lying to the public in order to defend its product and profits.

A Tale of Two Crises: Risk and DES as Object of Public Expertise

The FDA authorized DES for the US market in 1941. As reported by Susan Bell, the "synthetic estrogen" played an exemplary role in the history of the agency as it was one of the first drugs to be approved according to the procedures defined in the 1938 Food, Drug, and Cosmetic Act.[6] Approval was granted for basic gynecological indications—amenorrhea, menopausal symptoms, and infertility. However, DES was perceived as a potent estrogen rather than an aromatic compound with an ethylene side chain; consequently, "off label" uses rapidly surfaced in the 1950s. The most important of these—in terms of prescription numbers—was for managing the risk of spontaneous abortion during pregnancy. The idea was widely adopted that DES, a quasi-hormone, could replicate the changes in the concentration of sex steroids occurring during pregnancy. This justified the prescription of the drug as "replacement" therapy for a condition attributed to estrogen deficiency.

A decade later, in the 1950s, following pioneering work by animal nutritionists at Iowa State College under the leadership of Wise Burroughs, DES use was extended to agriculture. Burroughs and his colleagues discovered that DES given in minute amounts accelerated the growth of cattle.[7] The idea of adding DES to industrially prepared premixed livestock feed was patented by the university and licensed exclusively to the pharmaceutical firm Lilly. The process was a huge success. Lilly sublicensed it to a few dozen companies. As a result, within two years more than six million cattle were being fed with premixes containing DES. This transformation became a matter of concern and an issue of public debate in the 1960s. The trajectory of DES was rapidly affected by the emergence of critical voices associated with the consumer movement.

The enactment of what was known as the Delaney Clause in 1958 was a critical event in the polarization of US debates on the quality of food and the dangers it might pose to public health. The Delaney Clause crystallized the struggle in Congress on industrialized food for a decade after it was enacted. Farmers, industrial producers of animal feed, and drug companies joined forces in attempts to repeal the clause, arguing that it would necessitate banning all sorts of chemicals that could be used for the betterment, preservation, and conservation of food. They claimed that rodent carcinogenicity tests could not be trusted and that it would always be possible to find models and circumstances under which any chemical could become a carcinogen, and they explored ways of repealing the clause. In 1962, agricultural lobbying succeeded in inserting a short addition to the amended Food and Drug Act that stated that a substance could be authorized even if proved a carcinogen in the laboratory provided that it did not harm the livestock animals and left no residues in the meat for human consumption.[8]

Medical uses of DES, in contrast, remained basically unproblematic until the 1970s, even though feminist authors occasionally made the analogy to the contraceptive pill and targeted gynecological hormones for their purported carcinogenic properties.[9] The professional "warning" on DES—and the public controversy about its expertise—began only in 1970, when gynecologists at Massachusetts General Hospital reported a surprising series of vaginal tumors.[10] Considered very rare, this type of cancer had been diagnosed in rapid succession in very young women, a highly unusual feature. Following these early observations, Arthur Herbst and his colleagues reinforced their argument that this was a serious public health issue by concentrating on epidemiological evidence. More precisely, they mobilized a technology that had recently gained acceptance among physicians: risk-factor analysis. They selected a retrospective control group by looking at the records of women admitted the same day as their patients, matching age and social groups. The statistical comparison of their records revealed one single significant difference: the girls suffering from vaginal cancer were born from women who had been treated with DES during their pregnancy. The correlation was published in April 1971 in the *New England Journal of Medicine*, after Herbst had sent his data to the FDA.[11] The report alerted the New York Department of Health, and its cancer control bureau started to look for similar cases. Having found another five, also correlated with "DES mothers," they issued a general warning to the state's physicians.[12] It did not take long for the problem to find its way into the mainstream media.

In less than six months, the professional alert had become a national scandal, relayed by public health authorities and local gynecologists and widely discussed in the news, a topic of congressional hearings, and an ongoing cause for concern within the FDA. This alert had two effects. First, it gave very concrete meaning to the animal-human translation that had been at the core of all the arguments around the carcinogenic risks of the drug. If the correlation evidenced by Herbst and reinforced by the National Cancer Institute (NCI) survey was accepted, this meant that DES given in therapeutic dosage was inducing tumors in humans as well as in laboratory animals. As a consequence, the presence of DES residues in meat following its use as a growth enhancer could become a significant hazard. Second, the medical crisis reinforced doubts about the policy the FDA had implemented since DES had been put on the market. For many observers, the pharmacology division of the agency, blind to the doubts raised by a number of physicians regarding the clinical value of the pregnancy indication, had allied itself with a segment of the gynecological elite and the DES-producing chemical and pharmaceutical firms, choosing not to raise the problem of off-label prescription. Following the same logic, critics suspected that the FDA's Bureau of Veterinary Medicine's early defense of controlled DES in agriculture, a position it defended until 1973, originated in a similar alliance with nutritionists, agricultural scientists, and feed producers.

The two crises—on medical DES and on agricultural DES—actually reinforced each other. Throughout the 1970s, after the FDA had reacted by contraindicating pregnancy and announcing a ban of DES in animals, other medical uses of DES were questioned. In parallel with its campaigning on the adverse effects of the contraceptive pill, the nascent women's health movement succeeded in drawing attention to the prescription of DES as a morning-after pill. The pill was then regulated to include mandatory risk warning leaflets in the packages. A second problematic medical application was the use of DES to inhibit lactation and prevent breast engorgement after childbirth. The FDA plainly and simply contraindicated it in the late 1970s. This conjunction of a medical and an agricultural crisis was all the more powerful as links between DES uses in medicine and agriculture were at work at various levels outside the FDA, mobilizing texts, objects, and persons. The most important arenas for this public conjunction were, however, congressional hearings and—what are at the core of this chapter—the courts.

When the FDA decided to ban the use of implants in all farm animals in 1973, this policy was immediately challenged in court by a coalition of feed producers (the makers of DES followed at a close distance) with the result that the US Court of Appeals for the District of Columbia Circuit invalidated the FDA decision in 1973 as it had been taken without granting the interested parties the benefit of formal preliminary administrative hearings. The court thus mandated a judicial hearing. After lengthy negotiations between the parties, the public proceedings started in January 1977, with a final ruling from Judge Davidson, who was in charge of the case, in September 1978.

Sheila Jasanoff has insisted that an "adversarial" type of expertise characterized the mobilization of science in the US judicial system.[13] The DES trial thus provides an especially telling arena for the scientific evaluation of agricultural chemical risks, since (1) experts were proffered by and associated with the parties, (2) the corpus of acceptable documents defining the facts to be taken into account was negotiated between the judge and the parties' lawyers before the hearings, (3) each testifying expert was cross-examined by the opposite party with a right to follow up for the party putting forth the expert, and (4) one of the aims of the hearings was to include or exclude specific elements from the body of evidence that the judge would take into account in his final ruling. This procedure turned the trial into a stage for an open controversy on the properties, effects, and risks of agricultural DES, which resulted in major discussions about experimental and investigation data.

Judge Davidson's final ruling granted the FDA the right to ban all agricultural uses of DES, recognizing the existence of a public health emergency evidenced in the medical dimension of the DES crisis.[14] The Delaney Clause was actually difficult to invoke because of the 1962 amendment on the absence of residues. During the hearings, the controversy on the presence and the detection of residues had left open the question of the possible disappearance of the artificial hormone within the 120-day quarantine already mandated by the FDA and therefore the possibility that there was actually no carcinogenic DES in meat. Davidson grounded his evaluation in three elements: (1) the chemical and physiological specificity of DES, which he considered different from natural estrogens; (2) its demonstrated toxicity and carcinogenic potency in humans; and (3) the high probability of low-dose dangerous effects as demonstrated in animal experiments.

How can the role of the industry during these events be best understood? Alternative frameworks accounting for industrial influence on expertise have been promoted over the last decades within the social studies of science in general and the historiography of sanitary crises and risk management in particular—the production of ignorance, conflicts of interest, and capture among them.[15] The social phenomena these frameworks capture partly overlap, but they have been developed in distinct historical and intellectual contexts, and they convey different conceptions of power, and different understandings of the relationships between science, the state, the citizens, and economic interests.

The rest of this chapter attempts to understand and compare the benefits and drawbacks of these frameworks, using the DES affair and, more specifically, the documents of the 1978 trial to examine how the industry reacted and challenged the FDA ban.[16]

Knowledge on the Table: Meat, Cancer, and the Medicated Food Industry

Thomas Jukes, DES, and the Defense of (Industrial) Progress

In November 1977, *Feedstuffs*, then the main journal of the US feeding industry, published a long article on the DES affair whose appearance followed the hearings of the DES trial but preceded Judge Davidson's ruling.[17] The author was Thomas Jukes, a well-known biochemist, nutritionist, and molecular biologist, and at that time a professor emeritus at University of California, Berkeley. Best remembered for his contributions to the neutral theory of molecular evolution, Jukes was also an industrial researcher. For twenty years he worked at the drug company Lederle, where he investigated the metabolic role of folic acids and the growth-enhancing properties of antibiotics, contributing to their development as food additives for use in livestock.

Jukes's article was a strong plea against the ban of agricultural DES. It recapitulated the arguments presented by a palette of experts on the industry side during the hearings, including Jukes's own calculations. This paper's assessment of "what is known" about DES provided scientific justification of its uses in animal feeding in four main points: (1) The risk for consumers appears extremely small if one takes as reference the doses of DES administered to pregnant women; trying to find how much meat people would have to eat in order to ingest similar—cancer-inducing—quantities of DES and using the quantities of DES residues the US Department of Agriculture had found in its post-1973 surveys of meat, Jukes obtained the fantastic number of five hundred tons of beef. (2) The quantities of estrogens already ingested by US citizens independently of DES in meat are massive since sex hormones are widely used for medical purposes in women (e.g., the contraceptive pill and the hormonal therapies for menopause being targeted), while estrogens are naturally present in numerous vegetables consumed by humans, as well as in the natural grass eaten by cattle. (3) The main experiment mobilized during the hearings to show that DES is carcinogenic even at relatively low doses (the so-called Gass experiment conducted on mice) was flawed since (a) the mice used are quite abnormal animals genetically selected to display numerous mammary tumors when aging, (b) a fire destroyed some of the tests and controls during the course of the experiment, (c) the shape of the dose/effect curve is quite abnormal since it has a U shape with more tumors induced at low and high dosage. (4) As Herbst himself admitted in 1977, the variability of responses among the pregnant women who had been treated with DES implied that DES was not the only culprit—that is, that it "is not a complete carcinogen."

All in all, Jukes found the situation quite clear: the decision made by the FDA, recently approved in court, was not rational but political; there was no scientific basis for a ban, whose main effect would be detrimental to the consumer, who would no longer benefit from the increase in productivity that feeding DES to livestock had made possible.

In 1977, Thomas Jukes was not a newcomer in the DES debates. Before being selected to testify during the trial, he had regularly been advocating for medicated animal food. One reason for his inclusion among the close circle of scientists the industry councils mandated was the critical use they made of another DES-related calculation Jukes had produced. In 1976, in the newly funded academic journal *Preventive Medicine*, he had proposed a cost-benefit analysis of DES risks, which concluded that the ban of agricultural DES would cost consumers $500 million for the benefit of avoiding just one case of cancer in 133 years in the entire US population. The latter calculation of risk, which was the object of numerous comments during the hearings, was as follows:

> The lowest daily dosage reported by Herbst was 1.5 mg of DES and the highest was 225 mg. To be as conservative as possible, the lowest figure of 1.5 mg may be used. From this, we can estimate the average risk from beef consumption, using the following observations: (a) Linear relation between dose and risk. (b) Average per capita consumption of beef liver 0.7 kg annually; of beef muscle meats is 52 kg annually. DES disappears from tissues of animals fed DES when a withdrawal period is used, but not so rapidly when DES is implanted. DES in the livers of implanted steers was 0.12 ppb in the "radioactive test." None was detected in beef muscle. It has been argued that there is no DES in such meat, but to be conservative, we estimate the amount

as 10 percent of that in liver. The average daily intake, assuming all beef cattle are implanted with DES, is calculated as 1.9 ng (1 mg = 1 million ng) as follows:

0.7 kg beef liver per year x 0.12 ppb = 84 ng,
52 kg beef muscle per year x 0.012 ppb = 624 ng,
Total = 708 ng per year.
708 t 365 = 1.9 ng.

Female infant births per year in U.S.A. = 1.5 million. Cancer risk to female offspring from 1.5 mg DES per day given to pregnant women = less than four cases per 1,000 (25), which is equivalent, assuming ingestion of 1.9 ng per day and dose-effect linearity, to fewer than five cases per billion female births. In the United States, with 1.5 million female births per year, this is equivalent to a risk of one case of DES-induced genital tract cancer every 133 years.[18]

It is not necessary to delve deeply into this estimation to perceive the ways in which it aligned with the views of the feeding industry and therefore to acknowledge Jukes's important role in the deployment of arguments to save agricultural DES from strict regulation.

Ten years before, in the late 1960s, Jukes had similarly advocated against the ban of DDT and for the irreplaceable role of pesticides in the growth of agricultural output. In December 1969, he wrote his fellow nutritionists and biochemists Emil Mrak and William Darby, then heading the commission the State of California had created on the issue, to complain about the conclusions of their report on DDT's putative carcinogenicity—namely, that "the field of pesticide technology exemplifies the absurdity of a situation in which 200 million Americans are undergoing lifelong exposure, yet our knowledge of what is happening to them is at least fragmentary and for the most indirect and inferential."[19] In Jukes's eyes this was an extremely poor understanding of the political situation and of what experts should do:

Under "ordinary" circumstances, the report of your commission would have been received in an atmosphere which would have led to orderly reduction of the use of DDT and, hopefully, to the repeal of the Delaney clause. But, instead, the report will be seized on to further the current witch hunt. . . . With this background, to suggest in the absence of incontrovertible evidence that DDT is carcinogenic becomes the prelude to a major social cataclysm. . . . Fertilizers are under attack in a specious campaign by Barry Commoner. Antibiotics are being accused, without evidence, of endangering public health by producing transferable resistance. The EDF promises further attacks on pesticides, including malathion. 2,4-D, it is hinted, is teratogen—sentence first, verdict afterwards. I predict that by the end of 1970, we shall have a food shortage in this country produced not by the increasing demand of a burgeoning population, but by anti-scientific attacks on the technology used in agriculture.[20]

What motived Jukes in engaging in such expert activism? We lack any evidence of direct payments from the animal feed industry, although they might have existed.[21] However, the overall development of Jukes's career and the construction of his domain of expertise may be sufficient to account for his views. His personal ties with the industrial leaders and some of the companies that would suffer the most from the ban of DES reinforced a vision of the world that placed industrial agriculture, its use of chemicals, and the question of productivity center stage.

Jukes earned his PhD at the University of Toronto and then went to the University of California, first for a postdoctoral fellowship at Berkeley, followed by an instructorship and assistant professorship at Davis, the state system's preeminent agricultural school. Working at Lederle in the 1940s and 1950s, Jukes had been one of the first scientists to identify the growth-enhancing effect of antibiotics. Hired on the basis of his work on the B-complex vitamins and their effects on animal growth and diseases, he was initially investigating, with Robert Stokstad, the effects of Lederle's vitamin B_{12} on the growth of chickens. Late in 1948, they fed a deficiency-producing diet eventually supplemented with sterilized extracts of *Streptomyces aureofaciens* (the organism Lederle used to produce vitamin B_{12} as well as antibiotics such as the recently discovered aureomycin). They found that the animals eating the extracts grew faster than those not supplemented or supplemented with purified vitamin B_{12}.[22] Taken within the context of the industry's increasing search for new markets in order to valorize a booming production of antibiotics with its mounting piles of fermentation byproducts, the experiment became the cornerstone of a quick diversification of Lederle's sales and, for Jukes, of a decade-long in-house research program. Jukes thus became one of the key experts in the construction of the agricultural market for food additives. As such, he endorsed the industry vision of the nation's agriculture and its future.

The arguments he used in defense of DES were actually similar to those employed in the same circles to defend the role of animal food supplemented with antibiotics.[23] Jukes's public engagement increased in the wake of the DES debate. In 1978, he was one of the founding directors of a new industrial think tank, the American Council on Science and Health (ACSH), whose administrative director, Elizabeth Whelan, was a former research associate of Harvard School of Public Health. The ACSH positioned itself as providing expert information on food chemicals. In its first year of operation, ACSH released five reports, all but one dealing with the (non)carcinogenicity of additives or drugs. In 1981, Jukes served as a reviewer of an ACSH position paper on the Delaney Clause whose main plea was for an update of a law that had become scientifically "outdated" and created more regulatory problems than it helped solve. The organization explained their support for the attempted passage of a new bill modifying the clause:

> In the Hatch-Wampler bill, the Delaney Clause would apply only to those potentially carcinogenic food additives judged to be of *significant risks* to *humans*. This means that banning would no longer be the only regulatory option available for substances (1) that are found to be carcinogenic in animals but not in humans; (2) that may be very weak carcinogenic in man or (3) that are rarely used.[24]

In addition, this modification would rationalize a situation of chaos originating in the fact that many substances had been exempted from implementation of the clause through their inclusion in the list of "generally recognized as safe" (GRAS) additives.[25]

Greta Bunin of the citizens' organization Center for Science in the Public Interest (CSPI) questioned the ACSH's claim to represent "objective" science.[26] As she pointed out, ACSH's director Whelan's Harvard mentor Frederick Stare held numerous contractual ties with the sugar and cereal industries. Jukes's advocacy of chemical food additives did not escape scrutiny. The views of Whelan, Stare, and Jukes "have not been distinguishable from the industry point of view," as the director of Ralph Nader's Health Research Group, Sidney Wolfe, put it.[27]

CAST and the (Industrial) Production of Expert Reports

On 14 June 1974, a memo regarding the preparation of a "DES factsheet" left the premises of another industry think tank, the Animal Health Institute, whose activities included organizing a "DES task force," an ad hoc and unofficial group of representatives from the feed industry, plus their lawyers and advisors, which had been set up in order to coordinate the industry response to the FDA's ban of DES.[28] The group's main objective was to prepare for debates in court through the crafting of legal arguments (it thus produced an important report on the Delaney clause, its exemptions, jurisprudence, and the ways they could be used), but also through the gathering and active production of scientific expertise. This was what the factsheet memo was about. It insisted on the need to mobilize the entire food additive sector to support (1) the preparation of the above-mentioned scientific assessment of DES by the Council for Agricultural Science and Technology (CAST) expert committee, (2) the launch of new research, and (3) the broadening of issues in order to grant visibility to the nutritional, economic, and social benefits of agricultural DES.

CAST was founded in 1972 as another body of distinguished scientists concerned with what they perceived as the many biases in the US public debate on industrial agriculture.[29] Officially sponsored by an impressive array of professional societies, two-thirds of the very significant budget of the organization actually came from industry. CAST specialized in the preparation and circulation of scientific documents for use in the regulatory arenas and the media. To this effect, the council organized the documentation of cases and set up ad hoc panels whose memos, drafts, and oral advices were turned into formal reports by the council's staff. Depending on the collective assembled, the process could result in significant tensions. In contrast to the DES report, whose preparation proved remarkably smooth, the CAST 1979 report on the use of antibiotics as growth promoters turned into a matter of open controversy, with six members of the panel resigning and contesting the revisions of their writings.[30]

The CAST report on DES was issued late in 1976, shortly before the beginning of the court hearings. It was the most widely circulated product of the industry-based strategy of knowledge production and dissemination.[31] Its preparation mobilized two dozen specialists, whose domains of expertise included physiology, pharmacology, animal science, nutrition, food engineering, statistics, economics, and law. Its aim was nothing less than a comprehensive account of DES properties, effects, and uses in order to provide a balancing and rational assessment of risks and benefits.

The tone of the report was therefore less to stage "uncertainties" than to recall known "evidence." Its collective defense of DES ran parallel to Jukes's articles but in a much more quantified, technical, and encompassing way. The CAST DES expert committee, for instance, delivered a thorough and highly critical evaluation of the notion that carcinogenesis was a "no threshold" phenomena.[32] Following radiation studies,[33] cancer epidemiologists then increasingly discussed the possibility that one single mutagenic molecule could elicit changes in cellular DNA resulting in proliferation and in the formation of a tumor.[34]

By contrast, the CAST report insisted on the fact that all biological effects were bound to probabilities and therefore to effects of concentration. This implied that comparisons of dosages were central to any evaluation of risks and that thresholds associated with zones of "no risk at all" or of "negligible and acceptable risks" could be defined for all carcinogens. This epistemic choice grounded the critical role CAST experts attributed to issues of DES dosages, which were approached from two sides: estrogens and residues. The problem of DES was thus to be enlarged by taking into account all sources of estrogens in the body, whether related to food or to medical practices. CAST insisted on the wide use of contraceptives and the apparent incoherence of the FDA, which wanted to ban agricultural DES but "said nothing" about the risk of cancer associated with the pill and its high concentration of estrogens.

Regarding residues the text was even more assertive. Building on the measurements conducted by the USDA between 1973 and 1976, it concluded that there "are no residues of any significant importance" in DES-fed meat, since three years of monitoring beef livers had revealed only a few instances of DES in quantities amounting to a few ppb. The claim was central in shaping the core legal argument of the report, which was to use the amended Delaney law stating that a ban was necessary only if the carcinogen targeted was present in consumed products—presence being interpreted here in terms of "significant" amount.

The contribution of the CAST report most specific to framing the DES debate was to insist, in direct contradiction to the choice made by the FDA and its lawyers, on the need for a broad discussion balancing risks and benefits with the latter side of the equation, including nutrition and cost.[35] An entire chapter on meat and its nutritive value gathered references to studies documenting the improved quality of meat in DES-supplemented cattle as well as the benefits to consumers' health of a food regime enriched in animal products—the background to such nutritional progress being the higher productivity of animal production, with US consumers' savings estimated to reach $0.5 billion per year.

Building on the CAST report, the DES task force members did not consider that more economic or legal research was necessary. Priority was given to the production of new data directly relevant to the arguments the FDA would make during the hearings. They alternatively focused on the modeling of carcinogenicity and the status of the Gass experiment on mice.[36] The first issue was to get information regarding the new series of tests the congressional DES hearing committee had asked the National Toxicology Research Center (NTRC) to conduct on the same—criticized—C3H mice Gass had employed in his reference protocol. Hearings would become more difficult if any effect at low dosage could be confirmed. The task force felt relieved when one of its members was able to get the principal investigator on the phone and learned that NTRC had not tested doses as low as Gass had, and that preliminary results suggested the smallest concentrations checked (10 ppb) were not inducing more mammary tumors than appeared in the untreated controls.[37]

In parallel to the launch of the NTCR replication of his work, following the first congressional hearings, Gass received public funding for a new study of the effects of small doses of DES and estradiol in both C3H mice and typical white mice of the Swiss strain.[38] In July 1974, the draft of the protocol was transmitted to the industry lawyers and therefore to the task force, which initiated a strategy of involvement and amendment. The perspective was to supplement the palette of tests with two new series of feeding experiments: the first one with increasing quantities of processed liver from cattle treated with DES implants, the second one with increasing doses of what was then considered the main derivative of DES when it is metabolized in the body, its glucuronide-conjugate. In both instances, the task force hoped that the added tests would show no effect and could undermine the 1961 results. A meeting was thus organized with Gass himself, who proved quite willing to enlarge the protocol, accepting the help and financial support of the task force to provide for the necessary logistics, beginning with the numerous implanted steers needed, with the promise that there would be no industrial interference in the evaluation procedure, which would remain the sole affair of the researchers.

By the time of the hearings, however, only the preliminary results of the NTRC trial were available.[39] Those of the new Gass experiment were not. As a consequence, the debates focused on the 1961 curve and its interpretation, with most of the industry witnesses following the physiologist David Jones in his critical assessment:

> Jones. In my opinion there is no projectile [*sic*] risk to human beings from exposure to DES in food as far as the mammary cancer is concerned; and to understand my answer I would have to explain the dose-response relationship.

> Judge Davidson. Go ahead then, you've finally got your chance.

> Jones. Thank you. May I refer to G-22, the studies by Gass, Coates, and Graham? The point of the relationship in mammary cancer from exposure of animals to DES is the dose exposure to DES, which was given in the food and continued over lifetime. I'm sure the audience is generally familiar with this paper, so let me go then to the quantification that have been made by a number of us, including in the first place Gass and his colleagues, in which he showed a tumor response measured in a scale showing probits, which is to be seen on page 973, and a log dose response in the diet, this being a form of dose-response relationship that is customarily shown by toxicologists. Now, in the relationship there is a rather extraordinary response at the 6.25 parts per billion level, which is not significant in elevation from the controls by ordinarily statistical comparison; but nonetheless, it does catch the eye, and it catches the fancy of anyone interested in dose-response relationship as to whether this is an effect or not. The difference is great enough so that one should search in the experiment design for an alternative explanation, other than just random statistical variation although random statistical variation can account for it. In the Gass study, and I refer to you to his table 1, involving the time that tumors appear. There is a relationship between dose of DES and latent time: the bigger the dose, the shorter the latent time of the appearance of tumors. Before the experiment was concluded, a large fraction of the animals, all the surviving animals at that time, were killed in a fire. Gass reported his tabulation of these animals as though these animals had in fact been tested for tumorigenic response. In this 55 animals out of the 121

animals in the control group perished in the fire, whereas, in the succeeding intervals of dose there were only 17, 17, 18, and 18 respectively. One way of proceeding with an analysis is to take these animals out of the study, because in fact they weren't tested. When one does this, you can see that there is a perfectly reasonable explanation as to why the control animals were as low as they were.[40]

CAST's knowledge-making activities in favor of the medicated food industry did not stop with the ruling of Judge Davidson and the FDA confirmation of the ban. In the early 1980s, with the election of Ronald Reagan as president, the council felt times were changing. In a short time, it issued a very unusual evaluation of a National Academy of Sciences (NAS) document on the food supply. The impact of the latter had become a matter of concern since the NAS committee admitted that the US dominant food regimen contributed to the rising incidence of cancer in the country and therefore needed some reform. Members of the US Senate then asked CAST for an external evaluation. CAST sent the report for comments to forty-seven experts, among whom many had already served on the DES task force, including Jukes and Whelan. Given the absence of direct conversation between the members, there was no attempt to produce a consensus paper. CAST's staff juxtaposed the highly diverse texts collected and wrote its own summary. As reported there, a majority of reviewers were critical of the overall NAS approach, finding the evidence that fat and some additives were presumably carcinogenic too thin and suggesting that the changes in the American diet proposed by the NAS could result in "an opposite effect to the desired decrease in cancer incidence" because of their economic costs for the consumer.[41]

Two contributors focused on DES and the carcinogenicity of food additives. One was Jukes, who reiterated his general argument about the benefits of hormones in meat. The second one was the biochemist and specialist of estrogen metabolism Elwood Jansen, who had also testified in the DES 1976 hearings. Jansen's critique was straightforward:

In the minds of many leading scientists in the field of estrogenic hormones, the banning of DES by the FDA as a growth promoter represents not only economic masochism and a travesty of justice but also a triumph of emotion over scientific reason. Much of the FDA's case was based on the arguments that DES has cancer-inducing properties not shown by natural steroid estrogens and that there is no threshold or "no-effect" level of its action. Since both of these premises are scientifically untenable, I think it is important that a report bearing the imprimatur of the National Academy of Sciences present a balanced picture of the facts rather than superficial statements that could be interpreted as support for irrational arguments concerning imaginary hazards.[42]

The main target was therefore—again—the status granted to Gass's experiment, with Jansen insisting on two facts: (1) the absurdity of having the lowest dose of 6.25 ppb causing more cancer than the higher 12.5 and 25 ppb dosages—as it contradicted all toxicological reasoning, this result was obviously an artifact and its use by the FDA politically motivated; (2) the impossibility of replicating Gass's demonstration of carcinogenicity at low dosage in the course of the NTCR experiment since the lowest dosage tested (10 ppb rather than 6.25 ppb) "had a lower incidence of tumors than the control." In the eyes of Jansen, molecular and toxicological reason had thus been restored: "There is a 'no-effect' level as *rational consideration would predict for a phenomenon that depends on reversible interaction of hormone with receptor.*"[43]

The involvement of industrial think tanks was not limited to the production of counterfacts on DES. General principles of regulatory science were at stake too. For instance, in 1973, when the FDA shifted gear by endorsing the ban policy, the Animal Health Institute (a sister organization of CAST) started to circulate a report on the Delaney Clause that had been generated by its toxicology task force.[44] The motivation for this report could clearly be traced back to the agency's new positioning:

> Since the Delaney Clause provides such a strong regulatory tool for the FDA to use in determining which additives reach the food supply, it should be applied only with prudence and forethought; certainly this was the intent of Congress when the Clause was adopted. This should preclude prohibition of useful substances following the appearance or increased incidence of malignant tumors in limited animal studies. The public interest is not served when the banning of socially useful marketed additives or the prevention of the introduction of new additives increases the hazards of contamination or decreases the amount of food available for consumption.[45]

The main objective of the report was therefore, in parallel with listing instances of unreasonable regulatory intervention from the FDA, to propose a series of scientific guiding principles in order to define good modeling practices—from the absence of "exaggerated dosages" to special attention for "species-dependent toxicity," whose implementation would simply put an end to any regulatory use of Gass's experiment and its like.[46]

The FDA Bureau of Veterinary Medicine, or DES as Institutional Double Bind

On 28 February 1972, Senator Fountain, who was at the time leading the US Senate committee organizing the first congressional hearing on DES, wrote the head of toxicology at the FDA, Dr. Leo Friedman.[47] The questions Fountain sent him came after a first round of testimonies given by FDA officials with the goal of reassuring the committee about the policy followed by the agency, and more specifically about the soundness of the existing regulation of DES in agriculture. FDA officials were still thinking that a ban was not necessary since the chemical analysis tests used to detect DES residues had a sensitivity far below the concentrations of the molecule the Bureau of Veterinary Medicine (BVM) estimated as "safe," with the consequence that all "non negligible" DES residues could be detected. In the wake of the medical affair, the agency was nonetheless considering a move to strengthen DES regulation. The idea was to modify the residue policy in order to better enforce the amendment to the Delaney Clause, which stated that additives suspected of being carcinogenic would not be banned if proven absent from agricultural products for human consumption. The BVM was then thinking of changing the reference method for assessing the presence of DES, shifting from animal testing to chemical methods to diminish the detection threshold and extend the mandated withdrawal period during which animals were no longer given medicated feeds before their slaughtering. Adaptation rather than ban was thus still on the agenda.

Fountain was coming back to the agency because in the interim his staff had received (from unnamed FDA personnel) an internal report signed by Adrian Gross from the pharmacology division on the Gass experiment and the evaluation of safety levels based on a new model circulating within the agency:

> The memorandum of Adrian Gross is concerned with an animal study which revealed the carcinogenicity of DES at the extremely low level of 6.25 ppb. The evaluation of these results and their application in a determination of the safe level of DES for humans appears to have been the problem to which Dr. Gross addressed himself. In so doing he used the Mantel and Ryan concepts as applied to safety testing of carcinogenic agents. . . . Are the concepts of Mantel and Bryan as employed by Dr. Gross in his memorandum of December 5th 1971 appropriate for the evaluation and analysis of the specific study, which showed carcinogenicity at the level of 6.25 ppb? If "yes" did Dr. Gross properly employ these concepts?"

Mailed on 9 March, Friedman's response was carefully balanced.[48] On the one hand, he fully agreed with the application of the Mantel and Bryan model by Gross but immediately added that this very conservative calculus leading to a so-called "virtual safety level" of "a fraction of a part per trillion," which was several orders of magnitude lower than the concentration of residues detected in some meat samples and much lower than the sensitivity of existing tests, was not appropriate: "I cannot accept the concepts of Mantel and Bryan as crystallized in their specific procedure for estimating a 'virtual safe dose' for the same reason I indicated earlier" (during his testimony). Building on the comparison between normal levels of estrogens, estimates of the quantities found in meat, and the Gass data, Friedmann concluded he considered a safe level to be in the order of 0.01 ppb adding that "this is some three or four orders of magnitude *greater* than the estimates of a virtually safe level according to the procedure of Mantel and Bryan as calculated by Dr. Gross.[49]

In his letter, Friedman also considered another point raised by Senator Fountain. Friedman strongly opposed the "no threshold" model of carcinogenesis, defending the central paradigm of toxicology, the dose/effect proportionality: "I am of the firm opinion that for every carcinogen that we know of, even the most potent, there is a finite level that will definitely not produce a cancer in a human being or an experimental animal."

The Senate committee could appreciate the level of tensions and controversies at play within the FDA regarding these questions of low levels of exposure, probability of carcinogenesis, and safety threshold, when, a few days later, it received the copy of another four-page memo from the same Adrian Gross.[50] Sent to the head of pharmacology and to Dr. Edwards, the FDA commissioner, the second Gross memo recapitulated the ways in which Dr. Lehman, head of the BVM, had, in order to reassess the soundness of the above-mentioned policy of safe use, caricatured—both internally and in public—Gross's calculations, the Mantel and Bryan procedure, and all concerns regarding the existence of a "real" safe level.

One year later, relenting in the face of mounting demands and political pressure for a ban, Dr. Edwards reversed the FDA's position, abandoning safe use of DES in feeds using the "no threshold" framework Gross had mobilized in his calculations of risk levels. The FDA defense of a ban thus included the three ingredients of critical importance during the trial: Gass's experiment on low-level carcinogenicity in C3H mice, the "no threshold" idea, and the probabilistic modeling of carcinogenesis Mantel and Bryan had developed in order to compute "virtual safety level"—that is to say, levels of exposure resulting in less than one case of cancer per year in the entire US population.

The situation then seemed paradoxical. The Bureau of Medicine opposed the BVM, backing the commissioner's ban, whose vision of carcinogenesis challenged the FDA's long-term commitment to the dominant paradigm of regulatory toxicology, which had long granted centrality to concepts such as "no-effect threshold," "linear dose/effect relationship," and risk-benefit balancing in the administrative evaluation of drugs and additives. Expecting such a powerful and complex agency to be homogeneous would be a mistake. The oscillation between acceptance and rejection of both DES low-dosage carcinogenicity and the Mantel-Bryan modeling echoed other tensions within the agency, above all those underlying the parallel debates about antibiotics in animal feeds.

As Claas Kirchhelle recounts in *Pyrrhic Progress*, under the leadership of Van Houweling in the late 1960s and 1970s, the BVM repeatedly resisted attempts to link the problem of antibiotic resistance in humans and the non-prescription use of antibiotics as growth promoters and means of disease prevention.[51] Leaving out the possibility of horizontal transfer of resistance between microorganisms, BVM's vision of toxicity implied that medicated feeds could not be considered a primary hazard until antibiotics were shown to circulate from animals to humans through the meat. Prevention was therefore a matter of residue policing—namely a matter of threshold, withdrawal period, and monitoring—just like in the case of DES.

By the early 1970s, however, pressure from consumers' organizations as well as physicians was mounting, using the UK ban on antibiotics in feeds as precedent. Soon after his nomination in 1970, Commissioner Edwards set up an FDA task force on the question. After a fact-finding mission in the United Kingdom, preparation of the report proved highly controversial. One section of the task force recommended restrictions of use for those antibiotics important for human medicine through a status of prescription-only medication unless their producer could prove safety. BVM narrowed the scope of the review to focus on resistance in gram-negative bacteria, then the most visible problems of induced resistance. When manufacturers' data came in, they were considered so poor that FDA officials did not dare evaluate them. Following BVM's critique of British evidence and its opinion that the case for a ban of penicillin and tetracycline in feeds was legally too fragile, the agency dropped its initiative. Tightening the regulation of antibiotics was taken up again only in 1977 with the arrival of Commissioner Kennedy. This time the FDA faced such massive direct external pressures from farmers and Congress members that the Carter administration put the FDA action on hold, leaving as the only way out what BVM argued for—namely, engaging the agency in the definition of a more thorough and "definite" study of the resistance problem.

By 1977–78, defending the ban on DES could thus appear a decisive show of power and independence since the FDA was simultaneously losing another major initiative at food regulation. The FDA's defeat on the antibiotics front, however, was not simply the consequence of outside political pressure. It was also an effect of the agency's internal divisions and double bind since the BVM systematically balanced the defense of public health with that of industrial agricultural and farmers' interests.

Three Categories for Analyzing Industrial Influence

The manifestations of industrial influence made visible by the paper trail of the DES trial are far from specific to the case. Different ways of thinking about the problem of influence have been proposed in the historiography of public health over the last decades; conflict of interest, ignorance, and capture figure most prominently within this theoretical armamentarium.[52] These categories are sometimes used either implicitly or indiscriminately, but they developed in distinct historical and intellectual contexts, and they convey different conceptions of the relationships between expertise, regulation, and economic interests.

Conflict of Interest, or How to Follow Individuals Affiliated with the Industry

Conflict of interest is the most widely used category to discuss industrial influence on expertise and risk analysis. Conflict of interest, as the case of Thomas Jukes illustrates, focuses on the financial and, more generally, the personal ties between experts and the industries whose activities and/or products they contribute to evaluate. Technically, conflict of interest as category emerged in the 1950s United States as a portmanteau term used to designate the common target of different statutes dealing with the private activities and revenues of employees in the federal government.[53] Its use started to spread in 1963 when the Bribery, Graft, and Conflict of Interest Law was adopted, with the direct involvement of President John F. Kennedy. This law created a special status for external consultants and experts that exempted them from the rules banning private gainful activities imposed on ordinary government employees but required disclosure of their financial interests.[54] The FDA took advantage of this legal interpretation of conflict of interest to hire experts with industry connections during the 1970s. By that time, the notion of conflict of interest had emerged as an applied concept.

From the 1980s onward, conflict of interest increasingly became the focus of normative and conceptual debates in law and ethics with a strong emphasis on regulation.[55] In 1984, the most prestigious medical journal, the *New England Journal of Medicine*, decided to ask its authors to disclose any commercial interests that could potentially affect the opinions or results reported in their articles. This innovation was progressively taken up by all major scientific journals in biomedicine. From the late 1990s onward, the generalization of disclosure statements in scientific articles and the analysis of documents obtained through litigations fostered the development in medical journals of a quasi-subfield of research on conflict of interest.[56] The majority of these articles aimed at demonstrating that collaborations and financial ties with industries impact scientific practices and result in biased evaluation. Scholars and policy analysts have accordingly tried to substantiate links statistically between the source of funding (private/academic) and the claims made about the risks or benefits of products. A majority of the cases investigated dealt with pharmaceuticals,[57] but some also targeted food products.[58]

Social scientists have generally been suspicious of the notion. Sergio Sismondo, for instance, explained that the term "conflict of interest" "is well established and is best retained, but is misleading since it suggests that researchers act inappropriately to further their own interests."[59] Conflict of interest disclosure policies have also been criticized for their permissiveness and limited efficacy.[60] For instance, Sarah Wadmann has suggested that the focus on financial conflict of interest problems in physician-industry collaborations can lead to "blame games" that focus public debates on morality while leaving unaddressed more important issues such as political priorities or the role of research infrastructures.[61]

The popularity of conflict of interest as a key category stems from its normative and legal status but also from its vague and broad meaning. The concept at the same time (1) designates a wide range of problematic situations without assigning them to notions that, like bribery or fraud, are legally well defined and therefore more difficult to demonstrate, and (2) relies on an apparently simple psychological mechanism, which everyone has experienced, invoking tensions between competing motivations on a person's behavior.

Yet, the situations analyzed through this lens are diverse: some imply direct conflicts between interests (when a researcher is forbidden to publish bad results from a financing firm); some reflect persuasion (when a researcher is enrolled in a promissory claim about a drug through a collaboration); others (as in the case of Jukes) point to the convergence of views when a researcher's understanding of issues is informed by long-term collaboration with industrial patrons. As a descriptive tool, conflict of interest is therefore mainly cartographic: pointing to the existence of links but saying nothing about their specific nature and the modalities of interaction involved. In addition, from a more normative point of view, the denunciation of conflict of interest is infused with a Mertonian understanding of the norms governing science (disinterestedness, communalism).[62] These characteristics can explain why the notion has been less discussed in history and social sciences as in legal and ethical studies.

Ignorance, or the Industrial Making of "Undone Science"

In contrast to conflict of interest, the question of the "production of ignorance" has gained increasing visibility in the history and sociology of science during the past decade, with studies emphasizing the role of economic actors in the deliberate production of ignorance or in the slowing down of potentially inconvenient research.[63]

The litigation against the tobacco industry has played a structuring role in the emergence of the theme in the historiography. In a manner parallel to the use of FDA archives in this chapter, the trial archives mobilized as part of judicial decisions and settlements have given scholars the opportunity to conduct detailed investigations of the industry strategies and provided a reference case for understanding the role of firms in the production of "false controversies" through the promotion of "doubt and uncertainties" regarding the connection between cigarette smoke and lung cancer.[64] The "ignorance" approach is a highly political reading of the conflicts between, on the one hand, firms seeking to format scientific knowledge in order to minimize its negative consequences for their own interests and, on the other hand, government services or regulatory agencies, especially at federal level, seeking to thwart these strategies in order to regulate dangerous activities.

Besides tobacco, emblematic cases are that of asbestos, which several historians, journalists, and sociologists in several countries have studied,[65] and the lead and vinyl chloride industries,[66] which reveal the many situations in which firms have refused to publish certain results, have encouraged research that fitted their expectations, have concentrated research on subjects that did not directly challenge their economic interests, or have financed scientists to publicly criticize results that endangered the consumption of their products. The activities of CAST, mentioned above, and related industry think tanks during the DES trial fit perfectly in this landscape of influence through the production of expertise for public consumption.

Such studies of industrial expertise tend to emphasize the contrast between "ignorance" and "truth," leaving out uncertainty as a delusory category reinforcing the industry grip on the debates. As a consequence, this literature asserts the intentional and manipulative nature of this production of ignorance—a state of affairs clearly documented in the cases mentioned but far from being easy to generalize. Moreover, because ignorance tends to overshadow the more structural dimensions in the industrial production of knowledge, authors such as Scott Frickel have proposed to broaden the picture with the notion of "undone science," with the aim of taking account of, beside the direct pressure exerted by industry, many other factors explaining the unequal development of scientific knowledge within a given social, political, and economic configuration.[67]

Speaking of undone science is therefore a way to approach the structural inequalities between the groups that are mobilized to denounce a danger and the companies that produce that danger.[68] Knowledge is produced and enriched in path-dependent ways, which also generate—both intentionally and unintentionally—areas of nonknowledge.[69] Actions by interested parties (often from industry) range from direct suppression and undermining of information to disinvestment and targeted funding; the result is a *structural imbalance* between what is known and what is not known (and considered by actors other than the industry as needed). In contrast to conflict of interest studies, the level of analysis concentrates on research organizations and dominating epistemic frameworks rather than individual researchers and their choices. Studies of ignorance thus tend to put forward a strategic—eventually conspiring—reading of social relations and "undone science."

Capture, or Industrial Influence as Ever-Elusive Target

The notion of "capture" emerged in political science and public administration in the 1950s as a response to concerns that the functioning of federal commissions and agencies, because of the technicality of the laws they implement or flaws in their design, could tend to favor the interests of the regulated industry.[70] The 1970s and 1980s represent the "golden age" of regulatory capture as economists and journalists started to use the notion to analyze situations in which public regulatory bodies seemed "captured" by industry interests.

The idea of "captured" federal bodies was initially formulated to discuss the working of trade, finance, transportation, and energy commissions.[71] In the 1950s, Marver Bernstein, who had previously worked for the government, made it his project to evaluate critically the role and functioning of seven agencies. He explored the idea of "captive" agencies by describing their life cycle as they go through four stages. The "gestation" stage usually ends with the passing of ambiguous statutes. In the "youthful" stage, an agency relies on technical procedures and litigation to achieve its goals, a dynamic that benefits the industry. It then enters a "maturity" phase during which it becomes more and more attentive to the interests of the regulated industry. In the final, "old age" phase, it has become "retrogressive, lethargic, sluggish" and "surrenders," in other words, to being "captive."

The work of economist George Stigler is perhaps the most cited writing on the notion.[72] Stigler proposed to consider the implications that governments and industry operate in a regulatory marketplace. Governments supply products such as subsidies, control over competitive entry, products' regulation, or mandatory prices. Obtaining some of these products can then be a way to use them against one's competitors and dominate the market by setting up entry barriers. Stigler's work did not focus on agencies overseeing food, health, and environmental protection, but his general theory has inspired extensive research on the existence of capture as a main mechanism of influence in these sectors.[73]

Academic uses of "capture" have become less frequent after 2000. As far as food and public health is concerned, Daniel Carpenter is one of the few scholars who discussed the concept at length, if only to contest it.[74] His argument in *Reputation and Power*, a study of the FDA, is that the patterns of influence, and the power relationships at play in the context of American pharmaceutical regulation, do not follow the model of capture: companies have arguably not mobilized for specific regulation but resisted them, and big firms are not systematically favored over their smaller competitors. In other words, capture, according to Stigler, is not there. The discussions that followed led to the publication of an edited volume defending the idea that capture is always both preventable and manageable because it is a weak phenomenon, a "corrosive capture" whose aim is only to limit costly rules that would reduce profits and therefore rarely has systemic or structural dimensions.[75]

A striking dimension of this literature is the model-oriented nature of the approach and its reliance on only a very limited number of investigations into specific cases and the activities of industrial actors. The dynamics of interest are those taking place at the institutional level only. Those writing about capture from the administrative sciences and economics, given their normative orientation, aim to offer a practical assessment: can we consider that a given federal agency or commission is "captive" and what should we do about it?

Conclusion: Hegemony as an Additional Category

Conflict of interest, ignorance, and capture are not mutually incompatible categories. They overlap in many ways and tend to designate a common set of issues regarding the relationship between science, industry, and the regulation of markets. They all emphasize the ability of powerful economic actors to shape the law as well as the ways in which professionals and consumers use commodities. The case of the agricultural DES crisis shows that situations analyzed in terms of conflicts of interest often exhibit processes of expertise impacted by less visible and more structural forms of industrial influence, such as those operating on the production of knowledge, which the perspectives of undone science or institutional capture seek to illuminate. One may therefore consider these categories to be simply complementary tools, exploring different aspects of the same phenomenon.

Nonetheless, the invocation of these categories uncovers striking differences in the very definition of the problems. Conflict of interest and capture appear as notions with strong normative functions, widely used by the actors themselves. In contrast, undone science is a category elaborated by social scientists in order to point to the general, systemic nature of influence.

A better understanding of industrial influence should therefore imply a choice or at least a clear hierarchy between these analytical tools. This is not the main plea of this chapter. It is rather to say that, although there is no simple and straightforward complementarity, a comprehensive analysis of influence requires an articulation of categories as means of both mapping (objectifying) individual trajectories and links and exploring the asymmetries and power gradients at stake in the production and nonproduction of knowledge. However, in order to take into account not only the systemic industrial influence on regulatory science but also the ways in which industrial knowledge resonates with the processes of market construction and the diverse political economies at play in the logic of influence, one needs additional tools. The historiography and social studies of pharmacy have recently offered a variant of Gramsci's concept of hegemony to this effect.

Hegemony did not play any significant role during the first two waves of science studies development: the controversy studies in the 1980s and the network/coproduction analysis in the 1990s to early 2000s. Hegemony has thus gained visibility in studies of science, technology, and medicine only in the 2010s in relation to analysis of the rise of biotechnology and with studies of pharmaceutical capitalism and its role in the production of knowledge.[76] Sergio Sismondo, for instance, apprehends hegemony in a rather straightforward manner—that is, as the domination of capitalistic firms in the making and diffusion of pharmaceutical knowledge.[77] His core argument focuses on the industry-based system of knowledge legitimation and dissemination with its cortege of contract research organizations, ghost writing, sales representatives, and collaborating key opinion leaders. Sismondo thus echoes a significant body of historical and sociological work on the industry, its in-house research practices, and its links to practitioners. Kaushik Sunder Rajan refers to Gramsci and hegemony in a more specific and theoretical way.[78] His recent *Pharmocracy* explores the dynamics of pharmaceutical innovation, capital formation, and regulation in India, following two configurations of political conflicts over health and its appropriation: the mobilizations about the HPV vaccine and its adverse effects, and the denial of patent rights for Novartis's anticancer drug Gleevec by Indian courts. Hegemony is central to Kaushik's view, enabling him to understand the dynamics of capital in biomedicine beyond machinations or mere greed.

Hegemony is a very broad notion introduced to reflect upon the historical dynamics of capitalism and upon the complex, systemic, and dialectical relationship between the economical and the political. Given the existing corpus linking hegemony, knowledge, and state action, one may suggest that hegemony can be used to explore "pervasive powers" and "influence" in studies of science, food, and risks as a "structural" approach, one that places capital (rather than industrial production) and the construction of markets at the center of action while recognizing the critical role that knowledge and its production play. As a consequence, one should investigate not only the various forms of knowledge but also the ways in which they are practically appropriated by capital. This has been powerfully demonstrated by the historiography of the pharmaceutical industry, with its detailed deciphering of the intimate linkages between clinical trials and scientific marketing.[79]

Jean-Paul Gaudillière is a historian at Cermes3 and professor at École des Hautes Études en Sciences Sociales, Paris. His research explores the history of the life sciences and medicine during the twentieth century. His recent work focuses on the history of pharmaceutical innovation and the uses of drugs, and the dynamics of health globalization after World War II. He is coordinator of the European Research Council project From International to Global: Knowledge, Diseases, and the Post-War Government of Health.

Notes

1. "DES Livestock Implants."
2. Apfel and Fisher, *To Do No Harm*; Meyers, *D.E.S.*; Dutton, *Worse than the Disease*; Pfeffer, "Lessons from History."
3. Bell, "The Synthetic Compound Diethylstilbestrol"; Marks, *Sexual Chemistry*; Morgen, *Into Our Own Hands*.
4. Schell, *Modern Meat*; Marcus, *Cancer from Beef*; Rifkin, *Beyond Beef*; Gaudillière, "DES, Cancer, and Endocrine Disruptors."
5. Langston, *Toxic Bodies*.
6. Bell, "The Synthetic Compound Diethylstilbestrol."
7. Marcus, *Cancer from Beef*.
8. Temin, *Taking Your Medicine*; Daemmrich, *Pharmacopolitics*.
9. Seaman, *The Doctors' Case*.
10. Bonah and Gaudillière, "Faute, accident ou risque iatrogène?"
11. Herbst, Ulfelder, and Poskanzer, "Adenocarcinoma of the Vagina."
12. Greenwald et al., "Vaginal Cancer."
13. Sheila Jasanoff, *Science at the Bar*.
14. Daniel J. Davidson, Administrative Law Judge, Initial Decision, *Proposal to withdraw approval of new animal drug application for diethylstilbestrol*, 21 September 1978, 76N-002, FDA.
15. For a general discussion of the origins, uses, and limitations of these categories see a forthcoming review by Boris Hauray, Henri Boullier, Emmanuel Henry, and Jean-Paul Gaudillière, "Conflicts of Interests." This chapter owes much to the collective discussions that took place within the context of an ANR research project led by Boris Hauray on the history and dynamics of conflict of interests within the pharmaceutical sector.
16. The archives used for this chapter are the records the FDA assembled for the 1977–78 court hearings, which includes the corpus of scientific papers, reports and internal documents the agency submitted, those submitted by the councils of the industry, those mandated by the court, the testimonies and biographies of experts, as well as the transcripts of the proceedings. They have been complemented by documents on DES included in the Food Industry Documents database set up by the University of California San Francisco (UCSF); see https://www.industrydocuments.ucsf.edu/food/.
17. Jukes, "The DES Hearings." Jukes's article was included in the FDA documentation for the case. However, in all probability, it was used only internally in order to prepare the final memo the agency sent to the judge since there are no traces of it being used publicly in the transcripts of the hearings.

18. Jukes, "Diethylstilbestrol in Beef Production."
19. Thomas H. Jukes to Emil M. Mrak and William J. Darby, 1 December 1969, William F. Darby Papers, document ID snvb0228, FID.
20. Ibid., 2.
21. Jukes's papers, deposited at the Bancroft Library, are unprocessed and unavailable to scholars.
22. Kirchhelle, *Pyrrhic Progress.*
23. See Kirchhelle, this volume.
24. American Council on Science and Health (ACSH), *The U.S. Food Safety Laws: Time for Change?* (ACSH: New York, 1982), document ID kkdd0229, FID.
25. See Maffini and Vogel, this volume.
26. Greta Bunin, "ACSH—Another 'Objective' Group," *Nutrition Action*, February 1979, 12–13, document ID pycd0229, FID.
27. Ibid., 12.
28. The letter was appended material in a circular letter sent on 18 June 1974 from Dawes Laboratories Inc. to the members of the DES Task Force, 76N-002, FDA.
29. Finlay and Marcus, "'Consumerist Terrorists.'"
30. Kirchhelle, *Pyrrhic Progress*, 152.
31. Council for Agricultural Science and Technology (CAST), *Hormonally Active Substances in Food: A Safety Evaluation*, report 66, February 1977, 76N-0002, FDA.
32. Ibid., 31–40.
33. See de Chadarevian, this volume.
34. See Stoff and Creager, this volume.
35. CAST, *Hormonally Active Substances in Food: A Safety Evaluation*, report 66, February 1977, 59–69, 76N-0002, FDA.
36. G. H. Gass investigated DES carcinogenesis in the early 1960s at the University of Illinois with C3H mice. His results showed unexpectedly high levels of carcinogenesis (when compared with the control group) at low concentrations of DES, thus giving the dose response curve an unusual V- or U-shape. Gass's modeling was judged to be inconclusive by many experts in toxicology, who expected a linear dose/response curve.
37. Internal memo from Rhodia Division Hess & Clark to DES Ad Hoc Group, 28 January 1974, 76N-002, FDA.
38. Summary of meeting of industry representation and Dr. G. H. Gass on DES at American Society for Animal Science Meeting, 31 July 1974, 76N-002, FDA.
39. Department of Health, Education and Welfare, NTCR, January Progress Report Hormone Research Programs, 26 January 1976, 76N-002, FDA.
40. Transcript of the DES hearings, 2 November 1977, Testimony of D. Jones, 76N-002, FDA.
41. CAST, *Diet Nutrition and Cancer: A Critique*, Special Publication no. 12, October 1982, document ID rxnh0230, FID.
42. Ibid., 41.
43. Ibid., 42. Emphasis mine.
44. The Animal Health Institute acted as umbrella organization for the industry DES task force. In practice, however, its entire logistics was coming from the individual firms participating.

45. Animal Health Institute, *Scientific Application of the Delaney Clause. A Report by the Toxicology Task Force*, July 1973, 76N-002, FDA.

46. Ibid. 9 and 12.

47. L. H. Fountain to Leo Friedman (FDA Division of Toxicology), 28 February 1972, 76N-002, FDA. This section focuses on the relationship between the medicated feed industry and the FDA. In order to document the influence of the former on the latter, events before the DES ban policy are relevant. Therefore, this section moves back in time a bit to draw on FDA documentation dating from 1970–72, which was later included in the trial's background material.

48. Leo Friedman to Senator L. H. Fountain, 9 March 1972, 76N-002, FDA.

49. Ibid., 2.

50. FDA internal memo, Adrian Gross to William Wright, "Carcinogenicity of DES; Estimation of safe levels in the diet," 13 March 1972, 76N-002, FDA.

51. Kirchhelle, *Pyrrhic Progress*, 188–203

52. The following discussion is a modified version of a forthcoming paper by Boris Hauray, Henri Boullier, Emmanuel Henry, and Jean-Paul Gaudillière. I have adapted our ideas to the DES case and take responsibility for any interpretations that depart from our collective view.

53. Davis, "The Federal."

54. Perkins, "The New Federal."

55. Luebke, "Conflict of Interest"; Davis and Stark, *Conflict of Interest in the Professions*; Brody, "Clarifying."

56. Trost and Gash, *Conflict of Interest and Public Life*.

57. Lundh et al., "Industry Sponsorship."

58. Lesser et al., "Relationship."

59. Sismondo, "How Pharmaceutical Industry."

60. Rodwin, *Conflicts of Interest*; Sismondo, "Key Opinion Leaders"; Lexchin and O'Donovan, "Prohibiting or 'Managing.'"

61. Wadmann, "Physician-Industry Collaboration."

62. Merton, "The Normative Structure."

63. Proctor and Schiebinger, *Agnotology*; Gross and McGoey, *Routledge International Handbook*.

64. Glantz et al., *The Cigarette Papers*.

65. Proctor, *Cancer Wars*; Brodeur, *Expendable Americans*; Tweedale, *Magic Mineral*; Henry, *Amiante*; Henry, *Ignorance scientifique*.

66. Markowitz and Rosner, *Deceit and Denial*.

67. Frickel et al., "Undone Science."

68. Hess, "Undone Science and Social Movements."

69. Frickel and Edwards, "Untangling Ignorance."

70. Novak, "A Revisionist History."

71. Huntington, "The Marasmus"; Bernstein, *Regulating Business*.

72. Stigler, "The Theory."

73. Shapiro, "Blowout."

74. Carpenter, *Reputation and Power*.

75. Carpenter and Moss, *Preventing Regulatory Capture*.

76. Kleinman, "Untangling Context"; Salter, Zhou, and Datta, "Hegemony"; Holloway, "Normalizing Complaint."
77. Sismondo, "Hegemony of Knowledge."
78. Sunder Rajan, *Pharmocracy*.
79. Greene, *Prescribing by Numbers*; Greffion, "Faire Passer La Pilule"; Gaudillière and Thoms, *The Development*; Gaudillière, "Une manière industrielle."

Bibliography

Archives

Food Industry Documents (FID), hosted by University of California, San Francisco Library and Center for Knowledge Management
US Food and Drug Administration Archives (FDA), White Oak Campus, Silver Spring, Maryland; docket numbers provided in notes

Publications

Apfel, Roberta J., and Susan M. Fisher. *To Do No Harm: DES and the Dilemmas of Modern Medicine*. New Haven: Yale University Press, 1984.
Bell, Susan Elizabeth. "The Synthetic Compound Diethylstilbestrol (DES) 1938–1941: The Social Construction of a Medical Treatment." PhD diss., Brandeis University, 1981.
Bernstein, Marver H. *Regulating Business by Independent Commission*. Princeton: Princeton University Press, 1955.
Bonah, Christian, and Jean-Paul Gaudillière. "Faute, accident ou risque iatrogène: La régulation des événements indésirables du médicament à l'aune des affaires stalinon et distilbène." *Revue Francaise des Affaires Sociales*, no. 3–4 (2007): 123–51.
Brodeur, Paul. *Expendable Americans*. New York: Viking, 1974.
Brody, Howard. "Clarifying Conflict of Interest." *American Journal of Bioethics* 11, no. 1 (2011): 23–28.
Carpenter, Daniel P. *Reputation and Power: Organizational Image and Pharmaceutical Regulation at the FDA*. Princeton: Princeton University Press, 2010.
Carpenter, Daniel P., and David A. Moss, eds. *Preventing Regulatory Capture: Special Interest Influence and How to Limit It*. New York: Cambridge University Press, 2013.
Daemmrich, Arthur A. *Pharmacopolitics: Drug Regulation in the United States and Germany*. Chapel Hill: University of North Carolina Press, 2004.
Davis, Michael, and Andrew Stark, eds. *Conflict of Interest in the Professions*. Practical and Professional Ethics Series. New York: Oxford University Press, 2001.
Davis, Ross D. "The Federal Conflict of Interest Laws." *Columbia Law Review* 54, no. 6 (1954): 893–915.
"DES Livestock Implants Are Prohibited by FDA Because of Cancer Link." *Wall Street Journal*, 26 April 1973.
Dutton, Diana Barbara. *Worse than the Disease: Pitfalls of Medical Progress*. New York: Cambridge University Press, 1988.
Finlay, Mark R., and Alan I. Marcus. "'Consumerist Terrorists': Battles over Agricultural

Antibiotics in the United States and Western Europe." *Agricultural History* 90, no. 2 (2016): 146–72.

Frickel, Scott, and Michelle Edwards. "Untangling Ignorance in Environmental Risk Assessment." In *Powerless Science?: Science and Politics in a Toxic World*, edited by Soraya Boudia and Nathalie Jas, 215–33. New York: Berghahn Books, 2014.

Frickel, Scott, Sahra Gibbon, Jeff Howard, Joanna Kempner, Gwen Ottinger, and David J. Hess. "Undone Science: Charting Social Movement and Civil Society Challenges to Research Agenda Setting." *Science, Technology, & Human Values* 35, no. 4 (2010): 444–73.

Gaudillière, Jean-Paul. "DES, Cancer, and Endocrine Disruptors: Ways of Regulating, Chemical Risks, and Public Expertise in the United States." In *Powerless Science?: Science and Politics in a Toxic World*, edited by Soraya Boudia and Nathalie Jas, 65–94. New York: Berghahn Books, 2014.

———. "Une manière industrielle de savoir." In *Histoire des sciences et des savoirs 3—Le siècle des technosciences (dépuis 1914)*, edited by Christophe Bonneuil and Dominique Pestre, 85–105. Paris: Seuil, 2015.

Gaudillière, Jean-Paul, and Ulrike Thoms, eds. *The Development of Scientific Marketing in the Twentieth Century: Research for Sales in the Pharmaceutical Industry*. Studies for the Society for the Social History of Medicine. London: Pickering & Chatto, 2015.

Glantz, Stanton A., John Slade, Lisa A. Bero, Peter Hanauer, and Deborah E. Barnes, eds. *The Cigarette Papers*. Berkeley: University of California Press, 1996.

Greene, Jeremy A. *Prescribing by Numbers: Drugs and the Definition of Disease*. Baltimore: Johns Hopkins University Press, 2007.

Greenwald, Peter, Joseph J. Barlow, Philip C. Nasca, and William S. Burnett. "Vaginal Cancer after Maternal Treatment with Synthetic Estrogens." *New England Journal of Medicine* 285, no. 7 (1971): 390–92.

Greffion, Jérôme. "Faire passer la pilule: Visiteurs médicaux et entreprises pharmaceutiques face aux médecins : Une relation socio-economique sous tensions privées et publiques (1905–2014)." Doctoral thesis, EHESS, Paris, 2014.

Gross, Matthias, and Linsey McGoey, eds. *Routledge International Handbook of Ignorance Studies*. Routledge International Handbooks. New York: Routledge, 2015.

Hauray, Boris, Henri Boullier, Jean-Paul Gaudillière, and Emmanuel Henry, "Conflict of Interests, Capture, Making of Ignorance, Hegemony: Conceptualizing the Influence of Economic Interests on Public Health." In *Conflict of Interest and Medicine: Knowledge, Practices and Mobilizations*, edited by Boris Hauray, Henri Boullier, Jean-Paul Gaudilliere, and Hélène Michel. London: Routledge, 2021.

Henry, Emmanuel. *Amiante, Un scandale improbable: Sociologie d'un problème public*. Rennes: Presses Universitaires de Rennes, 2007.

———. *Ignorance scientifique & inaction publique: Les politiques de santé au travail*. Paris: Presses de la Fondation Nationale des Sciences Politiques, 2017.

Herbst, Arthur L., Howard Ulfelder, and David C. Poskanzer. "Adenocarcinoma of the Vagina: Association of Maternal Stilbestrol Therapy with Tumor Appearance in Young Women." *New England Journal of Medicine* 284, no. 16 (1971): 878–81.

Hess, David J. "Undone Science and Social Movements: A Review and Typology." In *Routledge International Handbook of Ignorance Studies*, edited by Matthias Gross and Linsey McGoey, 141–54. Routledge International Handbooks. New York: Routledge, 2015.

Holloway, Kelly Joslin. "Normalizing Complaint: Scientists and the Challenge of Commercialization." *Science, Technology, & Human Values* 40, no. 5 (2015): 744–65.

Huntington, Samuel P. "The Marasmus of the ICC: The Commission, the Railroads, and the Public Interest." *Yale Law Journal* 61, no. 4 (1952): 467–509.

Jasanoff, Sheila. *Science at the Bar: Science and Technology in American Law.* Cambridge: Harvard University Press, 1997.

Jukes, Thomas. "The DES Hearings: What Was Learned?" *Feedstuffs*, November 1977.

———. "Diethylstilbestrol in Beef Production: What Is the Risk to Consumers?" *Preventive Medicine* 5, no. 3 (1976): 438–53.

Kirchhelle, Claas. *Pyrrhic Progress: The History of Antibiotics in Anglo-American Food Production.* New Brunswick: Rutgers University Press, 2020.

Kleinman, Daniel Lee. "Untangling Context: Understanding a University Laboratory in the Commercial World." *Science, Technology, & Human Values* 23, no. 3 (1998): 285–314.

Langston, Nancy. *Toxic Bodies: Hormone Disruptors and the Legacy of DES.* New Haven: Yale University Press, 2010.

Lesser, Lenard I., Cara B. Ebbeling, Merrill Goozner, David Wypij, and David S. Ludwig. "Relationship between Funding Source and Conclusion among Nutrition-Related Scientific Articles." *PLoS Medicine* 4, no. 1 (2007): e5.

Lexchin, Joel, and Orla O'Donovan. "Prohibiting or 'Managing' Conflict of Interest? A Review of Policies and Procedures in Three European Drug Regulation Agencies." *Social Science & Medicine* 70, no. 5 (2010): 643–47.

Luebke, Neil R. "Conflict of Interest as a Moral Category." *Business & Professional Ethics Journal* 6, no. 1 (1987): 66–81.

Lundh, Andreas, Joel Lexchin, Barbara Mintzes, Jeppe B. Schroll, and Lisa Bero. "Industry Sponsorship and Research Outcome." *Cochrane Database of Systematic Reviews*, no. 2 (2017): MR000033. doi: 10.1002/14651858.MR000033.pub3.

Marcus, Alan I. *Cancer from Beef: DES, Federal Food Regulation, and Consumer Confidence.* Baltimore: Johns Hopkins University Press, 1994.

Markowitz, Gerald E., and David Rosner. *Deceit and Denial: The Deadly Politics of Industrial Pollution.* Berkeley: University of California Press, 2002.

Marks, Lara. *Sexual Chemistry: A History of the Contraceptive Pill.* New Haven: Yale University Press, 2001.

Merton, Robert K. "The Normative Structure of Science." In *The Sociology of Science: Theoretical and Empirical Investigations*, edited by Norman W. Storer, 267–78. Chicago: University of Chicago Press, 1973.

Meyers, Robert. *D.E.S.: The Bitter Pill.* New York: Seaview/Putnam, 1983.

Morgen, Sandra. *Into Our Own Hands: The Women's Health Movement in the United States, 1969–1990.* New Brunswick: Rutgers University Press, 2002.

Novak, William J. "A Revisionist History of Regulatory Capture." In *Preventing Regulatory Capture: Special Interest Influence and How to Limit It*, edited by Daniel P. Carpenter and David A. Moss, 25–48. New York: Cambridge University Press, 2014.

Perkins, Roswell B. "The New Federal Conflict-of-Interest Law." *Harvard Law Review* 76, no. 6 (1963): 1113–69.

Pfeffer, Naomi. "Lessons from History: The Salutary Tale of Stilboestrol." In *Consent to Health Treatment and Research: Differing Perspectives, Report of the Social Science Re-*

search Unit. Consent Conference Series 1. London: University of London, Institute of Education, Social Science Research Unit, 1992.

Proctor, Robert. *Cancer Wars: How Politics Shapes What We Know and Don't Know about Cancer*. New York: Basic Books, 1995.

Proctor, Robert, and Londa L. Schiebinger, eds. *Agnotology: The Making and Unmaking of Ignorance*. Stanford: Stanford University Press, 2008.

Rifkin, Jeremy. *Beyond Beef: The Rise and Fall of the Cattle Culture*. New York: Dutton, 1992.

Rodwin, Marc A. *Conflicts of Interest and the Future of Medicine: The United States, France, and Japan*. Oxford; New York: Oxford University Press, 2011.

Salter, Brian, Yinhua Zhou, and Saheli Datta. "Hegemony in the Marketplace of Biomedical Innovation: Consumer Demand and Stem Cell Science." *Social Science & Medicine* 131 (2015): 156–63.

Schell, Orville. *Modern Meat*. New York: Random House, 1984.

Seaman, Barbara. *The Doctors' Case against the Pill*. New York: P. H. Wyden, 1969.

Shapiro, Sidney A. "Blowout: Legal Legacy of the Deepwater Horizon Catastrophe: The Complexity of Regulatory Capture: Diagnosis, Causality, and Remediation." *Roger Williams University Law Review* 17, no. 1 (2012): 27–42.

Sismondo, Sergio. "Hegemony of Knowledge and Pharmaceutical Industry Strategy." In *Philosophical Issues in Pharmaceutics: Development, Dispensing, and Use*, edited by Dien Ho, 47–63. Philosophy and Medicine. Dordrecht, Netherlands: Springer, 2017.

———. "How Pharmaceutical Industry Funding Affects Trial Outcomes: Causal Structures and Responses." *Social Science & Medicine* 66, no. 9 (2008): 1909–14.

———. "Key Opinion Leaders and the Corruption of Medical Knowledge: What the Sunshine Act Will and Won't Cast Light On." *Journal of Law, Medicine & Ethics* 41, no. 3 (2013): 635–43.

Stigler, George J. "The Theory of Economic Regulation." *Bell Journal of Economics and Management Science* 2, no. 1 (1971): 3–21.

Sunder Rajan, Kaushik. *Pharmocracy: Value, Politics & Knowledge in Global Biomedicine*. Experimental Futures. Durham: Duke University Press, 2017.

Temin, Peter. *Taking Your Medicine: Drug Regulation in the United States*. Cambridge: Harvard University Press, 1980.

Trost, Christine, and Alison L. Gash, eds. *Conflict of Interest and Public Life: Cross-National Perspectives*. Cambridge: Cambridge University Press, 2008.

Tweedale, Geoffrey. *Magic Mineral to Killer Dust: Turner & Newall and the Asbestos Hazard*. Oxford: Oxford University Press, 2000.

Wadmann, Sarah. "Physician–Industry Collaboration: Conflicts of Interest and the Imputation of Motive." *Social Studies of Science* 44, no. 4 (2014): 531–54.

 CHAPTER 10

Defining Food Additives
Origins and Shortfalls of the US Regulatory Framework

Maricel V. Maffini and Sarah Vogel

The post–World War II period marked the beginning of an unprecedented economic boom in the United States fueled in part by cheap oil and the concomitant expansion in petrochemicals that served as the foundation for a dramatic change in the consumer market. Food—how it was produced, processed, manufactured, and packaged—was transformed by new pesticides, preservatives, dyes, and plastics—from DDT to Saran Wrap and other polyethylene plastic films, Teflon pans, and plastic resins used to line metal cans. This rapid transformation of the food supply brought cheap food to the market and tables of many Americans and, in the process, introduced hundreds of new chemical substances into the everyday diet—from pesticides to synthetic hormones, to preservatives, to compounds used to make plastic packaging.

By the late 1940s, low levels of some of these chemical inputs into food production, notably pesticide residues and the synthetic estrogen diethylstilbestrol (DES), were being detected in food and milk supplies. James Delaney, a young congressman from New York, told then Speaker of the House Sam Rayburn (D-TX) the following story shared with him by a colleague: a fellow congressman sprayed his midwestern lakeside property with DDT, and, while out fishing not long afterward, he discovered a disturbing number of dead fish. What might be the impact of these pesticides on people eating the food sprayed with them, Delany asked Rayburn. In 1950, Rayburn established a Select Committee to Investigate Chemicals in Food Production to investigate that question and the larger impacts of new chemical inputs into food, and appointed Delaney as its chair.[1] This marked the beginning of eight years of hearings and legislative negotiations that led to the passage of major reforms—namely, the 1958 Food Additive Amendment to the Federal Food, Drug and Cosmetics Act.

What became evident through the investigation and many hearings was that the 1938 Federal Food, Drug and Cosmetics Act inadequately protected the public from the new and expanding uses of chemicals.[2] How to manage the risks of the modern food industry lay at the center of the congressional debates that took place for much of the 1950s. The reforms laid out in the Food Additives Amendment defined for the first time what substances would be regulated and those that would not, and outlined how safety would be defined and risks managed.

Congressional hearing records revealed that the US Food and Drug Administration (FDA), the regulatory agency responsible for food safety, and the scientific community feared that many of the chemicals being added to food had not been adequately tested to establish their safe use in food.[3] The main concern was with chemicals that might cause harmful effects after being consumed for months or years. Managing risk and determining safety would require considering the cumulative health effect of chemicals in the diet, safety factors to address issues of uncertainty, and an exclusion of carcinogenic substances—what would come to be referred to as the Delaney Clause.

This chapter provides the legal and regulatory definitions of food additives and explains how safety decisions have been and are currently made and who makes them. This overview is not a historical analysis of these concepts and interpretations of safety but is meant to clearly define the regulatory framework, which has experienced little to no modification since the law was enacted more than sixty years ago. We highlight specific contemporary examples of food chemicals associated with adverse human health outcomes in order to define the limitations and flaws in the food additives law and the current regulatory process. And we present an alternative framework for assessing risk and determining safety that is grounded in the law but breaks from either a zero tolerance or a *de minimis* conceptualization. This framework focuses on a consideration of the cumulative effects of classes of chemicals that are pharmacologically related, putting the health outcome—notably chronic conditions—at the center of the risk assessment. Finally, in the conclusion, we flag how emerging technologies in food today are creating an even greater strain on this system for managing risk.

Defining Food Additives

From 1950 to 1952 and again from 1956 to 1958, members of Congress held two sets of hearings on the emerging problem of chemicals in foods and the need to expand federal oversight of such substances. With consumer spending booming, food production rising, and new materials, products, and chemicals driving a postwar economic boom, the issue before members of Congress was how to balance this economic expansion and innovation with public and consumer safety. In the process of publicly debating chemical safety, these two sets of hearings revealed diverse concerns among scientists, farmers, gardeners, consumers, and pure food activists regarding the implications of the radical transformation of the production, storage, packaging, and transport of food. Reports of the pesticide DDT in cow's milk, illnesses associated with pesticide use among soldiers during World War II, depletion of soil nutrients from fertilizers and pesticides, the carcinogenicity of DES, and the unknown hazards of hundreds of chemicals finding their way into foods, including new plastic packaging materials used to store and transport food, were among the issues presented to members of Congress. Rising concerns about cancer fueled debates about whether some substances should be allowed for use at all. Growing pressure from an international scientific community called for complete restrictions of carcinogens in foods.[4]

The central question before Congress and the FDA charged with oversight of the food supply was how to maintain food safety in the face of a food production process increasingly dependent upon petrochemical inputs—pesticides, insecticides, and even plastic compounds leaching from packaging materials. To secure the nation's supply of cheap food, the prevailing narrative by the late 1950s, frequently articulated by representatives of the chemical industry, held that some contamination of food was inevitable and a necessary risk that could be managed for the greater benefit of society.

The 1958 Food Additives Amendment outlined a new regulatory system for chemicals used to produce, process, transport, or package food. The law sought to both protect the health of consumers by requiring manufacturers of food additives and food processors to pretest any potentially unsafe substances that would be added to food, and to advance food technology by permitting the use of food additives at "safe" levels. For the first time, the law defined a food additive and determined what would and would not be allowed in food.

A food additive is defined as "any substance the intended use of which results or may reasonably be expected to result, directly or indirectly, in its becoming a component or otherwise affecting the characteristics of any food."[5] This includes chemical ingredients intentionally added to foods, or *direct additives*, such as preservatives, flavors, sweeteners, and emulsifiers. It also includes chemicals unintentionally added to foods, such as chemicals used in producing, packaging, preparing, or transporting foods that may enter the food supply through migration or leaching. Examples of these *indirect additives* are bisphenol A (BPA), ortho-phthalates, perchlorate, and tetrafluoroethylene-hexafluoropropylene-vinylidene fluoride copolymers. Today, other chemicals that may be present in the diet (e.g., color additives, pesticide chemicals or pesticide residues, animal drugs, dietary supplements) are regulated under different laws. (Notably, the Food Quality Protection Act in 1996 consolidated the regulation of pesticides in foods—raw and processed—that had been previously split under distinct agencies and laws. As such, pesticide residues are now regulated by the Environmental Protection Agency [EPA]).[6] Unless otherwise specified, the terms *additives* and *food chemicals* are used as synonyms throughout this chapter.

Excluded from the legal definition of food additives are *generally recognized as safe* (GRAS) substances. A GRAS substance is "generally recognized, among experts qualified by scientific training and experience to evaluate its safety, as having been adequately shown through scientific procedures (or, in the case of a substance used in food before January 1, 1958, through scientific procedures or experience based on common use in food) to be safe under the conditions of its intended use."[7] The GRAS exemption was intended for common food ingredients such as vinegar, oil, salt, and sugar; Congress interpreted that the use of those substances did not necessitate an FDA review of their safety.[8]

According to congressional intent and the FDA's regulations, the amount and type of toxicity and exposure data needed to make a safety determination for a GRAS substance are comparable to that for a food additive. Unlike food additives, whose safety data can be unpublished, for a chemical use to be considered GRAS, its safety assessment must be published so that the scientific community can assert its general recognition. Unfortunately, as discussed later, this is not always the case. The GRAS exemption has, over time, become the rule that has left the FDA and the public in the dark.

Over the past several decades, it has become increasingly evident that the FDA and industry have a different interpretation of GRAS.[9] Currently, as long as a substance is determined to be GRAS by somebody—without independent review or oversight—there is no requirement to inform the FDA of its identity, safety assessment, and uses. This appears to hold even if the determination is made in secret with no review by or reporting to the FDA. In other words, it is legal for a company to make its own decision that a chemical is GRAS, market it as GRAS, and completely bypass the agency in charge of ensuring food safety and public health. This is exemplified on the right side of figure 10.1.

Alternatively, some companies may seek FDA review of their safety assessment by voluntarily submitting data to the agency, but this has not been required. In cases of voluntary submission, the FDA acts more as an auditor or a peer reviewer than as a regulator; while the agency will offer an opinion about whether it agrees or disagrees with the industry assessment, its assessment or opinion is not binding. If the company perceives that the FDA may disagree with the GRAS determination, it can withdraw the notification at any time, asking the FDA to stop its review.

Regardless of potential concerns expressed by the FDA about the safe use of a voluntarily submitted chemical, companies are still allowed to market and sell the substance as GRAS. Although the FDA indicates in its website that a GRAS notification has been withdrawn, it does not publish the reasons. If the FDA does not have any concerns and agrees with the company's GRAS

Figure 10.1. Schematic representation of the decision-making process for chemicals and genetically engineered plants allowed in food. FDA = Food and Drug Administration; GRAS = generally recognized as safe. Shaded boxes indicate points at which companies are allowed to market and sell their products. Dashed lines indicate voluntary actions by companies. Adapted from Thomas G. Neltner, "Food Innovations: Innovative Foods and Ingredients. Institute of Science for Global Policy (ISGP), 2019," http://scienceforglobalpolicy.org/conference/innovative-foods-ingredients-ifi/.

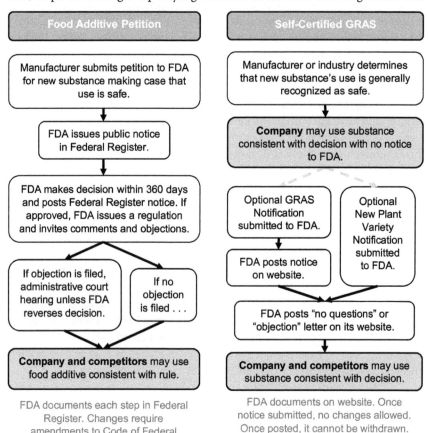

determination, it expresses that opinion in a letter (see figure 10.1, right side). Because it is only an opinion, it cannot be challenged in court.

There are also 120 chemicals used in food that were grandfathered in through a "prior sanctioned" process after the Food Additives Amendment was passed.[10] The use of these chemicals was explicitly approved by the FDA or by the US Department of Agriculture before 6 September 1958.[11] The uses with prior sanction are also exempted from the food additive regulations.

In the final negotiations of the Food Additives Amendment, Congressman Delaney managed to secure a single clause in the law that effectively restricted cancer-causing substances from the food supply. What came to be referred to as the Delaney Clause states that "no additive shall be deemed to be safe if it is found to induce cancer when ingested by man or animal, or if it is found, after tests which are appropriate for the evaluation of safety of food additives, to induce cancer in man or animals."[12]

Whether the Delaney Clause required the absolute restriction of a carcinogen (zero tolerance) or allowed levels below a detectible threshold (a *de minimis* interpretation of the law) has long been a source of conflict. As the regulated industry worked to respond to the 1958 law, the leading industry attorney and expert on the new law confirmed with the FDA that they had no plans for a zero-tolerance approach to carcinogens. Indeed, in 1962, the FDA passed the DES proviso that allowed the use of the synthetic estrogen—despite evidence of its carcinogenicity—provided the residue levels were below a detectible level.[13] This approach to the regulation of carcinogenic compounds as "not residue" has dominated the agency's approach. The detection of DES in the livers of cattle, therefore, required the agency to issue a limited restriction of the hormone as a feed additive for cattle in 1972.[14] In the negotiations that led to the passage of the Food Quality Protection Act, the Delaney Clause was eliminated in exchange for a lower safety standard for all pesticides—a one-in-a-million cancer risk standard.[15] The agency has, however, used the clause to restrict cyclamates, an artificial sweetener, in 1969, and since the Delaney Clause remains in the law for food additives, the agency restricted seven artificial flavors in 2018.[16]

When the legislative reforms were signed into law in 1958, the FDA gave the following tally: "Approximately 842 chemicals are used, have been used, or have been suggested for use in foods. Of this total, it was estimated that 704 are employed today."[17] Today, the universe of food chemicals is roughly ten thousand, with an approximately even split between direct and indirect additives.[18] It is unknown how many of these additives are in use currently or have been used in the past, or at what volume. An estimated 10 percent of the ten thousand chemicals have been declared GRAS by manufacturers with no review or disclosure to the FDA.[19] This absence of information results from the lack of requirements in the law for chemical manufacturers to report uses to the FDA.

The substantial increase in the number of food additives over the past sixty years reflects the continued transformation of food. Many of the foods and beverages found in a mainstream grocery store are highly engineered products in terms of the food itself and its packaging. According to the most recent data, almost 60 percent of the American diet is composed of ultra-processed foods.[20] A new food classification system based on the extent and purpose of industrial food processing defines an ultra-processed food as a "formulation of several ingredients, which, besides salt, sugar, oils and fats, include food substances not used in culinary preparations, in particular, flavors, colors, sweeteners, emulsifiers, and other additives used to imitate sensorial qualities of unprocessed or minimally processed foods and their culinary preparations or to disguise undesirable qualities of the final product."[21]

The proliferation of food chemicals and the incredible variety of foods in the marketplace suggest that the law has allowed food technology to advance at a rapid pace. However, in the face of rapid innovation and change, has the law adequately protected consumer health?

Safety of Food Chemicals

The Food Additive Amendment of 1958 fundamentally changed the way chemicals were supposed to come into use in or on food. Under the Federal Food, Drug, and Cosmetics Act of 1938, no testing of chemicals was required. In 1952, the FDA described to members of Congress that it knew that roughly one half of the chemicals in use (428 of the 842) were "known to be safe." The new law mandated that industry test "any potentially unsafe substances that are to be added to food" in order to provide the FDA with the information needed to make safety decisions.[22]

Although the law did not define *safe* or *safety*, the FDA has used language from the legislative history to define safe or safety to mean "a reasonable certainty in the minds of competent scientists that the substance is not harmful under the intended conditions of use."[23] This definition applies to all food additives and GRAS substances. In other words, a chemical is considered unsafe until proved safe.

Another important term that was not defined in the law or by the FDA is *harm*; this lack of clarity has been the source of many controversies about the safety of chemicals, including debates about the safety of chemicals that have been shown to have hormonal activity and may cause chronic health effects.[24]

The concern about chronic health effects of exposure to food chemicals was foremost in the congressional investigation in the early 1950s. The surgeon general pointed out that "the lack of adequate information on the chronic health effects" of chemicals precluded understanding the extent of the public health impact. Similar statements about the unknown impact of chemicals used in production and processing of foods on chronic health effects were made by the American Public Health Association and the American Medical Association.[25]

Despite the law's original intent to promote safety testing for food additives, there are significant information gaps, and the FDA has made minimal effort to conduct another review of additive safety, even in the face of new scientific evidence. The percentage of chemicals tested has decreased since the 1950s.

A 2013 report by the Pew Charitable Trusts on toxicity testing of chemicals allowed in foods showed that information sufficient to estimate a safe level of exposure is available for a mere 21 percent of direct additives.[26] An even more disturbing fact is that fewer than 7 percent of those additives have been tested for developmental or reproductive toxicity. If in vitro and computer-based (e.g., in silico) studies are taken into account, more than 50 percent of direct additives still lack information. Overall, the 2013 report found that fewer than 38 percent of FDA-regulated additives (i.e., direct and indirect additives and GRAS substances) have been investigated in published studies that included feeding the chemical to laboratory animals and evaluating for health effects.

There are several explanations for the lack of safety data based on laboratory animal feeding studies.[27] First, the GRAS exemption allows food manufacturers to determine that a chemical use is GRAS based on common uses before 1958 or on scientific procedures (e.g., toxicity testing, exposure estimates). A claim of common uses releases the manufacturer from performing a safety assessment. Second, the FDA approved thousands of chemicals before it defined safety or established how a safety determination should be conducted. An example is BPA, which was approved in 1963 together with hundreds of other can-coating chemicals with few or no data.[28] Third, the industry and FDA rely heavily on estimated (not experimentally determined) thresholds of exposure below which few (mostly *in vitro*) or no data are generated. The assumption is that at very low exposures, chemicals cannot present a risk.[29]

This is a very common practice for indirect additives and flavors.[30] For example, in 2005, the FDA approved, without any toxicity testing, the chemical perchlorate (a well-known thyroid disruptor) as a component of plastics designed to hold dry food. The manufacturer claimed that the dietary exposure would be below the threshold of 0.5 parts per billion (ppb) and therefore would be considered toxicologically insignificant.[31] The manufacturer invoked a 1995 regulation that exempts chemicals from regulation as food additives if they are in the diet at levels below 0.5 ppb and the chemical has not been shown to be a carcinogen.[32]

The fourth explanation for the lack of safety studies is that academic researchers are unlikely to conduct research aimed at filling data gaps, particularly in the absence of existing data indicating that there may be hazardous characteristics to explore, and there is no significant available funding for such work. Finally, industry has little or no incentive to generate additional information, particularly given that the FDA does not systematically review additive safety.[33]

The lack of review by the FDA of previous decisions about additive safety, even in the presence of new data indicating the existence of new hazards or exposures exceeding safe amounts, is particularly troubling. For instance, diethylhexyl phthalate (DEHP) is an industrial chemical commonly used to make plastics flexible. DEHP has been used as an indirect food additive since before 1958, and when the law was passed, it was grandfathered in as a "prior sanctioned" chemical for use as a plasticizer in food packaging for foods of high water content only.[34] It is unclear what type, if any, of safety data was available at that time. However, over the past fifteen years, DEHP has been well characterized as an endocrine disruptor with antiandrogenic properties, and its use has been effectively banned in children's toys and care articles due to adverse health effects associated with exposure.[35] Despite the FDA's knowledge of its toxic effects in 1973 and the additional evidence concerning human and animal health effects, DEHP is still allowed to be used in the manufacture of materials in contact with food without any limitation on the quantities that could migrate into food.[36]

There are a few notable examples of FDA reassessment of the safety of food chemicals. In 1969, President Richard Nixon issued a directive that ordered the agency to perform a full review of food additives, with a focus on the safety of chemicals that the FDA had designated as GRAS.[37] Specifically, the FDA was directed to reexamine the structure and procedures used by the agency to approve chemical additives to ensure they were "fully adequate."[38] In response, the FDA established the Select Committee on GRAS Substances (SCOGS) and charged it with the task of "evaluating the safety" of 468 substances (422 for direct addition to food and 46 as components of packaging materials).[39] The committee, made up of experts in biochemistry, pharmacology, and medicine, conducted its analysis from 1972 to 1982 and published its decisions in 115 reports.[40] It agreed that 92 percent of the chemicals were safe for direct use in foods. However, it stated that 35 chemicals required either "that safe conditions of use be established" or that, in the absence of relevant data submission, the FDA should move to "rescind GRAS status."[41] To our knowledge, the GRAS status of those substances has not been revoked.[42]

In 2009, a University of Illinois professor sued the FDA after its failure to respond to his citizen petition requesting a ban on the use of partially hydrogenated oils (the main source of artificial trans fats) due to their association with heart disease.[43] In 2015, after reviewing the evidence, the FDA determined that partially hydrogenated oils were not "generally recognized as safe for use in human food."[44]

The FDA took its most recent action in January of 2016, banning the use of long-chain perfluorinated compounds (PFCs) as indirect food additives because the chemicals are unsafe (i.e., they bioaccumulate and cause developmental and reproductive toxicity).[45] The decision was made in response to a food additive petition submitted by a group of public interest organizations.[46] These recent efforts to direct the agency to again review the safety of food additives suggest that the agency has the authority and the capacity to better manage these chemicals.

Who Determines the Safety of Food Chemicals?

According to the law, a chemical must be shown to be safe before it is used in or on food. The FDA states that it is the manufacturer's responsibility to demonstrate the safe use of its chemical. After a safety assessment is conducted, the manufacturer can submit a food additive petition to the FDA (figure 10.1, left side). The agency reviews the safety assessment and publishes its decision, the public submits comments, and the FDA eventually approves the chemical's safe uses. In the process Congress intended, the FDA reviews and approves the safety of chemicals and codifies its decision in the Code of Federal Regulations (CFR).

Alternatively, a manufacturer can submit a notification informing the agency of its safety determination and asking for its review (figure 10.1, right side). If the FDA agrees with the manufacturer, the agency publishes its decision on its website. This is not an approval but rather an opinion, and as such, it lacks the same legal weight. Figure 10.2 shows changes in the FDA's role over time.

However, because the notification system is voluntary, the manufacturer can make a GRAS determination and immediately market the chemical (Figure 10.1, upper right). The FDA is completely bypassed and therefore remains unaware of the chemical's identity and presence in the food supply. At least a thousand substances have been estimated by their manufacturers to be GRAS without notification to the FDA.[47] As the use—or arguably abuse—of this GRAS loophole has increased in recent decades, the filing of petitions has waned.

Figure 10.2. Trends in filings of petitions and notifications for food additives and GRAS substances directly added to human food submitted to the FDA from 1990 to 2010.

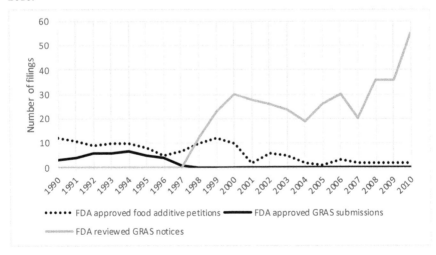

These GRAS self-determinations are riddled with conflicts of interest. When researchers analyzed more than 450 GRAS determinations voluntarily submitted to the FDA, they found that 22 percent of the safety assessments were made by an employee of the chemical manufacturer, 13 percent by an employee of a consulting firm hired by the company, and 65 percent by an expert panel selected by the manufacturer or its consulting firm.[48] The lack of independent review in GRAS determinations raises concerns about the integrity of the process and the safety of the chemicals, especially when the manufacturers do not notify the FDA. In the past two decades, the private sector has made more safety decisions than the FDA (see figure 10.2).

A 2014 report by the Natural Resources Defense Council (NRDC) identified 275 chemicals from fifty-six companies that were marketed for use in food based on undisclosed GRAS safety determinations.[49] In other words, the investigators found no evidence that the FDA had reviewed and cleared their use in food. Most of the GRAS chemicals the NRDC examined were primarily marketed as active ingredients in dietary supplements, suggesting that the market for such substances has expanded into conventional foods with claims that they make food "better for you." Many of the chemicals were extracts of plants or highly purified or synthetic versions of biologically active chemicals in plants, such as antioxidants, that are purported to have health benefits.[50]

While it is fair to assume that it is not in a company's best interest to commercialize an unsafe or questionable GRAS substance, there are a number of examples of GRAS substances that raise considerable concerns about the effectiveness of this kind of self-policing. The following are examples of self-determined GRAS substances submitted to the FDA for review that were found to have serious safety issues but are nevertheless used in and on food.

- *Epigallocatechin-3-gallate (EGCG)*. EGCG is a highly purified chemical from green tea extract; it can be labeled as EGCG or with the more appealing name of green tea extract. The FDA had serious concerns about its safety due to evidence that it might cause fetal leukemia based on in vitro studies using newborn and adult human cells.[51] The FDA pointed out the lack of explanation for potentially dangerous interactions with sodium nitrite, a common preservative, or with acetaminophen, the active ingredient in Tylenol and many other over-the-counter pain killers. EGCG was declared to be GRAS for use in beverages, including teas, sport drinks, and juices.
- *Sweet lupin protein, fiber, and flour*. Lupin is a legume belonging to the same plant family as peanuts. Lupin has been shown to cause allergic reactions similar to those from peanuts.[52] In its review, the FDA raised concerns that people with peanut allergies would be exposed to a similar allergen without any warning labels on the products.[53] Lupin products are

commonly used in gluten-free baked goods. They were declared GRAS also for use in dairy products, gelatin, meats, and candy.[54]

- *γ-Amino butyric acid (GABA).* GABA is a neurotransmitter that is naturally present in the central nervous system and inhibits brain signals. The FDA's chief concern was that the company's estimated exposure, based on its intended uses, was well in excess of what it had considered safe.[55] If estimated exposure levels exceed the estimated safe level, the chemical cannot be used safely. The company also failed to consider the contribution from natural sources of GABA (e.g., meats), which could increase exposure levels even further beyond a safe amount. The company also relied on unpublished data. Nevertheless, GABA was declared GRAS for use in beverages, chewing gum, coffee, tea, and candy.[56]

Although it is fairly certain that there is high degree of confidence that all ingredients listed on a label are legal for use in food under the current laws, the safety in many cases appears uncertain.[57] For many food chemicals, safety largely boils down to blind trust in the manufacturer. Investigation into the chronic, long-term exposures to chemicals in the diet has, as a result, been sidelined, despite the explicit requirement to consider these kinds of impacts under the law.

Cumulative Effects of Chemicals and Human Health

In its *Global Status Report on Noncommunicable Diseases* (NCDs), the World Health Organization (WHO) recognized that chronic diseases have become one of the "major health and developmental challenges of the 21st century in terms of the human suffering they cause and the harm they inflict on the socioeconomic fabric of countries."[58] NCDs are the leading cause of death globally. *Noncommunicable* means that the disease, usually chronic and with slow progression, is not transmittable from person to person. Examples are diabetes, cancer, obesity, and cardiovascular and chronic respiratory diseases.

The incidence of NCDs has increased in the past forty years due to environmental changes—not only lifestyle, diet, and behavioral changes but also environmental chemical exposures and, in particular, exposures that occur during fetal development.[59] Food is a constant source of environmental exposures ranging from fats, salts, and sugar to DEHP and other endocrine-disrupting compounds present in the diet. Exposure to chemicals in food is ongoing and occurs throughout all life stages, including the highly susceptible period of prenatal development.

When Congress passed the food additives law in 1958, it intended that the FDA determine the safety of chemicals in food. Discussions of lawmakers at the time reflect an understanding that some chemicals could pose potential health hazards and that co-exposure to chemicals is the norm because many additives are present in the diet. The law mandates that the FDA require three factors in determining the safety of a food chemical.[60] The first factor is the probable consumption of the substance and of any substance formed in or on food because of its use. To determine the risk of the chemical, there needs to be evidence of exposure (risk = exposure x health hazard). The second factor deals with the uncertainty of data and requires the use of safety factors that in the opinion of experts are appropriate. A ten-, hundred-, or thousand-fold safety factor is applied to the exposure level that does not cause harm, with the goal of lowering the risk. The third factor in a safety assessment states that the "cumulative effect of the substance in the diet, taking into account any chemically or pharmacologically related substance or substances in such diet" must be considered. In other words, chemicals that are structurally similar, such as ortho-phthalates (e.g., DEHP, benzyl butyl phthalate [BBP], di-isononyl phthalate [DINP]) should be assessed for safety together as a class.[61]

The reasoning behind this third factor is that structurally similar substances may have similar toxic effects. By assessing them cumulatively, the logic goes, a more accurate safe level for the class can be determined and the public can be better protected. Although the FDA has not defined *pharmacologically related* substances, it appears the agency interprets this concept as substances that have similar toxicity. For instance, in 21 CFR 170.18, the FDA describes how to set tolerances for related food additives and states that "food additives that cause similar or related pharmacological effects will be regarded as a class, and in the absence of evidence to the contrary, as having additive toxic effects and will be considered as related food additives." In a first-of-its-kind decision on the unsafe use of long-chain PFCs in food contact materials, the agency defined a class of chemicals, identified members of the class with toxicity data, and assumed that the same toxicity was expected for other PFCs within the class for which the agency had no data.[62]

The implications for considering the cumulative effects of pharmacologically related substances are significant. If this is how the agency should be interpreting the safety of food additives, it could be argued that even structurally unrelated chemicals should be assessed together.

Take the example of perchlorate (used in plastic packaging), nitrate (used in processed meats), and thiocyanate (food contaminant). All three of these chemicals found in the diet share a common toxic effect on the thyroid gland: inhibition of iodine uptake, which leads to decreased levels of thyroid hormone.[63] Decreased levels of thyroid hormone can present a health risk, and pregnant women are particularly vulnerable due to the permanent effects of low thyroid hormone levels on fetal brain development. One half of pregnant women in the United States already have inadequate iodine intake according to the WHO, and almost 16 percent are clinically deficient in iodine.[64] The risk of perchlorate, nitrate, and thiocyanate exposure is particularly high for these women and their fetuses. Thyroid hormone is fundamental for normal brain development, and long-term hormone deficiency, either maternal or after birth due to low iodine intake, is known to cause neurologic impairment.[65] Additional additives with thyroid-adverse effects in the FDA's toxicology database include flavors (e.g., brominated vegetable oil, butyl alcohol, ethyl isovalerate), synthetic colors (e.g., FD&C Red No. 3), preservatives (e.g., heptyl paraben, sodium nitrite), nutrients (e.g., tocopherols, vitamin D_3), and dough conditioners (e.g., potassium bromate).

Direct adverse effects on the brain or hypothalamus in animal studies are also documented for many food chemicals in the FDA's toxicology database. They include preservatives (e.g., butylated hydroxyanisole [BHA]), artificial sweeteners (e.g., aspartame, neotame), flavors (e.g., taurine [a common ingredient of energy drinks], theobromine, styrene, pulegone, peppermint oil), and the widespread ingredients monosodium glutamate (MSG) and caffeine.[66] Contemporary data indicate that BPA and some ortho-phthalates also adversely affect the developing brain.[67]

There are more examples of additives adversely affecting the same organs or systems. However, it is unclear whether the cumulative effect of pharmacologically related substances in the diet was ever considered when any of the chemicals mentioned previously were approved for uses in, or in contact with, food. It appears that this fundamental requirement has been ignored.

What has emerged is the single chemical–centric approach to the risk assessment process. The agency evaluates a single chemical and how it affects many organs.[68] Although this is an important step in understanding how the body reacts to a particular chemical in isolation and the potential health effects resulting from such exposure, it is arguably not equivalent to the safety assessment requirements mandated by Congress.

In the safety assessment process, a chemical's contribution to the health outcome of concern is to take into account the cumulative biologic effects of multiple chemicals on the same organ or system. This health outcome–centric approach to assessing chemical safety was proposed by the National Research Council in its report on phthalates.[69] Its Committee on Health Risks of Phthalates stated that "multiple pathways can lead to a common outcome" and that "the chemicals that should be considered for cumulative risk assessment should be ones that cause the same health outcomes or the same type of health outcomes."[70]

Consideration of the cumulative effects of pharmacologically related substances can better inform the social and financial costs of exposure. For example, Bellinger estimated that exposures to neurotoxic chemicals (e.g., methylmercury, organophosphate pesticides, lead) alone accounted for more than forty million lost full-scale IQ points in a population of 25.5 million

American children up to five years of age.[71] Other researchers have estimated the cost of disease and dysfunction (e.g., intellectual disability, autism, attention deficit hyperactivity disorder, childhood and adult obesity, endometriosis, male infertility, adult diabetes) associated with endocrine-disrupting chemicals at €163 billion ($180 billion) per year.[72]

A better understanding of the contribution of chemical exposures to the rising incidence of chronic diseases demands moving beyond a chemical-centric focus to consider the cumulative biologic effects on organs and their role in the onset of chronic diseases. For chemicals in food, this was the approach Congress intended in 1958.

The Triumph of Innovation over Safety

Today, a wave of innovations together with an increasingly competitive food market threatens to completely overwhelm the FDA. New sources of protein and fat (e.g., crickets, mung beans, algae, water lentils), new ingredients usually seen in dietary supplements (e.g., herbal extracts, CBD), and new foods produced using novel biotechnology methods (e.g., plant-based meat alternatives, cell-based meats) are examples of novel products coming onto the market at a frantic pace. Packaging has also evolved with new materials and added technology with the goals of improving sustainability and reducing food waste (e.g., bio-based polymers, intelligent materials that prolong shelf life, and sensors that detect spoiling). In a race to bring to market more sustainable and "healthy" options and to address food waste, new products and materials are altering how and what food is produced, packaged, and processed.

The agency appears increasingly concerned about the health consequences of the rapid influx of innovative substances onto the market, particularly through the GRAS process, and the insufficient resources and capacity available to assess the safety of these new food ingredients. In its budget request for fiscal year 2021, the agency included an innovative food products user fee of $26.1 million to provide for fifty-two full-time employees to carry out the FDA's "public health protection mission by assisting industry to meet its statutory responsibilities as it develops and implements new technologies in food, including cosmetics and biotechnology products."[73]

The agency also recently acknowledged the need for more regulatory oversight, flagging the GRAS loophole as a significant problem. In 2019, the agency worked to convene an invitation-only conference on innovative food ingredients. The multistakeholder discussions centered on how to connect "options based on evidence-based, credible scientific understanding with commercially realistic decisions concerning how plant- and microbial-derived foods and ingredients can be safely offered to the public under FDA's statutory framework."[74] While the agency touted instances where it has been able to review some innovative GRAS substances, such as mung bean protein isolate used in vegan egg products or algal fat intended to be used as butter-like product in bakery and spreads, it also recognized the market pressure to use the GRAS loophole and bypass the agency entirely.

The uneven approach taken by new food companies are exemplified by two products: alternative meats and cannabidiol. *Impossible Foods*, a plant-based meat alternative on the market, produces a soy protein called leghemoglobin that is similar to animal muscle protein. When leghemoglobin is added to a plant-based mixture and cooked, it produces a "bleeding" effect and provides a flavor similar to animal meat. The company voluntarily submitted a GRAS notification to the FDA. After an initial unfavorable review by the agency, the company produced additional information and later received a letter of no objection from the agency. The FDA's major concerns related to potential allergic reactions given that humans have never before been exposed to leghemoglobin. (The protein is present in parts of the soybean plant that are not edible.) In this case, the company chose a voluntary approach to working with the agency.

The situation is far different in the case of cannabidiol—the oil known as CBD—the use of which has spiked since Congress legalized industrial hemp production that contains no more than 0.3 percent of tetrahydrocannabinol or THC, the psychoactive component of marijuana in 2018.[75] Removing hemp-derived products from its Schedule I status under the Controlled Substances Act effectively legalized CBD. Foods containing CBD for humans and pets, as well as dietary supplements, have flooded the market and left the FDA with little to no control over its use. Most companies are using the GRAS loophole bypassing FDA review. However, the agency stated in warning letters to CBD companies that it "is not aware of any basis to conclude that CBD is GRAS among qualified experts for its use in human or animal food," and added that "there also is no food additive regulation which authorizes the use of CBD as an ingredient in human food or animal food, and the agency is not aware of any other exemption from the food additive definition that would apply to CBD. CBD is therefore an unapproved food additive, and its use in human or animal food violates the FD&C Act."[76] However, the agency has struggled to enforce the law and to gather safety information.[77]

In 2018, Susan Mayne, director of the FDA's Center for Food Safety and Applied Nutrition, gave a speech at the "Future Food-Tech" summit in San Francisco, CA, where she stressed to food innovators that the agency was there to help them innovate. She went on to emphasize the relationship between innovation and safety, noting that "even where pre-market authorization may not be required, if you don't consult with FDA, you can put your investments at risk because things can go wrong. It's really important to understand that people can become seriously ill if you haven't followed all the required steps."[78]

Yet, despite these sentiments and the agency's efforts to publish draft guidelines on advances in biotechnology, such as CRISPR-based gene editing, the GRAS loophole looms large over any attempt to safeguard the public's health.

Conclusions

The American diet is dramatically different today than in 1958, when Congress passed the Food Additives Amendment. Our food supply is more diverse, more processed, and in the case of new proteins, entirely novel. Innovations in processing, preserving, and packaging have made food more affordable, convenient, and available to all, but this transformation has brought with it thousands of new chemicals that have become an integral part of our daily diet. The unsettling aspect of this change is that the long-term, chronic effects of the use of these chemicals have rarely been studied or considered in the assessment of their safe use in food.

Today there are more than ten thousand additives allowed in food. The regulatory system designed to evaluate these additives for safety is plagued with problems. As a result, if a food additive is causing serious chronic health problems short of immediate serious injury, it is unlikely that the FDA would detect the problem unless the food industry alerted the agency. GRAS exemptions, lack of safety assessments that consider cumulative effects of chemically and pharmacologically related substances, and industry-determined safety assessments have created considerable uncertainty about the safety of food additives.

When Congress passed the Food Additive Amendment in 1958, it was in response to a rapidly changing food system and rising public and scientific concerns about the potential health risks of the use of new chemicals.[79] In the more than sixty years since that time, the number of chemicals used in food processing, packaging, transport, and handling has risen dramatically. During this same period, chronic diseases have become the leading cause of mortality and morbidity globally. Scientific knowledge about the impacts of chemical exposures—even very low-level exposures considered to be inconsequential by corporations and the FDA—on normal development and disease processes has raised new questions and concerns about chronic exposures to chemicals.[80]

We are faced with a regulatory system that has fallen short of being fully implemented and is weakened by decades of limited resources. The GRAS exemption has been exploited to the point that it is almost swallowing the law—a situation that in the face of new technologies in food is threatening to completely overwhelm the FDA. In response, some—including organizations we represent—have outlined necessary changes to the regulatory process. In the end, any reform will need to balance the pressure for speed to market with the public's need for independent oversight and an approach to assessing safety that improves our ability to manage chronic disease risks.

Maricel V. Maffini is an independent consultant in Washington, DC, where she works on public health advocacy through regulation and oversight of chemical exposure.

Sarah Vogel is vice president for health at the Environmental Defense Fund. A specialist in the regulation of chemicals in food and food packaging, she works with a team of scientists and policy experts to protect health by reducing exposure to toxic chemicals.

Notes

1. Vogel, *Is It Safe?*, 19
2. Vogel, *Is it Safe?*
3. US Congress, *Investigation.*
4. Vogel, *Is it Safe?*, 33
5. 21 USC §321(s); Maffini et al., "Looking Back."
6. Wargo, *Our Children's Toxic Legacy.*
7. Ibid.
8. Neltner et al., "Navigating."
9. Neltner and Maffini, "Generally Recognized as Secret."
10. 21 CFR §181.
11. Public Law 85-929, 72 Stat. 1784, 21 USC §348. 1958.
12. Vogel, *Is it Safe?*, 38.
13. Ibid., 47.
14. Ibid., 57.
15. Wargo, *Our Children's Toxic Legacy.*
16. *Federal Register* 83, no. 195 (9 October 2018): 50490.
17. US Congress, *Investigation.*
18. Neltner et al., "Navigating."
19. Ibid.
20. Steele et al., "Ultra-Processed Foods."
21. Moubarac et al., "Food Classification Systems"; Steele et al., "Ultra-Processed Foods."
22. US Congress, *Food Additive Amendment.*
23. 21 CFR §170.3(i).
24. Maffini et al., "Enhancing FDA's Evaluation."
25. US Congress, *Investigation.*
26. Neltner et al., "Data Gaps."
27. Ibid.
28. Bellinger, "A Strategy."
29. 21 CFR §170.39. Threshold of Regulation for Substances Used in Food-Contact Articles.

30. US Food and Drug Administration (FDA), *Guidance for Industry*; Hallagan and Hall, "Under the Conditions."
31. Maffini et al., "Looking Back."
32. US FDA, "Threshold of Regulation (TOR) Exemptions."
33. Maffini et al., "Looking Back."
34. 21 CFR §181.27. Plasticizers.
35. National Academy of Sciences, *Phthalates*; Consumer Product Safety Improvement Act of 2008. Public Law 110–314, 122 Stat. 3016. 15 USC 2051. 2008; Swan et al., "Prenatal Phthalate Exposure"; Swan, "Environmental Phthalate Exposure."
36. Shibko and Blumenthal, "Toxicology"; *Report to the U.S. Consumer Product Safety Commission.*
37. Nixon, "Special Message."
38. Ibid.
39. US FDA, "GRAS Substances (SCOGS) Database"; Select Committee on GRAS Substances, "Appendix," 5.
40. US FDA, "History of the GRAS List and SCOGS Review."
41. Select Committee on GRAS Substances, "Appendix," 5.
42. Maffini et al., "Looking Back."
43. Watson, "Researcher Files Lawsuit."
44. US FDA, News Release, "The FDA Takes Steps."
45. US FDA, "Indirect Food Additives," 5.
46. National Resources Defense Council et al., "Filing of Food Additive Petition."
47. Neltner et al., "Navigating."
48. Neltner et al., "Conflicts of Interest."
49. Neltner and Maffini, "Generally Recognized as Secret."
50. Ibid.
51. US FDA, "FDA Response," 197.
52. Bansal et al., "Variably Severe."
53. Ballabio et al., "Characterization."
54. US FDA, "FDA Response," 218.
55. Ibid., 206.
56. US FDA, *GRAS Notices: GRN-257.*
57. Neltner and Maffini, "Generally Recognized as Secret."
58. World Health Organization, *Global Status Report.*
59. Barouki et al., "Developmental Origins."
60. 21 CFR §170.3(i).
61. National Academy of Sciences, *Phthalates.*
62. US FDA, Memorandum P. Rice to P. Honigfort.
63. Maffini, Trasande, and Neltner, "Perchlorate and Diet."
64. Caldwell et al., "Iodine Status"; World Health Organization, *Assessment of Iodine.*
65. Rose and Brown, "Update."
66. Maffini and Neltner, "Brain Drain."
67. Rubin et al., "Evidence"; Miodovnik et al., "Developmental Neurotoxicity."
68. Maffini and Neltner, "Brain Drain."
69. National Academy of Sciences, *Phthalates.*
70. Bellinger, "A Strategy."

71. Ibid.
72. Trasande et al., "Burden of Disease."
73. US FDA, "Justification of Estimates."
74. Institute on Science for Global Policy, "Innovative Foods and Ingredients."
75. 2018 Farm Bill, US, available at https://www.agriculture.senate.gov/2018-farm-bill.
76. US FDA, "Warning Letter to Apex Hemp Oil LLC."
77. Margarita Raycheva, "FDA Says CBD Not GRAS."
78. Susan Mayne, "FDA's Role in Supporting Innovation."
79. Vogel, *Is it Safe?*
80. Vandenberg, "Classic Toxicology."

Bibliography

Ballabio, Cinzia, Elena Peñas, Francesca Uberti, Alessandro Fiocchi, Marcello Duranti, Chiara Magni, and Patrizia Restani. "Characterization of the Sensitization Profile to Lupin in Peanut-Allergic Children and Assessment of Cross-Reactivity Risk." *Pediatric Allergy and Immunology* 24, no. 3 (2013): 270–75.

Bansal, Amolak S., Mihir M. Sanghvi, Rhea A. Bansal, and Grant R. Hayman. "Variably Severe Systemic Allergic Reactions after Consuming Foods with Unlabelled Lupin Flour: A Case Series." *Journal of Medical Case Reports* 8, no. 1 (2014): 55.

Barouki, Robert, Peter D. Gluckman, Philippe Grandjean, Mark Hanson, and Jerrold J. Heindel. "Developmental Origins of Non-communicable Disease: Implications for Research and Public Health." *Environmental Health: A Global Access Science Source* 11, no. 1 (2012): 42.

Bellinger, David C. "A Strategy for Comparing the Contributions of Environmental Chemicals and Other Risk Factors to Neurodevelopment of Children." *Environmental Health Perspectives* 120, no. 4 (2012): 501–7.

Caldwell, Kathleen L., Yi Pan, Mary E. Mortensen, Amir Makhmudov, Lori Merrill, and John Moye. "Iodine Status in Pregnant Women in the National Children's Study and in U.S. Women (15–44 Years), National Health and Nutrition Examination Survey 2005–2010." *Thyroid* 23, no. 8 (2013): 927–37.

Hallagan, John B., and Richard L. Hall. "Under the Conditions of Intended Use: New Developments in the FEMA GRAS Program and the Safety Assessment of Flavor Ingredients." *Food and Chemical Toxicology* 47, no. 2 (2009): 267–78.

Institute on Science for Global Policy. "Innovative Foods and Ingredients." 23–27 June 2019. Accessed 26 March 2020, http://scienceforglobalpolicy.org/conference/inno vative-foods-ingredients-ifi/.

Maffini, Maricel V., Heather M. Alger, Erin D. Bongard, and Thomas G. Neltner. "Enhancing FDA's Evaluation of Science to Ensure Chemicals Added to Human Food Are Safe: Workshop Proceedings." *Comprehensive Reviews in Food Science and Food Safety* 10, no. 6 (2011): 321–41.

Maffini, Maricel V., Heather M. Alger, Erik D. Olson, and Thomas G. Neltner. "Looking Back to Look Forward: A Review of FDA's Food Additives Safety Assessment and Recommendations for Modernizing Its Program." *Comprehensive Reviews in Food Science and Food Safety* 12, no. 4 (2013): 439–53.

Maffini, Maricel V., and Thomas G. Neltner. "Brain Drain: The Cost of Neglected Responsibilities in Evaluating Cumulative Effects of Environmental Chemicals." *Journal of Epidemiology and Community Health* 69, no. 5 (2015): 496–99.

Maffini, Maricel V., Leonardo Trasande, and Thomas G. Neltner. "Perchlorate and Diet: Human Exposures, Risks, and Mitigation Strategies." *Current Environmental Health Reports* 3, no. 2 (2016): 107–17.

Mayne, Susan. "FDA's Role in Supporting Innovation in Food Technology." 22 March 2018. Accessed 18 March 2020, https://www.fda.gov/about-fda/what-we-do-cfsan/fdas-role-supporting-innovation-food-technology.

Miodovnik, Amir, Andrea Edwards, David C. Bellinger, and Russ Hauser. "Developmental Neurotoxicity of Ortho-Phthalate Diesters: Review of Human and Experimental Evidence." *Neurotoxicology* 41 (March 2014): 112–22.

Moubarac, Jean-Claude, Diana C. Parra, Geoffrey Cannon, and Carlos A. Monteiro. "Food Classification Systems Based on Food Processing: Significance and Implications for Policies and Actions: A Systematic Literature Review and Assessment." *Current Obesity Reports* 3, no. 2 (2014): 256–72.

National Academy of Sciences, National Research Council (US) Committee on the Health Risks of Phthalates. *Phthalates and Cumulative Risk Assessment: The Tasks Ahead.* Washington, DC: National Academies Press, 2008.

National Resources Defense Council et al., "Filing of Food Additive Petition." 7 January 2015. Docket No. FDA-2015-F-0714. *Federal Register* 80 (2015): 13508.

Neltner, T., and M. V. Maffini. "Generally Recognized as Secret: Chemicals Added to Food in the United States." National Resources Defense Council Report, April 2014. Accessed 24 January 2020, https://www.nrdc.org/sites/default/files/safety-loophole-for-chemicals-in-food-report.pdf.

Neltner, Thomas G., Heather M. Alger, Jack E. Leonard, and Maricel V. Maffini. "Data Gaps in Toxicity Testing of Chemicals Allowed in Food in the United States." *Reproductive Toxicology* 42 (December 2013): 85–94.

Neltner, Thomas G., Heather M. Alger, James T. O'Reilly, Sheldon Krimsky, Lisa A. Bero, and Maricel V. Maffini. "Conflicts of Interest in Approvals of Additives to Food Determined to Be Generally Recognized as Safe: Out of Balance." *JAMA Internal Medicine* 173, no. 22 (2013): 2032–36.

Neltner, Thomas G., Neesha R. Kulkarni, Heather M. Alger, Maricel V. Maffini, Erin D. Bongard, Neal D. Fortin, and Erik D. Olson. "Navigating the U.S. Food Additive Regulatory Program." *Comprehensive Reviews in Food Science and Food Safety* 10, no. 6 (2011): 342–68.

Nixon, Richard. "Special Message to the Congress on Consumer Protection (October 30, 1969)." Posted online by Gerhard Peters and John T. Wooley, The American Presidency Project. Accessed 20 January 2020, https://www.presidency.ucsb.edu/node/240003.

Raycheva, Margarita. "FDA Says CBD Not GRAS, Stirs Concern among Hemp Producers." *Food Chemical News*, 26 November 2019. Accessed 18 March 2020, https://iegpolicy.agribusinessintelligence.informa.com/PL222320/FDA-says-CBD-not-GRAS-stirs-concern-among-hemp-producers.

Report to the US Consumer Product Safety Commission by the Chronic Hazard Advisory Panel on Phthalates and Phthalate Alternatives. Bethesda: US Consumer Product Safety Commission, 2008.

Rose, Susan R., and Rosalind S. Brown. "Update of Newborn Screening and Therapy for Congenital Hypothyroidism." *Pediatrics* 117, no. 6 (2006): 2290–303.

Rubin, Beverly S., Jenny R. Lenkowski, Cheryl M. Schaeberle, Laura N. Vandenberg, Paul M. Ronsheim, and Ana M. Soto. "Evidence of Altered Brain Sexual Differentiation in Mice Exposed Perinatally to Low, Environmentally Relevant Levels of Bisphenol A." *Endocrinology* 147, no. 8 (2006): 3681–91.

Select Committee on GRAS Substances (SCOGS). "Appendix" to *Insights on Food Safety Evaluation*. Springfield: National Technical Information Service, US Department of Commerce, 1982.

Shibko, S. I., and H. Blumenthal. "Toxicology of Phthalic Acid Esters Used in Food-Packaging Material." *Environmental Health Perspectives* 3 (January 1973): 131–37.

Steele, Eurídice Martínez, Larissa Galastri Baraldi, Maria Laura da Costa Louzada, Jean-Claude Moubarac, Dariush Mozaffarian, and Carlos Augusto Monteiro. "Ultra-Processed Foods and Added Sugars in the US Diet: Evidence from a Nationally Representative Cross-Sectional Study." *BMJ Open* 6, no. 3 (2016): e009892.

Swan, S. H., F. Liu, M. Hines, R. L. Kruse, C. Wang, J. B. Redmon, A. Sparks, and B. Weiss. "Prenatal Phthalate Exposure and Reduced Masculine Play in Boys." *International Journal of Andrology* 33, no. 2 (2010): 259–69.

Swan, Shanna H. "Environmental Phthalate Exposure in Relation to Reproductive Outcomes and Other Health Endpoints in Humans." *Environmental Research* 108, no. 2 (2008): 177–84.

Trasande, L., R. T. Zoeller, U. Hass, A. Kortenkamp, P. Grandjean, J. P. Myers, J. DiGangi, et al. "Burden of Disease and Costs of Exposure to Endocrine Disrupting Chemicals in the European Union: An Updated Analysis." *Andrology* 4, no. 4 (2016): 565–72.

US Congress, House of Representatives. *Food Additive Amendment of 1958, by John B. Williams, July 28, 1958*. 85th Congress, Report No. 2284. Washington, DC: US Government Printing Office, 1958.

———. *Investigation of the Use of Chemicals in Foods and Cosmetics, by James J. Delaney, June 30, 1952*. 82nd Congress, Report No. 2356. Washington, DC: US Government Printing Office, 1952.

US Food and Drug Administration, "FDA Response to Natural Resources Defense Council's October 2013 Freedom of Information Request No. 2013-8042." GRAS Notice (GRN) 225. US FDA, 2014.

———. "The FDA Takes Steps to Remove Artificial *Trans* Fats in Processed Foods." News Release, 16 June 2015. Science Daily website. Accessed 10 March 2020, https://www.sciencedaily.com/releases/2015/06/150616160256.htm.

———. "GRAS Substances (SCOGS) Database." 2018. Accessed 15 March 2020, http://www.fda.gov/food/ingredientspackaginglabeling/gras/scogs.

———. "Indirect Food Additives: Paper and Paperboard Components," *Federal Register* 81, no. 1 (2016): 5–8.

———. "Justification of Estimates for Appropriations Committee, Fiscal Year 2021." Accessed 4 March 2020, https://www.fda.gov/media/135078/download.

———. *GRAS Notices: GRN-257, Gamma-Amino Butyric Acid*. Silver Spring: US FDA, 2008. Accessed 4 March 2020, https://www.accessdata.fda.gov/scripts/fdcc/?set=GRASNotices&id=257&sort=GRN_No&order=DESC&startrow=1&type=basic&search=257.

———. *Guidance for Industry: Preparation of Food Contact Notifications for Food Contact Substances; Toxicology Recommendations.* Silver Spring: US FDA, 2002. Accessed 4 March 2020, https://www.fda.gov/regulatory-information/search-fda-guidance-documents/guidance-industry-preparation-food-contact-notifications-food-contact-substances-toxicology/.

———. "History of the GRAS List and SCOGS Review." 2018. Accessed 15 March 2020, http://www.fda.gov/Food/IngredientsPackagingLabeling/GRAS/SCOGS/ucm084142.htm.

———. Memorandum P. Rice to P. Honigfort, "re: Indirect Food Additives: Paper and Paper Board Components," 27 July 2015. Docket No. FDA-2015-F-0714. Accessed 15 March 2020, https://www.regulations.gov/#!documentDetail;D=FDA-2015-F-0714-0016.

———. "Threshold of Regulation (TOR) Exemptions. TOR No. 2005-006. Sodium perchlorate monohydrate." 2005. Accessed 2 February 2020, http://www.accessdata.fda.gov/scripts/fdcc/?set=TOR&id=2005-006.

———. "Warning Letter to Apex Hemp Oil." 22 November 2019. Accessed 2 February 2020, https://www.fda.gov/inspections-compliance-enforcement-and-criminal-investigations/warning-letters/apex-hemp-oil-llc-592691-11222019.

Vandenberg, Laura N. "Classic Toxicology vs. New Science: Unique Issues of Endocrine-Disrupting Chemicals." In *Integrative Environmental Medicine,* edited by Aly Cohen and Frederick S. Vom Saal, 279–303. New York: Oxford University Press, 2017.

Vogel, Sarah A. *Is It Safe?: BPA and the Struggle to Define the Safety of Chemicals.* Berkeley: University of California Press, 2013.

Wargo, John. *Our Children's Toxic Legacy: How Science and Law Fail to Protect Us from Pesticides.* New Haven: Yale University Press, 1996.

Watson, Elaine. "Researcher Files Lawsuit vs FDA after It Ignored His Petition Calling for Ban on Artificial Trans Fats." *foodnavigator-usa.com,* 12 August 2013. Accessed 13 August 2013, https://www.foodnavigator-usa.com/Article/2013/08/13/Researcher-files-lawsuit-vs-FDA-after-it-ignored-his-petition-calling-for-ban-on-artificial-trans-fats.

World Health Organization. *Assessment of Iodine Deficiency Disorders and Monitoring Their Elimination: A Guide for Programme Managers.* 3rd ed. Geneva: WHO, 2007.

———. *Global Status Report on Noncommunicable Diseases.* Geneva: WHO, 2014.

 CHAPTER 11

The Rise (and Fall) of the Food-Drug Line

Classification, Gatekeepers, and Spatial Mediation in Regulating US Food and Health Markets

Xaq Frohlich

What makes something a food, and what makes something a drug? While such distinctions are usually considered self-evident, the appearance in recent decades of so-called "functional foods," foods marketed with specific medical or health-promoting properties, has raised questions about how best to mark the boundaries between food and drug and what is at stake when products fall into grey zones. Eating omega-3-enriched eggs, touted as a "good" fat, became popular in the United States in the early 2000s as a way to reduce the risk of heart disease. Danone's "probiotic" yogurt-based drink, Actimel (DanActive in the US and Canada), which was promoted to improve gut flora and digestion, triggered questions in the EU in 2007 about what level of scientific proof was needed to permit labeling a food with explicit health claims. Ginseng, an important traditional food ingredient and medicinal product in South Korea, has been packaged, classified, and marketed as food (*han-shik*), biomedical pharmaceutical, *and* dietary supplement (*yak-sun*), contributing to over a decade of tensions between Korean medicine practitioners who feel marginalized and a biomedical healthcare system that seeks legitimacy through largely Western standards for modern medical practice. These different cases illustrate how the problem of classifying new health foods is global, involves a wide variety of products and ideas about healthy living, and reflects a complex mixture of scientific, market, and legal concerns that converge in the central questions raised by functional foods: how much should managing health be about diet, and who should be responsible for that?

Food studies and science and technology studies (STS) scholars have taken up these questions out of their concern with how such products destabilize significant cultural boundaries or institutional classifications. Perhaps the most vocal and widely cited of these food critics is Gyorgy Scrinis. He coined the term "nutritionism" to explain how focusing on nutritional properties of a food has led to a form of reductionism, ignoring other important features of what makes food good to eat and also making it easier for the food industry to confuse consumers about processed foods.[1] More broadly, food studies scholars see functional foods as a new chapter in a longer history of corporate advertisers appropriating nutrition science to market questionable new industrial foods on the authority of science.[2] STS scholars, in contrast, have tended to avoid the frame of nutrition reductionism because it presumes some essential true state of food; instead they have viewed functional foods as a problematic "hybrid" that blends the disciplining languages of medicine and health with the pleasurable languages of food and eating.[3] When critical of functional foods, STS scholars focus their critiques on whether these foods represent a medicalization of food or marketization of health. Despite these differences, both groups have approached questions about functional foods with a set of assumptions: that marketing foods for biomedical properties is novel, and that the idea of doing this is in some way socially transgressive, even a new form of exploitation of consumers.

The problem with these assumptions is that the existence of product crossovers between food and health markets is nothing new. If one surveys the longer history of dietetics and medical therapeutics, one discovers numerous examples of diets or foods used to counter illness or lengthen life and of food-like substances attributed with drug-like, health-restorative powers when taken as additives or supplements with a meal. Historians of early modern Europe have documented the porous boundaries in that period between food and medicine in the healing arts, for which it was routine to consider diet regimen and table habits when addressing common maladies.[4] The *Journal of History of Medicine and Allied Science* dedicated a recent special issue to the subject of "food as medicine, medicine as food," exploring how "food, medicine, and science are not fixed or self-evident historical categories," but instead reflect a "moving boundary" shaped by evolving cultural norms about the "division of professional versus lay discourse, of theoretical versus tacit knowledge, of experiment versus experience."[5] It is not just historians. Legal scholars specializing in food law also tend to characterize functional foods as merely the latest in a long history of legal challenges with classifying border products in US law.[6] Most examples come from internalist histories of food law, focused narrowly on the practical challenges that policing product categories posed for state institutions given particular statutory powers and the gamesmanship and fraud of market actors. However, Lewis Grossman, an important expert in American food law, has written about the "indeterminate nature of extralegal notions of 'food' and 'drug,'" and argued that it has often been the law itself that has sought to impose a clarity on this division that the broader consuming culture has usually lacked.[7]

This chapter provides a prehistory of functional foods. As I will show, the concept of a "food-drug line," the idea that there can be a clear demarcation of these two product categories, is a fairly recent conceptualization, broadly dating back to the turn of the twentieth century, but more concretely institutionalized by the Food and Drug Administration (FDA) in the United States beginning in the 1930s. This food-drug line, which functional foods have so problematically transgressed in recent years, was constructed over decades by medical and regulatory professionals. The question is therefore not "how do functional foods transgress essential cultural boundaries for food and medicine?," but rather, "what work does this food-drug line do for the consumers, regulators, medical practitioners and (even) corporate marketers who helped develop it?" And, then, what forces in the 1980s and 1990s eroded the food-drug distinction and made functional foods viable (once again)? More broadly, for scholars looking at food, risk, and regulation, this chapter demonstrates how a conceptual practice, product classification in law and markets, results in material changes to foods and drugs that can, in turn, enlist those objects and their users in perpetuating, circumventing, or undermining that very conceptual division. Regulation for borderline products, in other words, becomes an important arena for innovation where different types of actors—regulators, medical professionals, and manufacturers—invoke competing epistemological claims about products in order to frame markets for food, diet, and health.

Boundary Work and Regulatory Objects: Mediating Markets through Classification

Tracing the rise and fall of a food-drug line in twentieth-century legal and medical practice shows how legal, medical, and regulatory experts have sought to mold everyday social categories by regulating the way products circulate in markets. A key point, articulated by Lisa Haushofer in her study of an early twentieth-century dietetic product and public health tool, is that "the intersection between food and medicine is not an unchanging and self-evident spillover of one realm into another, but the result of a historically specific process of creation and management."[8] Here I show some of the specific institutional and organizational tactics that legal and medical professionals used to create and manage the product division between foods and drugs, as well as their motivations for doing so. There is a larger story to tell about how similar contests over food/drug classifications have taken place in other countries.[9] Here I focus on its history in the United States, because the reliance on a food-drug line played a particularly significant role given that one agency, the FDA, oversees regulation of both kinds of products. The FDA's preoccupation with classification is tied to its authority to oversee a diverse marketplace of food, drug, and cosmetic products, each with different production histories, material cultures, and moral valences. A study of the FDA's efforts to demarcate "foods" from "drugs" quickly reveals what has become an established tenet of science studies, that classification has real-world consequences, particularly when part of an expert system of regulating risk.[10]

Regulation is not a fixed, predetermined governmental domain, limited to legislation or its administrative implementation, but rather a nebulous mix of practices, both formal and informal, public and private, individual and collective, that range from the interpersonal interactions of a doctor managing a patient's regimen to the impersonal market interfaces of label claims or product placement. Huising and Silbey show how regulation is relational, shaped by the social status and negotiations between the regulator and regulated.[11] Here it is argued that the relational forms of regulating food and drugs have been varied and evolved dramatically over the twentieth century. As food and drug sales became mass markets built on anonymous relationships, both market and regulatory institutions developed new audit instruments and forms of what Ted Porter calls "mechanical trust" as a practical substitute for the personal trust that had characterized earlier local buyer-seller markets.[12] One approach has centered on building up certifying authorities, such as the FDA, as a kind of government brand whose management and use of its reputation serves as a powerful public check on market activities.[13] Manufacturers might invite this "outside" government "interference" in the market to add credibility to their brands and to reduce competition with cheaper competitors or fraudulent producers.[14] For the FDA, part of the pragmatics of regulating a wide field of products is determining when a product merits intensive agency resources and review, and when it should be left to other nonstate actors to carry the burden of preventing risk. This delegation of authority by the FDA often takes the form of "boundary work," where the agency defers to certain expert professionals (but not to others) on matters of risk, such as is the case with prescription drugs being restricted to board-licensed medical professionals.[15] Gaudillière and Hess show how pharmaceuticals have been subjected to multiple "ways of regulating," both governmental and nongovernmental.[16]

All of this points to one of the key questions at stake in classifying foods and drugs: who should do the work of regulating risk in food versus drug markets? Who should manage the responsibility for maintaining personal health, professionals and government institutions or private companies and individual consumers? Through this prehistory of functional foods I will show how the answer to this question would become very different for foods than for drugs. Classification worked as a tool in a larger machinery of regulatory tactics designed to mitigate risk, but also to allocate it to different actors in food and drug markets. For drugs, the FDA would position itself and medical practitioners as expert "gatekeepers" to stand between manufacturers and the consuming public. For foods, the agency wouldn't regulate consumers directly. Instead the FDA sought to regulate the everyday market spaces that shoppers passed through, by determining what products were sufficiently mundane and self-evident to end up on a supermarket shelf and not behind the pharmacist's counter. Drug-using patients would be told what to do, while the food-consuming public would be free to choose.

In this story, the objects regulated played an important role. Foods and drugs became "regulatory objects" as they were designed to conform to (or resist) the standards and rules the FDA developed to organize markets for both. Different regulatory styles would be embedded into their design, but also, conversely, regulatory objects could "take on different identities and be embedded in different social arrangements with different consequences."[17] To understand how this came to be so, it is helpful to look to the past and ask, what is a drug, what is a food, and how have these categories blurred in recent years? The three sections below explore these questions.

What Is a "Drug"? Managing Drug Dependence and the Calculus of Risk

Today it is taken for granted that certain substances are controlled by the state, either because they are dangerous or habit-forming to the consumer or because they are believed to induce behavior that could be dangerous to the community. Prescription drugs, for example, require a certified doctor to grant the consumer access to their dispensation by pharmacies. This approach to risky drug products was the culmination of over a century of efforts by medical associations, law enforcement agencies, in the United States principally the FDA, and large pharmaceutical manufacturers to establish regulatory mechanisms to manage risk in the marketplace. At the beginning of the twentieth century, by contrast, there was a prevalent culture espousing a "right to self-medication." This was a combination of an age-old philosophy of medical skepticism, embodied in the idea that an individual might know his or her body and ills better than a doctor, and a legal culture of *caveat emptor* centered on the ideals of consumer autonomy, personal judgment, and one's assumption of risk when one knowingly chooses to take a potentially dangerous treatment. This right of self-medication also reflected the more porous boundaries at the time between credentialed and reputable medical practitioners and more dubious health profiteers. Medical pluralism was the norm, and health consumers could shop around in mixed markets for medicines and health products that ranged from extreme and acute medical interventions requiring a degree of expert skill to more innocuous forms of dietary modification or herbal remedies dispensed at druggist shops.[18] The creation of a special legal category for prescription drugs, and by extension a more clearly defined legal classification between foods and drugs, would be a significant step toward linking the state's interest in managing socially problematic, illicit substances to the medical profession's efforts to define a limited monopoly over certain forms of healthcare.

Historians of medicine have written extensively about the central importance therapeutics played in the rising status of medical practitioners in the modern era. Charles Rosenberg argues that changes in the medical profession in nineteenth-century America helped to usher in a "therapeutic revolution" by the end of the century. For much of the nineteenth century, American doctors grappled with a tension in medical practice. Many therapeutics were decried as unnatural and unhelpful interventions in the natural course of illness, and many risky forms of popular "heroic" medicine, such as blood-letting and mercury, were subjected to rational, empirical testing and found to be wanting. Yet in everyday doctor-patient relationships, therapeutics formed a central part of routine practice and a key validation of why patients sought out expert medical care. How could doctors modernize by incorporating a new empiricism that was discrediting the usefulness of many traditional therapeutics, yet still meet their patients' demand to do something, to prescribe something?

By the turn of the twentieth century, medical professional associations, chief among them the American Medical Association (AMA), began to take measures to distinguish their expert professional authority from mere profiteering by alternative medical practitioners.[19] One of their chief challenges was the explosion in what were known as "patent medicines."[20] Entrepreneurial local and regional salesmen, or health "quacks" depending on one's perspective, found a ripe consumer market for proprietary cure-alls in a highly mobile society.[21] Interest in health product markets was not limited to small players. Several large chemical and drug firms began to formulate and successfully mass-market synthetic medications. One iconic example was the German multinational company Bayer, which in the 1890s synthesized acetylsalicylic acid and began marketing it under the branded name of Aspirin. Given the wide variety of medical therapeutics flooding the market of variable size, shape, packaging, and, more importantly, reliable quality and trustworthy origin, the question arose, what could be done to help consumers differentiate sound and credible medical treatments from the more fantastical, unscientific ones also in circulation? The American Medical Association began to identify medicines that met its safety and manufacturing standards and listed them in the US Pharmacopeia as "ethical drugs," in contrast to patent drugs. The AMA encouraged its members to recommend their patients use only the former and not the latter. It also implemented rules about advertising in its flagship journal, *JAMA* (*Journal of the American Medical Association*), refusing to advertise any company that marketed patent drugs. Doing so put pressure on big drug manufacturers, who were eager to market to the many doctors who subscribed to *JAMA*, to reject selling patent drugs, enlisting big manufacturers as allies in the AMA's campaign against them.

These measures in many ways would come to define the ideal type for the concept of "drug" that would be instituted in the food-drug line. In place of advertisements selling cure-alls, the new drug paradigm was built on a modern epistemology of "disease specificity," the idea that illnesses were caused by a specific, single disease-causing agent, and with it the "lock and key" metaphor that for each disease or symptom there was a specific remedy that might cure it.[22] Advances in physiology and pharmacology also led to a growing gap between the patients' understanding of illness and cure versus the expert physicians' model of disease and therapeutic. Before, patients and doctors shared an understanding of what a cure was doing to the body. As drugs increasingly targeted not only visible symptoms of a disease, but also those signs that only a doctor could measure (such as body temperature, blood pressure), patients had to take it on faith that the cure was working as the medical authority claimed. As Rosenberg notes, "In a sense, almost *all* drugs now act as placebo, for with the exception of certain classes of drugs such as diuretics, the patient experiences no perceptible physiological effects."[23] This faith was made possible by new forms of market regulation and evidence-based medicine that certified the treatment was scientifically proven to work, most famously the requirement of randomized clinical trials.[24] What emerged from these trends was a shift from the drug as a craft formula to the modern medical view of drug, characterized by at least three traits: (1) it was mass-produced, (2) it was composed of purified chemical components and an "active ingredient," and (3) its action would address only a narrow range of medical applications—that is, drug "specificity."[25]

A second front in this battle to restrict the right to self-medication was being fought over moral problems posed by drug dependency and risk-taking. When did the state have the authority to infringe on personal liberties to experiment with drugs if doing so became a public menace? This can be seen most clearly in regulatory movements to define licit versus illicit drugs and acceptable versus unacceptable recreational drugs. What made certain substances a matter of moral concern was that they were habit-forming, and, in the case of drugs like opium, could even lead to a self-destructive dependency. Moreover, some substances such as alcohol were seen to promote aggressive, even violent, behavior, which disturbed the public order. Participants in the temperance movement advocated controls on addictive products ranging from coffee and tobacco to alcohol and opium and pointed to cases of downward-spiraling substance abuse to campaign against so-called "self-medication."[26] The problem was how to differentiate between, first, acceptable medical uses of products like laudanum, a common home remedy that contained opium and was used to relieve pain or help one sleep, from, second, harmless recreational uses of popular goods such as coca beverages, which included "coca wines" like the early Coca-Cola that even had traces of cocaine, and, third, illegitimate and dangerous substances that should be kept out of the hands of ordinary consumers.

These questions about one's right to take risks with pleasurable substances, what would become a "licit-illicit line" for drugs, was also implicated in the problem of defining a food-drug boundary: to what extent was a substance "needed" for restoring health or simply being enjoyed for recreation? Many of these habit-forming plants and products, which today we think of as food ingredients or additives, were, at the start of the twentieth century, part and parcel of the diverse market for patent drugs. Even the use of spices and flavorings was often more strongly associated with medicines than foods.[27] Indeed, a 1947 US Supreme Court case, *United States v. Kordel*, illustrated how companies selling health tonics such as Kordel's "Kola" sarsaparilla sought to play with culturally fluid boundaries between food and drug and thereby skirt increasingly stricter drug laws by labeling their products as food. The prohibition of alcohol from 1920 to 1933 soon made visible the challenges of sacrificing personal liberties to the enforcement of public safety standards: should the fact that some people abused a substance mean that other legitimate uses also be restricted, or would restricting their use drive otherwise responsible people underground to seek them out more recklessly? A variety of lawsuits and legislative acts, from the 1912 International Opium Convention to the 1970 Comprehensive Drug Abuse Prevention and Control Act, established a boundary between legal versus illegal drugs and institutionalized the concept of "controlled substances."[28] It reinforced the emerging policy that situated medical practitioners as "gatekeepers" for certain special but risky drugs. Some substances, albeit habit-forming and potentially easy to abuse, held health-restoring benefits if used under the care of a trained and trustworthy physician. The moral threat of dependency on a drug, under this rationale, would be neutralized by a more socially acceptable kind of dependency on expert care.

In the 1950s and 1960s, the AMA and FDA consecrated this gatekeeper model of market regulation by implementing new rules about "prescription drugs." Congress passed the Durham-Humphrey Amendments in 1951, which required any drug that was habit-forming or potentially hazardous to be dispensed under the supervision of a physician. Following the thalidomide scandal over unanticipated "adverse reactions" from a pregnancy drug, the 1962 Kefauver-Harris Amendments required premarket testing and approval from the FDA for mass-marketed drugs. It increased standards of proof for pharmaceutical sales and raised the importance of randomized clinical trials. This new system of product regulation posed a substantial financial burden on pharmaceutical companies. Premarket trials were expensive and slowed down how quickly companies could bring new products to market. Requirements for physician prescriptions and premarket approval placed doctors and the FDA as important market gatekeepers between consumers and pharmaceutical companies.[29] Large pharmaceutical companies accepted these terms in part because any FDA approval served as an independent "third-party certification" of their products' safety and raised the barriers to market entry for competitors.

Companies also accepted these regulations because the FDA and AMA were investing substantial organizational resources in building up the credibility of prescription drugs while discrediting alternative medicine markets.[30] The Kefauver-Harris Amendments represented a substantial hardening of the food-drug line. A prominent feature of FDA Commissioner George Larrick's leadership from 1954 to 1962 was the Campaign Against Medical Quackery that his agency coordinated with the AMA, where they targeted for criticism many patent drugs and nutrition quackery claims made for diet foods. Both organizations believed these claims undermined the integrity of the new drug safety system. The campaign against medical quackery rested on the boundary work that some market actors, generally big companies that agreed to FDA standards, were legitimate, while others, especially small "hoax" patent drug producers, should be actively discredited. The AMA pressured its members to be wary of such discredited home remedies and to focus on prescribing the more credible, clinically tested pharmaceuticals the FDA approved. These efforts consecrated the food-drug line, since now any foods that made disease claims, such as the Kordel sarsaparilla "Kola" case mentioned above, would be treated as drugs that had not received FDA premarket approval.[31] By the 1960s, one can see an alliance of large pharmaceutical companies, physicians organizations, and the FDA that rested on what could be called a "liberal consensus" about the necessity for systematic regulatory controls, so long as those controls ensured safe and effective drugs would come to market.

Even at the height of this liberal consensus, when the FDA and AMA were most heavily committed to shoring up the food-drug line, there were significant cracks in the facade. One was the growing interest by pharmaceutical companies in expanding into diet food markets. Mass marketing diet foods, discussed below, undermined the FDA's efforts to restrict "special dietary foods" to unhealthy people who were supervised by a doctor, and to block all health-related claims from food labels thereby protecting "ordinary" (i.e., healthy) consumers from unwelcome health quackery. A second challenge were "lifestyle drugs." The best example was the birth control pill, first approved by the FDA as a contraceptive in 1960. It was a drug to be taken by otherwise healthy people, reflecting their personal lifestyle choices and not a specific medical motive. The FDA's problems with the pill illustrates how drug regulation in the 1960s and 1970s would be reshaped by consumer politics and anti-paternalist sentiment. Because of adverse side effects (and also broad moral concerns), the FDA classed the birth control pill as a prescription drug, meaning women would have to get their doctors to prescribe it to them. This led to awkward, even tense, negotiations between women and their doctors about what constitutes a "reasonable risk" for the "recreational benefits" the pill provided. When it was discovered that the FDA and Searle, the company that manufactured the first pill, Enovid, had glossed over certain side effects in the push to get it approved, there was a backlash to the doctor-centered prescription system. Critics, especially feminists and consumer safety advocates, protested that important safety information had been withheld, and argued that women had a right to that information.

In 1969, the birth control pill would become the first prescription drug to include an information package insert directed at the consumer and not the doctor.[32] Since it was women's bodies that faced the risk-benefit tradeoff between unwanted pregnancies and potential health side effects from the pill, women would be empowered with information about these risks, not just their doctors. In a speech that year on the relationship between doctor, patient, and the FDA, FDA commissioner Herbert Ley called for greater humility using a line from a poem by Alexander Pope: "Who shall decide when doctors disagree, And soundest casuists doubt, like you and me?"[33] The speech hinted at an end to the liberal consensus and collaborative spirit that had characterized FDA and AMA cooperation on rules for drugs the decade before. It would also reflect a broader "informational turn," discussed below, which eroded the role of doctors as gatekeepers and market intermediaries.

What Is a "Food"? Freedom to Choose and Freedom from Worry

In many respects, the same forces that were transforming drugs were transforming the idea of what is a "food." At the end of the nineteenth century, most foods in Americans' diet were relatively unprocessed and purchased at local shops, which displayed them to shoppers in bulk. Over the course of the twentieth century, more and more of the consumer's budget would go to "consumer packaged goods," or CPGs as the industry called them, and more and more of items once sold in bulk were now sold precooked, processed, and packaged for the emerging self-service economy of supermarkets. The industrialization of food meant that foods were increasingly mass-produced using synthetic and purified components, much like modern drugs. In short, by the 1960s many foods in the supermarket were formulated rather than grown. Yet, recognition of this change was slow, and one can see a gap emerge in the mid-twentieth century between regulations that imagine food qua food—time-honored recipes whose authority rested on familiarity and tradition—and the enforcement of these standards, which had to accommodate the reality that foods in the supermarket hardly met these ideals. Moreover, if drugs, as discussed above, were to become a special category of product for a special consumer in need of special market protections and gatekeepers, the consumer of foods by contrast ought to be free to choose. This would mean defining foods in a way that prevented any confusion with the newly established therapeutic authority of drugs. For these reasons, in the United States it is important to view legal efforts to define food as a foil to parallel efforts to define drugs.

Nowhere is the tension clearer between traditional ideas about food and the realities of modern industrial food than in the FDA's applications of "pure food" laws in the first half of the century and its implementation of food "standards of identity" in the second. The 1906 US Pure Food and Drug Act culminated decades of efforts by regulatory scientists, often calling themselves "official agricultural chemists," who advocated for better regulation of the food market through the establishment of scientific standards for defining and testing food ingredients. Chief among them was Harvey W. Wiley, who famously led the "Poison Squad" in the US Department of Agriculture's Division of Chemistry before the establishment of the FDA in the 1930s. The focus of their concern was with food fraud and the "adulteration" of foods—for example, when producers used certain hazardous chemicals to cover up the flavor of rotten foods or preserve them well past their natural date of freshness.[34] Fraud included not only cases where the adulteration resulted in direct harm, but also "economic adulteration": the use of cheap substitutes that cheated consumers out of the more wholesome authentic product. The frame of food "purity" had popular appeal, as it suggested that traditional "authentic" foods were whole, and thus wholesome, and industrial adulteration, by contrast, entailed a chemical corruption of that authentic purity. Chemical corruption would, in turn, represent a legal corruption that could be prosecuted.[35] In practice, however, establishing a legal concept of purity was problematic. For one, what kinds of food processing and adulteration constituted fraud and what industrial processing was meeting a legitimate consumer demand for shelf-stable, palatable packaged food? Once you established this defining line between fraudulent adulteration and legitimate product innovation, how do you go about enforcing it?

Efforts by Wiley and other regulatory scientists to defend food's integrity in this period were symptomatic of a broader crisis in how to establish food quality and safety in an industrialized food chain. Their solution, however, reflected their profession's tendency to reframe safety in chemists' terms. Traditional means of policing food focused on knowing the supply chain and knowing your supplier. Trust was relational, and the relation largely interpersonal, reflecting continuous interactions between buyer and seller. The scaling up of national, even global, supply chains and the greater complexity for processed foods meant more people of unknown reputation handling the food between producer and supplier. Branded national manufacturers during this period implemented a variety of management and logistical practices to assure consumers that their products would be uniformly fresh and safe, despite the long distances they traveled.[36] Regulatory scientists, however, had to deal with the problem of bad actors, which is why they focused on methods for testing the final product that would not depend on knowing the process or history of a product. Wiley, for example, sought to call out industry trickery through the use of analytic chemistry. This led to a shift in the meaning of the term "pure" because of what historians of science call the "proxy problem." Scientists develop a proxy measure to study a social problem, but over time they come to focus on the proxy in place of the original problem.[37] Unable to come up with a workable definition for most "pure" whole foods, regulatory chemists increasingly characterized food ingredients by the purity of their chemical components. Sugar content, for example, came to be levels of sucrose. The flavor vanilla came to be equated to the chemical vanillin. The emphasis on chemical purity even had the ironic consequence of helping certain products that were traditionally *not* food, such as cottonseed oil, be marketed as a "pure" and clean processed alternative to more traditional cooking fats.[38] While these reductionist rules for food components and substitutions would be disputed by many in the food industry as arbitrary or at odds with the diversity of customary traditions in America, they offered the possibility of establishing testable and thus enforceable standards for preventing food adulteration.

A second approach to reducing food fraud was to focus on how food itself was marketed, and especially what was said about it on the label or in advertising. Following the passage of the 1938 Food, Drug, and Cosmetic Act (FDCA), the FDA was authorized to implement premarket approval for fixed "standards of identity" for all mass-produced foods. Sociologist Lawrence Busch argues that "standards always incorporate a metaphor or simile, either implicitly or explicitly."[39] In the case of FDA food standards, the metaphor was the "time-honored" home recipe of housewives. The idea of establishing food standards was simple: regulators working with industry would set standard recipes matched to a common name and with certain agreed-upon ranges for ingredients. This would provide food manufacturers some room for developing their particular brand, but also make standard foods self-explanatory to the imagined target consumer, the average "housewife," such that the common name was enough to explain the product. Setting standards was motivated by two distinct and potentially conflicting ethos: was it primarily about coordinating markets, making them accountable through standard procedures and conventions, or was it about addressing substantive safety concerns? This tension was not evident to policymakers at the time. The idea of establishing food standards reflected their belief that most whole foods or traditional processed foods should be familiar to a consumer. In the words of one eminent food law scholar, reflecting on this system in the 1970s, "The FDA's prolonged adherence to its original recipe format" reflected the agency's "desire to preclude any modifications of basic food formulas that could contribute to consumer deception" and its "concern to restrain the growing use in food production of chemical additives whose safety had not been demonstrated."[40] The first standards for canned tomatoes, for example, removed benzoate of soda, one of Wiley's many maligned chemical adulterants, from the list of permitted ingredients. Nonstandard products could be subject to removal from the market, subject to stricter labeling requirements, and required to carry the "imitation" label widely recognized by consumers to mean an inferior, substandard quality.

The problem is, how do you establish what is a customary recipe when the foods you are making are new and industrial? "Recipe," after all, *was* just a metaphor. The consumer packaged goods the FDA was regulating were not prepared by housewives, but were being cooked and processed by industrial food manufacturers at scale. In a series of lawsuits in the 1940s and 1950s, courts largely upheld the FDA's right to use standards as a tool for seizing problematic products, such as a lower-fat creamy "Neufchâtel" cheese or an imitation jam with less percent fruit. However, standards hearings quickly became an entrenched battleground where manufacturers disputed each other on what ingredients or processes their customers believed were customary. The bread standards hearings, which ran from 1941 to 1943 and from 1948 to 1950, were particularly protracted and costly, as questions arose about the legitimacy of new emulsifiers used to help make sliced bread feel "fresh."[41] Rather than stemming the tide of industrial foods entering US markets after World War II, making foods standard ultimately facilitated their market acceptance. Much the way food safety had been transformed by "pure" food policies, standards of identity helped transform what was "food." Standards accommodated a wide variety of food additives "generally recognized as safe" and new ingredients and techniques to make processed foods. Once a standard recipe was set by the FDA, the only information the consumer was guaranteed by law was the common name.[42] Standards thus helped to disguise transformations taking place in the processing of familiar foods behind a simple common name. This was why consumer advocates who criticized the food standards system in the late 1960s would call them "silent labels."[43]

The standardization of food in the 1950s and 1960s, making food self-evident, involved several different forms of "boundary work" to keep consumers from confusing foods with drugs. Here I focus on three examples: identifying "good" versus "bad" market advice on healthy foods, linking product classification to the distinct retailing architectures for both, and associating the form of a product to its function.

The first of these entailed the boundary work of identifying bad actors who sought to profit off consumers' anxiety by promoting nonmedical goods as if they had magical health properties. The AMA's and FDA's campaign against "health quackery," described above, included a "campaign against nutrition

quackery" that targeted what regulators saw as "the usual suspects" of unfounded nutrition claims and products marketed to profit off them. The campaign identified four common nutrition "myths" used to promote alternative
health foods, including the beliefs that industrial fertilizers and overprocess-
ing stripped foods of minerals and that resulting "subclinical deficiencies" explained many ailments and illnesses.[44] The FDA and AMA also aggressively
targeted many alternative health products which, though harmless, undermined the credibility of officially sanctioned drug markets. One example was
the continued presence of many health tonics and drinks purporting health
benefits but not marketed as drugs. *United States v. Kordel* (1947) was one of
the earliest cases upholding the FDA's power to remove a food product from
the market because it was a "misbranded" drug. The FDA argued that even
though the sarsaparilla "Kola" type beverage in the case was sold as a food,
pamphlets marketed with it described the drink's ability to relieve stomach
pain, anemia, high blood pressure, and numerous other physical ailments.[45]
While colas could be "refreshing" and "rejuvenating," under the emerging
food-drug line, beverage advertisers had to be careful to avoid language that
suggested it could cure or promote good health.

A second axis of the FDA's boundary work on the food-drug line centered
on how the distribution channeling for a product might shape a consumer's
understanding of it. As described above, rules developed in the 1950s and
1960s made doctors and pharmacists market gatekeepers with whom consumers would have to consult before they could "buy" a drug. Foods had no
comparable gatekeeper, certainly not an expert one. Indeed, marketing literature during this period identified the "housewife" as the primary gatekeeper
for food purchases, since she was the consumer in charge of family household
food decisions. In the early decades of the twentieth century, local grocers were
also seen to be important influencers, often steering customers away from the
manufacturer's branded product to other similar ones.[46] The rise of supermarkets in the 1940s and 1950s, and with it a new self-service approach to food retailing, meant food marketers could worry less about these market gatekeepers
and instead focus more on branding, product placement, and, in the words of
one head of a chain of supermarket stores in the 1960s, "the increasing importance of the written word amongst an educated population."[47] The possibility
of convenient one-stop shopping was important to supermarkets' postwar expansion in a newly suburban America. But the supermarket came to symbolize more than just convenience. The everyday shopper's freedom to choose
from a wide assortment of diverse food items became a key proof of concept
in the Cold War struggle for ideological self-definition, where supermarkets
represented the successful culmination of American capitalism and hands-off
governance.[48] Key to this construction of a "consumer sovereign" was the consumer's free flow through the aisles of the supermarket, unencumbered with

the help of the shopping cart and free from opinionated, pushy vendors in this self-service paradigm.[49]

Yet butchers, bakers, and other mom-and-pop food retailers had served as personal guarantors of quality, available for consultation when consumers were confused and directly accountable when products failed. Trust was relational, based on how well you knew the vendor, not only about proof. Given the absence of these gatekeepers and market mediators, it was even more important to create standard foods that were self-explanatory. This was why the FDA tried to keep food product labels simple and to channel more complicated "special" products into separate special retailing spaces. A further challenge was raised by the way that supermarkets brought a wide variety of foods, drinks, and spices under one roof. Previously, specialized retailing geography functioned to segregate different classes of products, making it easier to police them. The new one-stop shopping paradigm led regulators to wonder how one prevents consumer confusion arising from ambiguous product placement. When assessing certain new low-cal diet foods in 1953, FDA commissioner Crawford worried that their mass marketing

> brings up a serious problem of keeping their distribution channeled to people who know what they are and want them, and to prevent such products from being supplied for the staple articles consumed by the bulk of the population who need the caloric intake they are getting, or who need to have their diet selected by experts. The problem is very much more difficult than the comparable one of keeping dangerously potent drugs channeled to those who need them. To obtain these, a physician's prescription is required and this can be filled only at drugstores who employ licensed pharmacists.[50]

Preserving the supermarket as a space for the free movement of consumers and consumer choice meant separating out problematic drug-like products, which demanded some form of expert mediation. One expected foods at the supermarket, but drugs didn't belong there.

One problem with this simple division was pill-like foods, such as vitamin supplements, and specially formulated diet foods, such as low-cal sodas.[51] These challenged a third front of the FDA's boundary work in separating food from drug: linking form to function so that the food's design aided in the division of foods and drugs. At the start of the twentieth century, the boundary between food and drug being more porous, many medicine makers used food as a vehicle for treatments.[52] Efforts by the AMA, FDA, and pharmaceutical manufacturers recounted above helped to ensure that the modern "drug" was most commonly taken in pill form. Yet some pills were not drugs per se. Vitamin supplements, initially sold as oils but over time marketed as pills, remained in a grey zone under the 1938 FDCA because they were not expressly mentioned in the legislation. They resembled drugs in shape, packaging, and health claims, but they were intended to supplement food, not medicine. From the 1940s to 1960s, the FDA choose to treat supplements as a food unless they made a specific disease claim, in which case they were regulated as a drug. The FDA focused on identifying bad faith actors who espoused what the agency saw to be "bad science." A famous example of this would be the vitamin C mega-dosing fad advocated by the nutritionist Adelle Davis throughout the sixties and seventies.[53] For the most part, these campaigns were seen to be directed at fringe elements of the health community.

Over the course of the 1960s, the agency's campaign against nutrition quackery began to draw the ire of big food and pharmaceutical companies who were increasingly interested in the markets for dietary supplements, enriched foods, and other "special dietary foods." The FDA's policy of not allowing enrichment on foods it deemed to be candy, what became known as the "jellybean rule," invited complaints from producers who felt agency staff used it arbitrarily to enforce their own standards of what was "good" or "bad" to eat. Carbonated beverage bottlers protested standards that allowed enrichment for fruit drinks like orange juice, but not for soft drinks. In the 1963 case *United States v. 119 Cases . . . New Dextra Brand Fortified Cane Sugar*, a district court noted that "the real basis of the Government's objection to the sale of fortified sugar is the notion that sugar is not a preferable vehicle for distributing vitamins and minerals." The court ruled against the FDA, arguing the FDCA "did not vest in [the FDA] the power to determine what foods should be included in the American diet; this is the function of the marketplace."[54] The FDA also initially classified artificially sweetened products as "special dietary foods" and thus subject to distribution and marketing restrictions. This ran contrary to dramatic changing market trends in low-cal diet foods in the 1960s. The pharmaceutical company Abbott Laboratories, for example, had branched into diet food markets with a new trademarked product, Sucaryl, a blend of cyclamate and saccharine. The evolution of Sucaryl's marketing mirrored the evolution of the diet market more broadly. Initially Sucaryl was sold in two forms: in barrels to the food industry for "technological use" at low doses to restore sweetness to processed foods, and in sachet packets to diabetics, which they could open and add to their foods and beverages. Over time, and despite repeated confrontations with the FDA, Abbott Laboratories began to broaden the uses for Sucaryl to promote it in mass-marketed diet products, most famously TaB diet cola.[55] The shift in the form of how artificial sweeteners were used reflected a growing opinion in the public, as well as among medical experts, that ordinary people also might have a legitimate interest in using specially formulated diet foods.

All of these contradictions would come to a head with the "diet-heart thesis." Much has been written about the history of the diet-heart thesis and "cholesterol controversy."[56] Here I will not tell that full story, but mention it for how it got tied to the FDA's food-drug challenges at the time. In the 1950s and 1960s, medical scientists began to argue that there was a link between future risk for cardiovascular disease, measurements of blood-serum cholesterol levels, and diets high in certain fatty foods. These scientists encouraged Americans to consider modifying their diet to reduce that risk. Food companies, particularly vegetable oils and margarines, capitalized on the popular interest in this by advertising the polyunsaturated-to-saturated fats or "P/S" levels on their product labels. As with artificially sweetened low-cal foods, "low-fat" foods became part of the shift to what historian Harvey Levenstein calls "negative nutrition," diet advice focused on eating less in contrast to an earlier public health focus on increasing nutrition.[57] However, unlike low-cal diets, which were seen to be an extension of vanity dieting, low-fat foods marketed on the diet-heart thesis were tied to an emerging medical paradigm that focused on identifying and managing environmental "risk factors," including diet, with the aim of preventing disease. This "preventive medicine" approach had strong advocates in the public health sector, but it also generated debates in the medical community and among regulators because it expanded the target of health campaigns to both the healthy and unhealthy alike. Everyone could potentially be "at risk" for diet-related heart disease.

Initially, the FDA treated health claims on low-fat margarines and vegetable oils much the way it had low-cal diet foods, as "misbranded" drugs and a marketing nuisance and diet fad that undermined the agency's policies separating food and drug markets. Yet there were contradictions in how the FDA handled the diet-heart thesis for food versus drug markets. In a letter written to the FDA in 1960, Jeremy Stamler, one of the researchers who helped establish the diet-heart link, noted a double standard in the FDA's policies. The agency publicly claimed, "The role of cholesterol in heart and artery disease has not been established" sufficiently to warrant the marketing of low-fat foods. Yet that same year the FDA approved the drug Triparanol (MER/29) for general medical distribution, a drug whose purpose was to lower blood serum cholesterol and thereby reduce the risk of heart attacks. To Stamler, this double standard, allowing doctors to use a drug to treat cholesterol levels in patients, but denying health messages to accomplish the same effect through diet, was both nonsensical and hazardous to the public's health. One can reconcile the FDA's inconsistent position here on food versus drug treatments because of its bureaucratic preoccupation with delegating risk and responsibility: the diet-heart thesis was adequately established for doctors-as-risk-gatekeepers to act on, but not a lay public.[58] By the end of the 1960s, the FDA had come to accept the legitimacy of the diet-heart thesis and therefore largely ignored the growing proliferation of low-fat foods with implied health claims. It did not, however, abandon its food standards approach and official stance prohibiting these kinds of health claims on food.

In a broad sense, the FDA's food standards system had been about maintaining the fiction that food was self-evident, safe, and "traditional," while drugs, by contrast, were "innovative" and worth the assumption of risk. Drugs were expected to be technological and complex, while the opposite was expected of food. That this expectation in the case of food was contradicted by reality was evident in the recurring debates over what novel ingredients would be allowed in standard foods, or in the regulation of food "additives," food ingredients that in many ways were treated with the kind of stricter regulatory scrutiny of risks associated with drugs, albeit without the consideration of benefits.[59] Mass-marketed diet foods, however, were opening up the idea of food as an engineered product with novel value-added payoffs. The old idea of food as whole (uncomplicated), wholesome, nonrisky, and self-evident was giving way to a new idea of food as something to be modified and designed to fit new needs and lifestyles.

Making Foods "Functional": Lifestyle Politics and the Informational Turn

"Americans, heal thyselves." This was the opening injunction of then secretary of health and human services Margaret M. Heckler in the preface to the 1984–85 edition of *Prevention*, a journal of the US Public Health Service. Heckler's play on the biblical warning "Physician, heal thyself" underscores the extent to which a new health movement had caught on in the United States that emphasized personal responsibility for health. Over the previous decade, a new health consciousness ranging from holistic medicine to popular self-care remedies "situated disease and health at the level of the individual." This cult of "healthism," as sociologist Robert Crawford called it,[60] would become a signature feature of American culture in the 1980s, and arguably still today, but it was rooted in the changing cultural and political winds of the seventies. Writing the decade before, amid a heated public debate about the rising costs of healthcare, prominent physician and Rockefeller Foundation president John H. Knowles wrote an essay, "The Responsibility of the Individual," arguing that greater affluence had changed the nature of disease and life expectancy. The solution to solving the new pattern of disease, according to Knowles, was greater emphasis on preventive health, in particular avoiding bad habits, which would require that individuals take greater personal responsibility for their wellbeing.[61] Healthism and this ethos of personal responsibility would become one important thread in the emergent fabric of late twentieth-century "lifestyle politics," a new axis of political identity linking personal lifestyle to self-definition, political mobilization, and, increasingly, checks and limits on government-curtailed liberties.[62]

The novel foods and new policy tools that appeared during this period reflected this new ethos. A key change took place in 1973 when the FDA chose to shift its style of governance of foods away from standards setting and toward informational regulation, requiring informative disclosures such as nutrition information and ingredients panels on health foods, and now allowing producers to market diet foods as ordinary foods and not specially regulated dietetic goods. Popular frustration with the FDA's time-consuming and onerous standards of identity hearings on special dietary foods in the early 1970s had provided some of the impetus for changing how the FDA approached the food-drug line with the new health food products that blurred that boundary. Many consumers were also frustrated with the FDA's handling of the 1969 ban on cyclamate. The ban heightened public anxiety about the risks of what had become a common diet food ingredient, but also generated backlash over what some saw as a heavy-handed use of the Delaney Clause, a zero-risk anticarcinogen rule on food additives. It denied consumers choice on their artificially sweetened diet products. Staffing changes at the FDA that came with a new commissioner reopened the possibility of changing the agency's approach to problem products, in particular low-fat foods and the agency's decade-long trouble handling health claims about diet, cholesterol, and heart disease. Peter Hutt, an industry lawyer who served as the FDA chief counsel from 1971 to 1975, represented the agency's greater openness to market-friendly reform. Starting in 1973, the FDA signaled that it would allow companies to create new foods, including mass-marketed diet recipes, without seeking a standard, so long as companies included both an ingredients panel and "nutrition information" label.

The changes in food labeling resulted in an "informational turn" in the governance of food markets.[63] The FDA's reliance on informative labeling, allowing companies to market health information about foods to consumers, circumvented the previous market relationship which posited health as a special product area requiring the consultation of expert gatekeepers. In other words, the medium of the information label was substituted for the earlier market intermediary of the expert doctor. At the time, most people celebrated the nutrition information and ingredient panels as a pro-public advance for the FDA. However, informative labeling registered a new political formulation of personal responsibility for health as well as a broad cultural rejection of the authority of the FDA and doctors to restrict the health options people sought for health foods and diet products. Anti-government sentiment across political parties in 1970s America fueled a variety of movements to explore private solutions to what had previously been seen as public affairs.[64] "Informational regulation"—regulating markets through information disclosure—would be an increasingly popular style of governance during this tumultuous political period because of its cross-partisan appeal.[65] The turn from standards to labeling was neither a simple dismantling of the state—deregulation favored by small-government conservatives and feared by the Left—nor the bureaucratic ramping up of direct state controls that might result in onerous procedures feared by the Right.

The popular push for self-help dieting and medicine was so great during the seventies that Congress passed not one but two separate pieces of legislation that effectively limited the FDA's power to regulate certain health products. In 1976, Senator William Proxmire succeeded in getting a Vitamin-Mineral Amendment passed, known subsequently as the Proxmire Amendment, which prohibited the FDA from setting limits on the strength of vitamin supplements or regulating them as "drugs" simply based on their potency. Proxmire saw the amendment as a counterbalance to the "hostility and prejudice" the FDA directed at small retailers of health products, and what he felt was the agency's collusion with Big Pharma.[66] The FDA also experienced legislative pushback on its policies on artificial sweeteners, what one historian has called a "saccharin rebellion." In 1977, the agency announced its intention to ban saccharine for much the same reasons it had banned cyclamate in 1969. Carolyn De la Peña described a flood of angry letters written by consumers to Congress and the FDA protesting the proposed ban. Consumers voiced their strong attachment to saccharine-sweetened products. The ban removed the only remaining artificial sweetener from the market and thus represented the FDA taking away from consumers their right to make their own calculus of risk given their personal lifestyles and dietary concerns.[67] Congress passed the Saccharine Study and Labeling Act imposing a two-year moratorium on any ban on the additive, and instead mandating a warning label for its risks. These two pieces of legislation reflected the widespread hostility toward the FDA "meddling" with individuals' food choices

Here I will provide two historical cases that illustrate how dramatically the informational turn and new lifestyle politics transformed what we mean by "food," presaging present-day challenges with functional foods. The first example is that of filled milk. Filled milk is any milk or cream product that has been reconstituted with fats from nondairy sources, most commonly vegetable oils. Filled milk, much like margarine, had long been the target of state regulation because of concerns that filled milk could be fraudulently marketed as a cheap alternative to the more expensive and natural original. Laws intended to prevent such "economic adulteration" banned any interstate trafficking of products made in imitation or semblance of milk unless it was exempted as a clearly labeled and packaged prepared food. A 1938 Supreme Court case, *United States v. Carolene Products Company*, for example, upheld a federal law used to seize packages of Milnut, later named Milnot, a condensed skim milk with coconut oil, in part under the rationale that there was substantial public-health evidence for such laws. The popularization of the diet-heart thesis in the 1960s turned this logic on its head. Vegetable oil companies aggressively advertised their products as better for your heart than many dairy and meat originals. Margarine and filled milk were potentially value-added health foods. Perhaps for this reason, a 1972 district court case, *Milnot Co. v. Richardson*, effectively reversed the 1938 decision and ended filled milk restrictions, arguing that such products were not an imitation but rather a legitimate healthy alternative. The FDA was quick to follow this turning tide. In its 1973 labeling reforms, it included a standard of identity for mellorine, an ice cream substitute using vegetable oils. Publishing the mellorine standard was a clear signal to industry that the agency was now encouraging untraditional, novel low-fat health foods. The changing status of filled milk from the 1930s case to the 1970s ruling registered a change in the idea of what is food—from food as whole and natural to food as component chemical ingredients. It also reflected a new idea of dietary health and risk—from whole foods being wholesome to foods engineered for health. This meant that foods, as with drugs, could now incorporate additives and "active ingredients" with health-promoting features.

The second example, legal disputes over fiber-enriched foods in the late 1980s, reiterates the form and function ambiguities of the food-drug line. Despite the FDA's turn to nutrition labeling and health claims in the seventies, the agency continued to take a strong stance against disease claims on foods: advertisements or label statements that specifically mentioned a disease, such as "prevents cancer," or "structure/function" claims that implied one (e.g., "calcium builds strong bones" or "fiber maintains bowel regularity"). This policy was undermined in 1984 by a move by Kellogg's to include an NIH National Cancer Institute statement about fiber and colon cancer on its All-Bran cereal boxes. The NIH statement on the food was technically a violation of the food-drug line, but the NIH was a peer institution to the FDA, so the claim was permitted. What followed was a flood of food products into supermarkets with fiber and disease claims. This direct-to-consumer advertising would birth the subsequent consumer "confusion," "information explosion," and category blurring that has characterized criticisms of functional foods ever since.

Kellogg's soon faced another product challenge. In 1989, Kellogg's introduced Heartwise cereal, which contained a substantial quantity of psyllium, the primary ingredient in many laxatives. Because the FDA had not determined psyllium to be GRAS, or "Generally Recognized as Safe," at the levels found in Heartwise, the state of Texas had it seized as an untested misbranded drug. In this case, a US district court ruled in favor of state enforcement, supporting the Texas attorney general's motion to remove Heartwise from the market. In the decision, the court noted that the cereal's labeling, which included a heart symbol and endorsements by a hospital and heart institute, suggested it was intended as a medical cure and therefore legitimately subject to scrutiny as a drug.[68] This case illustrates how purification of a thing is partly what makes it drug-like. Fiber, something wholesome in a food naturally, could also be purified into an "active ingredient" in a manner similar to many laxative drugs. And yet "pure" food additives had long been used by industry to enhance and modify properties of food. What made psyllium drug-like, while artificial sweeteners and flavors like vanillin were not?[69]

In 1990, the US Congress passed the Nutrition Labeling and Education Act (NLEA) which directed the FDA to rework food labels—specifically, to introduce mandatory "Nutrition Facts" labels for all food products and to determine which health and disease claims should be permitted for foods. The intent of the legislation was to clear up the legal confusion that had arisen with products such as Kellogg's All-Bran and Heartwise cereals, but also to lay the groundwork for more health-promoting foods. Nutrition labeling and what would soon be called functional foods were a politically appealing solution to the "nutrition transition," the shifting burden of disease and increase in chronic degenerative illnesses, like heart disease and cancer, that contributed to rising healthcare costs. Preventive care was cheaper than postdiagnosis care, and the new diet products suggested this could be outsourced to consumers as self-care.

Despite this political appeal and legislative mandate, the FDA resisted watering down the food-drug line and continued to restrict most disease claims on food. The agency also created a new category of products, "medical foods," or foods with medically significant properties designed to be used by doctors in treating patients. The borderline category "medical foods" served to protect the agency's continued commitment to differentiating marketing for foods and drugs, much as the earlier category of special dietary foods had. Frustrated by what some saw as the FDA's stranglehold on new modes of health marketing, diet supplement industry and advocacy organizations lobbied for a further change in the law. In 1994, Congress passed the Dietary Supplement Health and Education Act (DSHEA), which, for the first time since the 1976 Proxmire Amendment, once again limited the FDA's authority over certain marketed products, in this case giving a free hand to marketers of products classed as "dietary supplements." Dietary supplements, drug-like insofar as they were often in pill form, but food-like insofar as they were meant for routine, supplemental health maintenance and not disease treatment, became a new middle category straddling the food-drug line without erasing it.

Conclusion: The Food-Drug Line as a Legal Market-Making Device

The phrase "functional food" came into wide usage in the US food industry in the early 1990s right at the moment when food labeling laws were being rewritten by Congress and the FDA.[70] Food industry experts described the opportunity that "designer foods," functional foods, or "nutraceuticals" portended given the new laws for marketing nutrition information and the growing public interest in so-called superfoods and in modifying diets to slow aging. It was also at this moment when food policy consultants first coined the phrase "food-drug interface" or "food/drug line," which, they argued, posed the principal challenge and, as I will argue below, the main opportunity for industry seeking to promote novel foods in these new markets.[71] In an influential 1994 publication, the *Nutrition Labeling Handbook*, a guide for the food industry to the FDA's new Nutrition Facts labeling system, several authors described the categories and health claims rules in the 1990 NLEA, and those under development by the FDA, as an inflexible regulatory environment that "discourages rather than encourages investment in . . . important advances in functional food technology." They argued that, in the absence of clear and encouraging regulation, companies would need to carefully develop products to find strategic gaps or areas "outside the scope of NLEA health claims regulations."[72] The food-drug interface was thus characterized as a legal lacuna created by new food technologies that circumvented established but outdated legal categories for food versus drug. In short, they presented what STS scholars have criticized as a "law lag" narrative for functional foods.[73] Scholars in the late 1990s and 2000s who followed with concern the rise of these functional foods and lax regulation can be forgiven for largely accepting this narrative of functional foods as something new, boundary-pushing, and transgressive of traditional cultural categories separating eating and treating.

The food-drug line, suddenly under assault in the late 1980s and early 1990s, is, however, much older, and reflects decades-long efforts by FDA regulators and medical professionals to manage sensationalist and confusing marketing for health products. Looking at the longer history of the food-drug line, it is clear that there is nothing novel about seeking health in food or making medicine nourishing. In the twentieth century, new professional arrangements and institutional commitments led groups like the AMA and FDA to establish their authority over drug markets by marking boundaries between legitimate but risky and therefore highly regulated drugs, and less regulated and (ideally) carefree and self-evident foods. Foregrounding these professional and institutional stakes in the US history of classifying food, drug, and risk helps to clarify the normative concerns that float in the background of much of the social science literature on functional foods. Are consumers of functional foods "knowledge-able,"[74] or are they dupes? In some respects, the surge in sales of functional foods and dietary supplements in the last few decades, the popular interest in what Dumit calls "surplus health," can be seen as a revival of pre-twentieth-century markets for health tonics and other "just-in-case" medicines, now guaranteed not to hurt you.[75] One can imagine many consumers who buy them thinking, "Who knows if it will help, but why not?" Indeed, recent arguments in favor of the use of placebo medicines marks a dramatic reversal from early twentieth-century progressive campaigns, whose moral outrage rested on the falseness and deception of "health quackery" and "economic adulteration" in food-or-drug fraud.[76]

So what *is* new about functional foods? While I argue that there is nothing new in the boundary crossing, there are important ways that functional foods are different from their antecedents. The characterization of functional foods as "hybrid" does not arise out of essentialist claims about their mixed material composition, but rather from the different social functions they are put to and the mixed ideas about what is "health" and what is its relation to edible, nourishing, health-promoting things.[77] Functional foods are marketed on the authority of modern biomedicine, with its claims of specificity and mass-produced, standardized safety; yet they circumvent the expensive system of clinical testing for efficacy that characterize those markets. In other words, functional foods dance around the food-drug line rather than obliterate it. They free-ride on the prestige of drug testing and marketing without paying the cost of truth-testing restrictions. Woolgar and Neyland have asked the question, "Can things 'have politics?,'" and the answer here is a resounding *yes!* Companies who developed hybrid products such "special dietary foods" or "medical foods," or who marketed foods with health claims not supported by FDA policy, did so intending for the objects to be "made to 'do' political work."[78] This is why the food-drug line is not so much a problem created by a supposed "law lag" behind new food technology, but rather a profit engine for inspiring borderline products that thrive in a system of looser standards and ambiguous institutional control. It is an example of regulation as a driving force for product innovation, and also an example of how science, law, and markets coproduce one another.[79] Viewed in this light, the "food-drug line" is recast as a discursive market tool in which marketers, regulators, and consumers are invited to participate.[80] Ambiguity and hope are the product.

Xaq Frohlich is an assistant professor of history of technology at Auburn University. His research focuses on the intersection of science, law, and markets, and how the three have shaped our modern, everyday understanding of food, risk, and responsibility. He is currently completing a book, *From Label to Table: Regulating Food in the Information Age*, which explores the history of efforts by the US Food and Drug Administration to manage food markets and health risk through the regulation of food standards and informative labels.

Notes

1. Scrinis, "On the Ideology of Nutritionism"; Scrinis, "Nutritionism and Functional Foods"; Scrinis, *Nutritionism*.
2. Mudry, *Measured Meals*; Nestle, *Food Politics*.
3. Mol, "Good Taste"; Weiner, "Configuring Users"; Lezaun and Schneider, "Endless Qualifications"; Jauho and Niva, "Lay Understandings"; Azimont and Araujo, "Credible Qualifications"; Weiner and Will, "Materiality Matters"; Moore, "Food as Medicine."
4. Albala, *Food in Early Modern Europe*; Shapin, *Never Pure*; Gentilcore, *Food and Health*. For similar arguments in modern Asia, see Leung and Caldwell, *Moral Foods*.
5. Adelman and Haushofer, "Introduction: Food as Medicine."
6. Hutt, "Government Regulation," 3; Termini, "Product Classification," 1.
7. Grossman, "Food, Drugs, and Droods."
8. Haushofer, "Between Food and Medicine," 170.
9. Indeed, Carof and Nouguez have made very similar arguments in their parallel study of food and medicine markets and the genesis of functional foods in France. See Carof and Nouguez, "At the Boundaries." They see product classification as forming a part of what Michel Callon calls the "framing/overflowing" processes of innovation in market-making. Callon, "An Essay on Framing and Overflowing."
10. Star and Bowker, *Sorting Things Out*.
11. Huising and Silbey, "Governing the Gap."
12. Porter, *Trust in Numbers*.
13. Carpenter, *Reputation and Power*.
14. Stanziani, "Negotiating Innovation."
15. Gieryn, "Boundary-Work"; Jasanoff, *The Fifth Branch*.
16. They discuss *professional regulation*, where control of a product is delegated to a particular profession (what below I refer to as a gatekeeper model); *industrial regulation*, where the standardizing and systematic controls implemented by industrial firms have a kind of regulatory effect; *administrative regulation*, such as the FDA's process for pre-market approval; *public regulation*, meaning the regulating power that public interest groups and consumerism can have; and *juridical regulation*, which arises from the role that courts and court cases play in shaping drug policy. Gaudillière and Hess, *Ways of Regulating Drugs*. To this list, one could add others, such as *market regulation* based on the kinds of market-making practices economic sociologists study. See, for example, Cochoy, "A Sociology of Market-Things"; Chessel and Sophie Dubuisson-Quellier, "The Making of the Consumer."
17. Fisher, "Chemicals as Regulatory Objects," 163–64. It is not just regulation that gets embedded in objects' designs. Foods and drugs also embody market actors' "differentiated ontologies." See Cochoy and Mallard, "Another Consumer Culture Theory."
18. Kleinman, *Writing at the Margin*.
19. Starr, *The Social Transformation*.
20. "Patent" here referred to the fact that they were proprietary, heavily advertised, and promoted, and *not* to actual patents. Many of these medical elixirs would not have been patented because that would require their mysterious formulas be registered and made public.

21. Young, *The Toadstool Millionaires*.
22. Jauho and Niva, "Lay Understandings."
23. Vogel and Rosenberg, *The Therapeutic Revolution*.
24. Timmermans and Berg, *The Gold Standard*.
25. Grossman, "Food, Drugs, and Droods." Drug specificity would be given further validation with the rise of generics, which reinforced the idea of active ingredient essentialism and chemical reductionism, since carefully formulated branded drugs could be substituted for generic equivalents. See Greene, *Generic*.
26. Courtwright, *Forces of Habit*.
27. Grossman, "Food, Drugs, and Droods."
28. Courtwright, *Forces of Habit*.
29. Marks, *The Progress of Experiment*; Daemmrich, "Pharmacovigilance"; Greene, *Prescribing by Numbers*; Carpenter, *Reputation and Power*.
30. Companies may also have sought this preapproval system as a form of legal insurance against a broad shift taking place in American tort law for defective products, which involved the removal of privity as a requirement for seeking damages and a new standard of strict liability between branded manufacturer and end consumers. Friedman, *A History of American Law*.
31. The rising importance of a physician's prescription powers has also contributed to the increased division of labor in medical care between doctors and nurses. Much of the dietetic management important to nineteenth-century physicians was now left to nurses and considered secondary to medical treatment. Haushofer, "Between Food and Medicine," 182.
32. Watkins, *On the Pill*, 121.
33. By "doctors," Pope meant scholars, but Commissioner Ley in his presentation "The Doctor, the Patient, and the FDA" to the American College of Legal Medicine, 13 July 1969, was implicating uncertainty among physicians. Speech found in PBH/CBLF.
34. White, "Chemistry and Controversy."
35. Cohen, *Pure Adulteration*.
36. Freidberg, *Fresh*; Stoll, *The Fruits*.
37. I am grateful to Bruce Hunt for this insight.
38. On sugar and sucrose, see Singerman, "Inventing Purity." On vanilla and vanillin, see Berenstein, "Making a Global Sensation." On cottonseed oil and "pure" marketing, see Veit, "Eating Cotton."
39. Busch, *Standards*.
40. Merrill and Collier, "'Like Mother," 568.
41. Junod, "Food Standards"; Freidberg, *Fresh*.
42. The law also required certain declarations of weights and measures and a mailing address for the manufacturer.
43. Jacobson, *Eater's Digest*.
44. Apple, *Vitamania*.
45. Termini, "Product Classification."
46. Strasser, *Satisfaction Guaranteed*.
47. Bowlby, *Carried Away*, 194; Cochoy, *On the Origins*.
48. Hamilton, *Supermarket USA*.

49. This experience was constructed through what Franck Cochoy has called "faire laissez-faire" devices such as shopping carts, entry and exit barriers, and open-display shelving information tools. Cochoy, "A Sociology of Market-Things."

50. As quoted in Frohlich, "The Informational Turn," 154.

51. See Weiner and Will, "Materiality Matters."

52. Cross and Proctor, *Packaged Pleasures*.

53. Carstairs, "'Our Sickness Record.'"

54. *231 F. Supp. 551* (S.D. Fla. 1963), as quoted in Hutt, Merrill, and Grossman, *Food and Drug Law*, 233.

55. This change in Abbott Laboratories' marketing strategy can be seen in the Sucaryl business-to-business (B2B) advertisements that they ran in the trade journal *Food Technology* in the 1950s and 1960s.

56. Garrety, "Social Worlds," 750. Olszewski, "The Causal Conundrum"; Rothstein, *Public Health*; Aronowitz, *Making Sense of Illness*.

57. Levenstein, *Paradox of Plenty*.

58. "December 21, 1960 Letter to O. L. Kline, from Jeremiah Stamler" found in the binder "5.Polyunsaturates3-1965," PBH/CBLF. The Merck drug triparanol would soon be at the center of the scandal over industry drug safety testing and the balance of drug benefits against risks. In 1961, as the drug was coming on the market, Merck disclosed to the FDA that the drug had certain toxic side effects in lab tests on rats and dogs. The company would eventually be taken to court over the findings, and the drug removed from the market. The "triparanol fiasco," as one cardiologist would later call it, led many other companies to halt their research on cholesterol-lowering agents. Steinberg, "Thematic Review." On triparanol and the history of cholesterol-lowering drugs at this time, see also Greene, *Prescribing by Numbers*, 159–64.

59. See Stoff, this volume; Maffini and Vogel, this volume.

60. Crawford, "Healthism."

61. Knowles, "The Responsibility."

62. Giddens, *Modernity and Self Identity*.

63. Frohlich, "The Informational Turn."

64. Tuck, "Introduction: Reconsidering the 1970s," 617.

65. Sunstein, "Informational Regulation," 613.

66. Apple, *Vitamania*. In two separate cases in the 1970s, the FDA tried unsuccessfully to argue that premarket approval drug status should be applied to high-dose vitamin preparations. In the absence of a disease claim, courts rejected these arguments, drawing a distinction between nutritional use and therapeutic purposes. Grossman, "Food, Drugs, and Droods."

67. Peña, *Empty Pleasures*.

68. Termini, "Product Classification."

69. Psyllium was not a "natural" fiber to find in American foods. It came from a plant in India and had become popular in dietetic products and the food industry because its effects for relieving constipation and diarrhea were stronger than most naturally occurring fibers in Western grains. Much more could be said about the food-drug line in a global market for medicines. The globalization of alternative medicine markets and the intersections between traditional Eastern medicines and modern biomedi-

cal regulatory systems has often resulted in interesting transformations in how older medicines are framed in modern markets. See Kim, "Alternative Medicine's Encounter"; Gaudillière, "An Indian Path to Biocapital?"; Ma, "Join or Be Excluded." In 1992 the National Institutes of Health (NIH) established an Office of Alternative Medicine. When the NIH opened the National Center for Complementary and Integrative Health in 1998, it signaled the growing institutional interest in nonconventional therapeutics, including Eastern medicines. In the United States, these therapeutics have generally entered markets as dietary or health "supplements" to avoid the stricter premarket approval process on "drugs." Mason, "Drugs or Dietary Supplements." This adds an interesting cross-cultural dimension to debates about the lack of restrictions on such supplements.

70. Discussions of "functional foods" had appeared earlier in Japan in the 1980s, though by 1991 its ministry chose to refer to them as FOSHU, "Food for Specific Health Uses," because the word "function" implied such foods would legally be classed as drugs.

71. The first use of the phrase "food-drug interface" appears to be in Wrick et al., "Consumer Viewpoints," 94.

72. Hasler, Huston, and Caudill, "The Impact," 478–79, 485.

73. The belief that the law, as a conservative institution, naturally lags behind science and technology, which are always innovating and moving ahead. Jasanoff, "Making Order," 761–86.

74. "Knowledge-able" is Jasanoff's phrase for who administrative law imagines to be the ideal subject, someone presumably equipped to seek out and act rationally on information related to their safety and health. Jasanoff, "Governing Innovation."

75. Dumit, *Drugs for Life.*

76. Cohen, *Pure Adulteration.*

77. Much more could be said about the difference between "health" versus "medicine," the former broader and including health maintenance, the latter about treating illness. The murky boundaries between the two has prompted much of the category confusion highlighted by the food-drug line. A "healthy" label on a food could mean it is related to a biomedical condition or risk ("reduces risk of heart disease"), related to general health promotion (e.g. tonics, supplements), sustaining baseline health (nourishing, preventing malnutrition), "part of a balanced diet," or good for the body and soul (e.g. "wholesome" hearty, homemade soul food). These different kinds of "healthy" evoke different types of care work done by mothers, midwives, nurses, doctors, and others. The concept of "at risk" and preventive medical care have added to the blurring of boundaries between what is treating illness versus maintaining health. One can also review holistic medicine and its critiques of narrow biomedical definitions of health. See Greene, *Generic,* 11–13; Kleinman, *Writing at the Margin.*

78. Woolgar and Neyland, *Mundane Governance,* 16, 39. See also Barry, *Political Machines.*

79. Taylor, Rubin, Hounshell, "Regulation."

80. One can see this most clearly with products that carry labels that state that a supplement is "not yet FDA approved," or with the "off label" use of drugs. In some markets, the FDA's conservative status is itself a marketable feature for making claims about a treatment's cutting-edge value. This reveals the kinds of "ironies" and interpretative flexibility that shape the "mundane governance" of household items like foods and drugs. Woolgar and Neyland, *Mundane Governance.*

Bibliography

Archives

Personal archives of Peter Barton Hutt, Private library of Covington & Burling Law Firm (PBH/CBLF), Washington, DC

Publications

Adelman, Juliana, and Lisa Haushofer. "Introduction: Food as Medicine, Medicine as Food." *Journal of the History of Medicine and Allied Sciences* 73, no. 2 (2018): 127–34.

Albala, Ken. *Food in Early Modern Europe*. Westport: Greenwood Press, 2003.

Apple, Rima D. *Vitamania: Vitamins in American Culture*. New Brunswick: Rutgers University Press, 1996.

Aronowitz, Robert A. *Making Sense of Illness: Science, Society, and Disease*. Cambridge: Cambridge University Press, 1998.

Azimont, Frank, and Luis Araujo. "Credible Qualifications: The Case of Functional Foods." In *Concerned Markets: Economic Ordering for Multiple Values*, edited by Susi Geiger, Debbie Harrison, Hans Kjellberg, and Alexandre Mallard, 46–71. Cheltenham: Edward Elgar Publishing Limited, 2014.

Barry, Andrew. *Political Machines: Governing a Technological Society*. New York: A&C Black, 2001.

Berenstein, Nadia. "Making a Global Sensation: Vanilla Flavor, Synthetic Chemistry, and the Meanings of Purity." *History of Science* 54, no. 4 (2016): 399–424.

Bowlby, Rachel. *Carried Away: The Invention of Modern Shopping*. New York: Columbia University Press, 2001.

Busch, Lawrence. *Standards: Recipes for Reality*. Cambridge: MIT Press, 2011.

Callon, Michel. "An Essay on Framing and Overflowing: Economic Externalities Revisited by Sociology." *Sociological Review* 46, no. 1 (1998): 244–269.

Carof, Solenne, and Etienne Nouguez. "At the Boundaries of Food and Medicine: The Genesis and Transformation of the 'Functional Food' Markets in France and Europe." Sciences Po LIEPP Working Paper. Paris: Laboratoire interdisciplinaire d'évaluation des politiques publiques, September 2019.

Carpenter, Daniel P. *Reputation and Power: Organizational Image and Pharmaceutical Regulation at the FDA*. Princeton: Princeton University Press, 2010.

Carstairs, Catherine. "'Our Sickness Record Is a National Disgrace': Adelle Davis, Nutritional Determinism, and the Anxious 1970s." *Journal of the History of Medicine and Allied Sciences* 69, no. 3 (2014): 461–91.

Chessel, Marie-Emmanuelle, and Sophie Dubuisson-Quellier. "The Making of the Consumer: Historical and Sociological Perspectives." In *The SAGE Handbook of Consumer Culture*, edited by Olga Kravets, Pauline Maclaran, Steven Miles, and Alladi Venkatesh, 43–60. London: SAGE Publications, 2018.

Cochoy, Franck. *On the Origins of Self Service*. London: Routledge, 2015.

———. "A Sociology of Market-Things: On Tending the Garden of Choices in Mass Retailing." *Sociological Review* 55, no. 2 (2007): 109–29.

Cochoy, Franck, and Alexandre Mallard. "Another Consumer Culture Theory: An ANT Look at Consumption, or How 'Market-Things' Help 'Cultivate' Consumers." In *The*

SAGE Handbook of Consumer Culture, edited by Olga Kravets, Pauline Maclaran, Steven Miles, and Alladi Venkatesh, 384–403. London: SAGE Publications, 2018.

Cohen, Benjamin R. *Pure Adulteration: Cheating on Nature in the Age of Manufactured Food*. Chicago: University of Chicago Press, 2019.

Courtwright, David T. *Forces of Habit: Drugs and the Making of the Modern World*. Cambridge: Harvard University Press, 2001.

Crawford, Robert. "Healthism and the Medicalization of Everyday Life." *International Journal of Health Services* 10, no. 3 (1980): 365–88.

Daemmrich, Arthur. "Pharmacovigilance and the Missing Denominator: The Changing Context of Pharmaceutical Risk Mitigation." *Pharmacy in History* 49, no. 2 (2007): 61–75.

Dumit, Joseph. *Drugs for Life: How Pharmaceutical Companies Define Our Health*. Durham: Duke University Press, 2012.

Fisher, Elizabeth. "Chemicals as Regulatory Objects." *Review of European, Comparative & International Environmental Law* 23, no. 2 (2014): 163–71.

Freidberg, Susanne. *Fresh: A Perishable History*. Cambridge: Belknap Press of Harvard University Press, 2009.

Friedman, Lawrence M. *A History of American Law*. 3rd ed., rev. Touchstone ed. New York: Simon & Schuster, 2005.

Frohlich, Xaq. "The Informational Turn in Food Politics: The US FDA's Nutrition Label as Information Infrastructure." *Social Studies of Science* 47, no. 2 (2017): 145–71.

Garrety, Karin. "Social Worlds, Actor-Networks and Controversy: The Case of Cholesterol, Dietary Fat and Heart Disease." *Social Studies of Science* 27, no. 5 (1997): 727–73.

Gaudilliére, Jean-Paul. "An Indian Path to Biocapital?: The Traditional Knowledge Digital Library, Drug Patents, and the Reformulation Regime of Contemporary Ayurveda." *East Asian Science, Technology and Society: An International Journal* 8, no. 4 (2014): 391–415.

Gaudillière, Jean-Paul, and Volker Hess, eds. *Ways of Regulating Drugs in the 19th and 20th Centuries*. New York: Springer, 2012.

Gentilcore, David. *Food and Health in Early Modern Europe: Diet, Medicine and Society, 1450–1800*. New York: Bloomsbury Academic, 2016.

Giddens, Anthony. *Modernity and Self-Identity: Self and Society in the Late Modern Age*. Stanford: Stanford University Press, 1991.

Gieryn, Thomas F. "Boundary-Work and the Demarcation of Science from Non-Science: Strains and Interests in Professional Ideologies of Scientists." *American Sociological Review* 48, no. 6 (1983): 781–95.

Greene, Jeremy A. *Generic: The Unbranding of Modern Medicine*. Baltimore: Johns Hopkins University Press, 2014.

———. *Prescribing by Numbers: Drugs and the Definition of Disease*. Baltimore: Johns Hopkins University Press, 2007.

Grossman, Lewis A. "Food, Drugs, and Droods: A Historical Consideration of Definitions and Categories in American Food and Drug Law." *Cornell Law Review* 93, no. 5 (2008): 1091–148.

Hamilton, Shane. *Supermarket USA: Food and Power in the Cold War Farms Race*. New Haven: Yale University Press, 2018.

Hasler, C., R. L. Huston, and E. M. Caudill. "The Impact of the Nutrition Labeling and Education Act on Functional Foods." In *Nutrition Labeling Handbook*, edited by Ralph Shapiro. Boca Raton, FL: CRC Press, 1995.

Haushofer, Lisa. "Between Food and Medicine: Artificial Digestion, Sickness, and the Case of Benger's Food." *Journal of the History of Medicine and Allied Sciences* 73, no. 2 (2018): 168–87.

Huising, Ruthanne, and Susan S. Silbey. "Governing the Gap: Forging Safe Science through Relational Regulation." *Regulation & Governance* 5, no. 1 (2011): 14–42.

Hutt, Peter Barton. "Government Regulation of Health Claims in Food Labeling and Advertising Symposium: Health Claims in Food Labeling and Advertising." *Food, Drug, Cosmetic Law Journal* 41, no. 1 (1986): 3–73.

Hutt, Peter Barton, Richard A Merrill, and Lewis A. Grossman, eds. *Food and Drug Law: Cases and Materials*. New York: Foundation Press, 2007.

Jacobson, Michael F. *Eater's Digest: The Consumer's Factbook of Food Additives*. Garden City, NY: Doubleday, 1972.

Jasanoff, Sheila. *The Fifth Branch: Science Advisers as Policymakers*. Cambridge: Harvard University Press, 1990.

———. "Governing Innovation: The Social Contract and the Democratic Imagination." *Seminar* 597 (May 2009): 16–25.

———. "Making Order: Law and Science in Action." In *The New Handbook of Science and Technology Studies*, 3rd ed., edited by Edward J. Hackett, Olga Amsterdamska, Michael E. Lynch, and Judy Wajcman, 761–86. Cambridge: MIT Press, 2007.

Jauho, Mikko, and Mari Niva. "Lay Understandings of Functional Foods as Hybrids of Food and Medicine." *Food, Culture & Society* 16, no. 1 (2013): 43–63.

Junod, Suzanne White. "Food Standards in the United States: The Case of the Peanut Butter and Jelly Sandwich." In *Food, Science, Policy and Regulation in the Twentieth Century: International and Comparative Perspectives*, edited by David F. Smith and Jim Phillips. New York: Routledge, 2000.

Kim, Jongyoung. "Alternative Medicine's Encounter with Laboratory Science: The Scientific Construction of Korean Medicine in a Global Age." *Social Studies of Science* 37, no. 6 (2007): 855–80.

Kleinman, Arthur. *Writing at the Margin: Discourse between Anthropology and Medicine*. Berkeley: University of California Press, 1997.

Knowles, John H. "The Responsibility of the Individual." *Daedalus* 106, no. 1 (1977): 57–80.

Leung, Angela Ki Che, and Melissa L. Caldwell, eds. *Moral Foods: The Construction of Nutrition and Health in Modern Asia*. Honolulu: University of Hawaii Press, 2019.

Levenstein, Harvey A. *Paradox of Plenty: A Social History of Eating in Modern America*, rev. ed. Berkeley: University of California Press, 2003.

Lezaun, Javier, and Tanja Schneider. "Endless Qualifications, Restless Consumption: The Governance of Novel Foods in Europe." *Science as Culture* 21, no. 3 (2012): 365–91.

Ma, Eunjeong. "Join or Be Excluded from Biomedicine? JOINS and Post-Colonial Korea." *Anthropology & Medicine* 22, no. 1 (2015): 64–74.

Marks, Harry M. *The Progress of Experiment: Science and Therapeutic Reform in the United States, 1900–1990*. Cambridge: Cambridge University Press, 2000.

Mason, Marlys J. "Drugs or Dietary Supplements: FDA's Enforcement of DSHEA." *Journal of Public Policy & Marketing* 17, no. 2 (1998): 296–302.

Merrill, Richard A., and Earl M. Collier. "'Like Mother Used to Make': An Analysis of FDA Food Standards of Identity." *Columbia Law Review* 74, no. 4 (1974): 561–621.

Mol, Annemarie. "Good Taste." *Journal of Cultural Economy* 2, no. 3 (2009): 269–83.

Moore, Martin D. "Food As Medicine: Diet, Diabetes Management, and the Patient in Twentieth Century Britain." *Journal of the History of Medicine and Allied Sciences* 73, no. 2 (2018): 150–67.

Mudry, Jessica J. *Measured Meals: Nutrition in America.* Albany: SUNY Press, 2009.

Nestle, Marion. *Food Politics: How the Food Industry Influences Nutrition and Health.* Berkeley: University of California Press, 2002.

Olszewski, Todd M. "The Causal Conundrum: The Diet-Heart Debates and the Management of Uncertainty in American Medicine." *Journal of the History of Medicine and Allied Sciences* 70, no. 2 (2015): 218–49.

Peña, Carolyn Thomas de la. *Empty Pleasures: The Story of Artificial Sweeteners from Saccharin to Splenda.* Chapel Hill: University of North Carolina Press, 2010.

Porter, Theodore M. *Trust in Numbers: The Pursuit of Objectivity in Science and Public Life.* Princeton: Princeton University Press, 1995.

Proctor, Robert, and Gary S. Cross. *Packaged Pleasures: How Technology and Marketing Revolutionized Desire.* Chicago: University of Chicago Press, 2014.

Rothstein, William G. *Public Health and the Risk Factor: A History of an Uneven Medical Revolution.* Rochester: University of Rochester Press, 2003.

Scrinis, Gyorgy. "Nutritionism and Functional Foods." In *The Philosophy of Food,* 1st ed., edited by David M. Kaplan, 269–91. Berkeley: University of California Press, 2012.

———. *Nutritionism: The Science and Politics of Dietary Advice.* New York: Columbia University Press, 2013.

———. "On the Ideology of Nutritionism." *Gastronomica* 8, no. 1 (2008): 39–48.

Shapin, Steven. *Never Pure: Historical Studies of Science as If It Was Produced by People with Bodies, Situated in Time, Space, Culture, and Society, and Struggling for Credibility and Authority.* Baltimore: Johns Hopkins University Press, 2010.

Singerman, David Roth. "Inventing Purity in the Atlantic Sugar World, 1860–1930." *Enterprise & Society* 16, no. 4 (2015): 780–91.

Stanziani, Alessandro. "Negotiating Innovation in a Market Economy: Foodstuffs and Beverages Adulteration in Nineteenth-Century France." *Enterprise & Society* 8, no. 2 (2007): 375–412.

Star, Susan Leigh, and Geoffrey C. Bowker. *Sorting Things Out: Classification and Its Consequences.* Cambridge: MIT Press, 1999.

Starr, Paul. *The Social Transformation of American Medicine.* New York: Basic Books, 1982.

Steinberg, Daniel. "Thematic Review Series: The Pathogenesis of Atherosclerosis. An Interpretive History of the Cholesterol Controversy, Part V: The Discovery of the Statins and the End of the Controversy." *Journal of Lipid Research* 47, no. 7 (2006): 1339–51.

Stoll, Steven. *The Fruits of Natural Advantage: Making the Industrial Countryside in California.* Berkeley: University of California Press, 1998.

Strasser, Susan. *Satisfaction Guaranteed: The Making of the American Mass Market.* New York: Pantheon Books, 1989.

Sunstein, Cass R. "Informational Regulation and Informational Standing: Akins and Beyond." *University of Pennsylvania Law Review* 147, no. 3 (1999): 613–75.

Taylor, Margaret R., Edward S. Rubin, and David A. Hounshell. "Regulation as the Mother of Innovation: The Case of SO_2 Control." *Law & Policy* 27, no. 2 (2005): 348–78.

Termini, Roseann B. "Product Classification under the Federal Food Drug and Cosmetic Act: When a Food Becomes a Drug." *Journal of Pharmacy & Law* 2, no. 1 (1993): 1–14.

Timmermans, Stefan, and Marc Berg. *The Gold Standard: The Challenge of Evidence-Based Medicine and Standardization in Health Care.* Philadelphia: Temple University Press, 2003.

Tuck, Stephen. "Introduction: Reconsidering the 1970s; The 1960s to a Disco Beat?" *Journal of Contemporary History* 43, no. 4 (2008): 617–20.

Veit, Helen Zoe. "Eating Cotton: Cottonseed, Crisco, and Consumer Ignorance." *Journal of the Gilded Age and Progressive Era* 18, no. 4 (2019): 397–421.

Vogel, Morris J., and Charles E. Rosenberg, eds. *The Therapeutic Revolution: Essays in the Social History of American Medicine.* Philadelphia: University of Pennsylvania Press, 1979.

Watkins, Elizabeth Siegel. *On the Pill: A Social History of Oral Contraceptives, 1950–1970.* Baltimore: Johns Hopkins University Press, 1998.

Weiner, Kate. "Configuring Users of Cholesterol Lowering Foods: A Review of Biomedical Discourse." *Social Science & Medicine* 71, no. 9 (2010): 1541–7.

Weiner, Kate, and Catherine Will. "Materiality Matters: Blurred Boundaries and the Domestication of Functional Foods." *BioSocieties* 10, no. 2 (2015): 194–212.

White, Suzanne Rebecca. "Chemistry and Controversy: Regulating the Use of Chemicals in Foods, 1883–1959." PhD diss., Emory University, 1994.

Woolgar, Steve, and Daniel Neyland. *Mundane Governance: Ontology and Accountability.* Oxford: Oxford University Press, 2013.

Wrick, K. L., L. J. Friedman, J. K. Brewda, and J. J. Carroll. "Consumer Viewpoints on Designer Foods." *Food Technology* 47, no. 3 (1993): 94–104.

Young, James Harvey. *The Toadstool Millionaires: A Social History of Patent Medicines in America before Federal Regulation.* Princeton: Princeton University Press, 1961.

Afterword

Deborah Fitzgerald

As I write this in March 2020, my home in Boston is under quarantine. The coronavirus SARS-CoV-2 has infected over one million people worldwide and killed tens of thousands, has wreaked havoc with our financial systems, has closed schools and universities, has canceled all large social events, and has driven businesses to shutter their doors and people to hunker down in their homes, trying to comprehend how a virus has so profoundly transformed our day-to-day lives. Most people have never imagined such a dramatic phenomenon; it is the stuff of science fiction. We are not accustomed to this level of social anxiety or to the shockingly plain shortcomings of our previously admired medical system. We are unnerved that the scientific and technological accomplishments that have cushioned us from the harsher biological facts of life—germs, food shortages, epidemic disease—have let us down. Our confidence in the modern world is no longer quite as firm as it once was. We feel we are entering an age of profound uncertainty in which microbes call the shots. The emergence of COVID-19 seems to represent, as other pathogens have before this, that barriers between humans and animals are more porous than we had thought. Scientists initially suspected that the interaction of wild bats with other animals slaughtered in a Chinese wet market was the source of COVID-19, perhaps a warning that we interfere with wild animal ecosystems at our peril.[1]

As the essays in this volume attest, microbes have a great knack for finding their way into our bodies, our food supplies, and our animal populations. In the United States, aspects of the massive agricultural production system and low reporting of foodborne illness contribute to problems with contaminating bacteria.[2] Microbes are resourceful and wily, capable of mutating from one strain to another, sometimes able to resist our fiercest drugs. Some of the most virulent pathogens, such as *E. coli* O157: H7, are products of our industrialization of livestock, with its heavy use of antibiotic-laced animal feed. Whether concerning *E. coli*, SARS, *Salmonella*, MERS, Ebola, or COVID-19, the pathogens responsible for these outbreaks are difficult to predict and control.[3] Other kinds of food contaminants, from PCBs to pesticides, behave in unruly ways in our environments and food systems, ending up on our plates despite (or, in some cases, because of) existing regulations. And unlike disasters such as hurricanes or housing bubbles, microbial outbreaks and chemical residues endanger our very bodies as well as the many networks that bind us to one another and to our lifelines of food, shelter, money, and healthcare.[4]

At the heart of the essays here are the agricultural and food production systems that emerged in the mid to late nineteenth century in the United States. These innovations, each in its own way, profoundly transformed both production and consumption practices. The first was the mechanization of farm work, particularly the development of the reaper in the western United States and eventually the combined reaper/harvester, or combine, in the early twentieth century. These machines, which were designed to manage the vast flat landscapes of inland California and the high prairies of Montana, Wyoming, and Kansas, created an unprecedented amount of raw material, wheat. More than any other grain in the world, wheat is considered indispensable because of the centrality of bread in so many cultures. As wheat harvests grew, so too did the derivative by-products of wheat—cereal, crackers, desserts, and by the 1950s a host of new, highly processed foods, such as pancake mix. Over the next fifty to seventy-five years, most agricultural commodities became mass-produced trade goods, introducing an important microbial opportunity into the national food supply. But wheat had another important feature: its importance as a trade good. Particularly in times of war, the trade in wheat became a crucial factor in people's survival, as well as an indicator of national health or weakness.[5]

The second key development was the regional and transcontinental railroads that emerged in the 1860s. We tend to think that the importance of railroads was primarily in allowing people to travel much further and faster. But it was arguably as or more important as a food distribution machine. It moved not only wheat but perishables like fruits and vegetables, dairy products, and ultimately meat and fish. Never before had the diets of ordinary people been so divorced from local landscapes and climate; thanks to the food chains made possible by new modes of transport, diets became more varied, more interesting, and, potentially, more problematic in the years before robust inspections and regulations.[6]

The third innovation was the creation of a powerful national system of marketing and advertising, which ultimately provided standards by which to sort and grade raw materials such as wheat and pears and hams, as well as standards for packaging, processing, and mass production. By World War I, the system of producing agricultural goods and that of marketing and consuming such goods as food was well integrated.[7]

And finally the fourth development was the federal system overseeing both the production of food on farms (the United States Department of Agriculture) and the regulation of the food and pharmaceutical industry (the Food and Drug Administration). These two government agencies became sprawling organizations by the mid-twentieth century, often developing policies that were mutually contradictory and shortsighted. As several of these chapters make clear, these agencies have repeatedly fallen short in protecting the health of consumers.[8]

Another important transition occurred in World War II, when this provisioning system dramatically expanded to deal with the unique and unprecedented food requirements of the armed forces. The military had to locate and ship both raw food materials as well as prepared rations to over sixteen million military personnel in twenty-three different climatic theaters of combat during the war, an extremely demanding operation. To do this, the military worked very closely with the food industry, academic scientists and engineers, nutritionists, and farmers to ensure that every soldier had adequate and safe food at all times. The Quartermaster Subsistence Research Lab and its Food and Container Institute was located in Chicago, the epicenter of the meat and grain processors and one of the key nodes on the railroad lines crisscrossing the country. From here, the military coordinated scientific research happening across sectors, collaborating with others to produce the recipes, the packaging, and the protocols to make the whole system work successfully. Near the end of the war, the previously distinct military and civilian food markets had merged into a single market that privileged ever-ready food for the next military crisis—that is, food that was insensitive to sitting on shelves, or in bunkers, for months and years at a time. In an important sense, the purpose of time-insensitive food *is* to sit around waiting for a national crisis. As the food processors might ask, aren't you glad that during the current pandemic, you have all the shelf-stable food you could want? And it is hard to argue the point. But it has not come without a cost. World War II gave the food industry all the incentives it needed to create an extraordinary number of food additives— spray-on flavors, synthetic vitamins, stabilizers, radioactive preservation, and all the things that would allow foods to weirdly resist the passage of time. This obsession with nonspoiling food became the defining feature of the postwar American diet. The seemingly artificial nature of much of this food, I would argue, offered consumers a risky and false sense of security regarding dangers lying within either the food or its packaging. Highly processed food was seen as safer than raw food.[9]

Of course for many Americans, particularly scientists and engineers, experimentation and invention are themselves a social good, a way of pushing back against stale convention. Consumers seemed to welcome the postwar influx of new foods, which were sold as expressions of victory and a brighter, more space-age future. Foods that were sold as so-called convenience foods were often wartime surpluses transformed into novel new forms—crunchy flavored snacks, cheese spreads, sugary cereals in funny shapes, milk that could sit in a cupboard rather than the fridge, plus a lot of the new and intriguing packaging—that defined the 1950s and 1960s food revolution. One outcome of this shift was that slowly but surely, consumers began to lose track of what was actually in their food and where it originated. And, evidently, most of them thought that this was completely fine, the happy result of scientific supremacy. This disconnect between the raw material of foods, the origins and circumstances of its production, and our plates set the stage for the kinds of issues the authors here explore.[10]

The chapters in this volume offer a host of intriguing and crosscutting themes regarding food and health. One is the complexity of the food chains, and consumers' limited awareness and understanding of them.[11] The chains clearly reflect the fact that there are a number of differing social interests operating all along the chain, but none of them hold public health as their primary concern. Farmers, who have been notoriously unsuccessful in organizing themselves behind a unified economic or social plan, grow those commodities that the market will support, in large quantities and as cheaply as possible. For over fifty years this has centered on a heavy usage of chemicals to control pests and plant disease, a reliance on antibiotics to bring livestock to maturity faster, and a persistent disregard for soil and water quality beyond the farm itself. While farmers have not intended to contaminate food and water, or cause antibiotic resistance, their relation to the market, regulators, and the food industry has made them culpable nonetheless. Food processors, for their part, have often taken a minimalist approach to food safety. In many cases, rather than insist upon rigorous testing of ingredients in food products, they have fought regulations intended to protect consumers against unknown (often low-dose) effects, preferring a dubious practice of following threshold allowances. Food firms operate in an intensely competitive atmosphere, in which creating neverending novelty is the norm. Combined with an obsession with the bottom line, this has impacted their ability to patiently test and evaluate each new ingredient or process. Rather than find out the facts, processors have taken a GRAS approach ("generally recognized as safe" until proven otherwise), which puts all consumers at risk. And scientists who work inside processing companies, as well as at the USDA and FDA, often likewise have been inclined to err on the side of industry rather than consumers. Many scientists are trained to compartmentalize their research, eschewing information that is not strictly based on laboratory results. Scientists have also been inclined to take an uncritical view of innovation, tending to assume that new solutions are a good thing until proven otherwise.[12]

A second important theme in these papers is the complex and sometimes contradictory role of regulators. Particularly in the Food and Drug Administration, regulators have tried to strike a balance between protecting consumers from unknown or unproven ingredients in foods and protecting the food and drug industries from unfounded and libelous complaints. Our sense that the government agencies are primarily looking out for consumers turns out to be rather misguided. In their attempts to be fair, and following the presumption that most toxic effects from ingredients result not from the absolute toxic properties of a substance but from incorrect dosages, government scientists have not protected the public. Obviously, getting an X-ray examination is not as dangerous as entering the Fukushima reactor as it melts down, yet the idea that it is all about dosages remains controversial when applied to what humans ingest on a daily basis. Thus, regulators developed a threshold approach that delimits a range beyond which something might be dangerous, an approach that allows industry to have more leeway in selecting ingredients and amounts. But it also pushes the burden of dosage onto consumers, who are expected to educate themselves about the risks and benefits of each thing they ingest.[13] When consumers fail to do this, they are considered at fault. The practice of individual risk management, then, encourages a kind of intentional blindness among regulators, who may choose to define the potential health problems as mostly outside some specified norm. The fact that some consumers get sick and die after eating contaminated industrial food is evidently less important than the statistically low numbers of consumers involved.[14]

A third theme running through these essays is the sociological dimension of expertise. Again and again, we see scientists at work in the food and drug industries and in federal and state agencies seeming to ignore the basic questions about chemical exposure from eating. If card-carrying scientists have abandoned the Mertonian norms of objectivity, truth-seeking, and universalism, then whom are we to trust with our health? The sociology of expertise tells us that young physicians in training imbibe an intense dedication to the Hippocratic Oath, and scientists in general are trained to adhere to strict and clear values and behaviors. Because of this training they are inclined to trust each other and avoid the kind of collegial criticism that can characterize other professions, in which intellectual and personal commitments are not so uniformly anchored. But these tight alliances at the personal and professional level can be fraught and dangerous if the price of stepping out of line and becoming a whistleblower amounts to professional death.

Another aspect of professional hegemony is momentum. When studying technological systems, Thomas P. Hughes noted that as systems expand and become ever more complex, they achieve an internal momentum that is very difficult to adjust or stop. As these systems develop, scientists and engineers make thousands of decisions that foreclose some options and open up others, creating layers of pathways that depend on each other. At some point these systems can seem barely comprehensible to the experts themselves, as Charles Perrow has shown with the nuclear and aerospace industries and many others. But it is not just the technologies that become highly complex, interdependent, and often opaque—the human relationships that create and maintain such systems similarly take on a momentum that is challenging to interrupt. Whether it is devising the protocols that govern relationships between industry and the government, or the protocols of laboratory testing and clinical trials, or the belief that industry can regulate itself when introducing new drugs or additives, these relationships develop an internal momentum that maintains business as usual.[15]

Finally, the inherent instability of food contaminants, whether it is the liveliness of microbes or the unpredictable reactions of chemicals, is very much at odds with our insistence upon order, predictability, and linear progression of cause and effect.[16] The social systems we have worked so hard to create—clean water distribution, growing and stockpiling healthy foods, medical schools, clinics, and hospitals, sophisticated financial and transportation systems—none of these well-planned networks matter to microbes exquisitely developed to live in our guts and perhaps kill us. A viral strain can morph more quickly than our clinical trials can change gears, and a bacterial strain can survive even meticulous precautions. The question is, how can the historical incidents laid out in this volume inform our procedures and expectations going forward? Can we shift our professional norms to accommodate scientific serendipity? Can we halt the momentum in scientific organizations and their funding agencies to recalibrate our relationships with pathogens? And can we offload the hubris that characterizes so much of our leadership's response to outbreaks? Time will tell if policymakers can take to heart the lessons of the past.

Deborah Fitzgerald is the Cutten Professor of the History of Technology in the Science, Technology and Society Program at MIT. She specializes in the history of food and agriculture and is the author of *The Business of Breeding: Hybrid Corn in Illinois* (Cornell,1990) and *Every Farm a Factory: The Industrial Ideal in American Agriculture* (Yale, 2003), as well as various articles, essays, and opinion pieces. She is currently working on a book that explores the industrialization of food during World War II. She also writes and speaks on the importance of humanities and social science education, particularly for students of STEM.

Notes

1. For current numbers of recorded infections and deaths (much higher than at the time of writing this chapter), see "Corona Map: Tracking the Global Outbreak," *New York Times*, https://www.nytimes.com/interactive/2020/world/coronavirus-maps.html. On the connections between COVID-19 and modern industrial systems of animal production see, e.g., Dutkiewica, Taylor, and Vettese, "Covid-19 Pandemic."

2. Nestle, "The Politics of Foodborne Illness."
3. Kirchhelle, *Pyrrhic Progress*.
4. For an early treatment of microbes as actors, see Latour, *Pasteurization of France*.
5. On the rise of industrial farming, see Fitzgerald, *Every Farm a Factory*; on wheat, see Cronon, *Nature's Metropolis*. On worldwide food shortages, see Collingham, *Taste of War*.
6. On fruit see Stoll, *Fruits of Natural Advantage*; on the history and culture of preservation see Freidberg, *Fresh*.
7. On the rise of consumer culture see Strasser, *Satisfaction Guaranteed*; on advertising and marketing see Tedlow, *New and Improved*.
8. On the history of the United States Department of Agriculture, see Baker, *Century of Service*. On the Food and Drug Administration, see Hilts, *Protecting America's Health*.
9. See Fitzgerald, "World War II."
10. Levenstein, *Revolution at the Table*; and Levenstein, *Paradox of Plenty*.
11. Belasco and Horowitz, *Food Chains*.
12. On agricultural chemicals, see Vail, *Chemical Lands*. On livestock, see Marcus, *Cancer from Beef*. On the FDA's GRAS provision and its consequences, see Maffini and Vogel, this volume.
13. Frohlich, "Informational Turn."
14. On scientists, see Oreskes and Conway, *Merchants of Doubt*.
15. Hughes, *Networks of Power*; Perrow, *Normal Accidents*.
16. For a provocative and useful analysis of this see Boudia et al., "Residues."

Bibliography

Baker, Gladys. *A Century of Service: The First 100 years of the Department of Agriculture.* Washington, DC: Government Printing Office, 1963.

Belasco, Warren, and Roger Horowitz, eds. *Food Chains: From Farmyard to Shopping Cart.* Philadelphia: University of Pennsylvania Press, 2009.

Boudia, Soraya, Angela N. H. Creager, Scott Frickel, Emmanuel Henry, Nathalie Jas, Carsten Reinhart, and Jody A. Roberts. "Residues: Rethinking Chemical Environments." *Engaging Science, Technology, and Society* 4 (2018): 165–78.

Collingham, Lizzie. *A Taste of War.* London: Allen Lane, 2011.

Cronon, William. *Nature's Metropolis: Chicago and the Great West.* New York: Norton, 1991.

Dutkiewica, Jan, Astra Taylor, and Troy Vettese. "The Covid-19 Pandemic Shows We Must Transform the Global Food System." *The Guardian*, 16 April 2020.

Fitzgerald, Deborah. *Every Farm a Factory: The Industrial Ideal in American Agriculture.* New Haven, CT: Yale University Press, 2003.

———. "World War II and the Quest for Time Insensitive Foods." In *Food Matters: Critical Histories of the Food Sciences*, edited by Emma Spary and Anya Zilberstein, *Osiris* 35 (2020).

Frohlich, Xaq. "The Informational Turn in Food Politics: The U.S. FDA's Nutrition Label as Information Infrastructure." *Social Studies of Science* 47 (2017): 145–71.

Freidberg, Susanne. *Fresh: A Perishable History.* Cambridge, MA: Harvard University Press, 2009.

Hilts, Philip J. *Protecting America's Health: The FDA, Business, and 100 Years of Regulation.* New York: Knopf, 2003.

Hughes, Thomas Park. *Networks of Power: Electrification in Western Society, 1880–1930.* Baltimore, MD: Johns Hopkins University Press, 1983.

Kirchhelle, Claas. *Pyrrhic Progress: The History of Antibiotics in Anglo-American Food. Production.* New Brunswick, NJ: Rutgers University Press, 2020.

Latour, Bruno. *The Pasteurization of France,* trans. Alan Sheridan and John Law. Cambridge, MA: Harvard University Press, 1993.

Levenstein, Harvey. *Paradox of Plenty: A Social History of Eating in Modern America.* New York: Oxford University Press, 1993.

———. *Revolution at the Table.* New York: Oxford University Press, 1988.

Marcus, Alan. *Cancer from Beef: DES, Federal Food Regulation, and Consumer Confidence.* Baltimore, MD: Johns Hopkins University Press, 1994.

Nestle, Marion. "The Politics of Foodborne Illness: Issues and Origins." In *Safe Food: The Politics of Food Safety,* 33–61. Berkeley: University of California Press, 2010.

Oreskes, Naomi, and Erik Conway. *Merchants of Doubt: How a Handful of Scientists Obscured the Truth on Issues from Tobacco Smoke to Global Warming.* New York: Bloomsbury, 2010.

Perrow, Charles. *Normal Accidents: Living with High-Risk Technologies.* Princeton, NJ: Princeton University Press, 1999.

Stoll, Steven. *The Fruits of Natural Advantage: Making the Industrial Countryside in California.* Berkeley: University of California Press, 1998.

Strasser, Susan. *Satisfaction Guaranteed: The Making of the American Mass Market.* New York: Pantheon, 1989.

Tedlow, Richard. *New and Improved: The Story of Mass Marketing in America.* New York: Basic Books, 1990.

Vail, David. *Chemical Lands: Pesticides, Aerial Spraying, and Health in North America's Grasslands Since 1945.* Tuscaloosa: University of Alabama Press, 2018.

Index